Tectonic, Magmatic, Hydrothermal and Biological Segmentation of Mid-Ocean Ridges

Geological Society Special Publications
Series Editor A.J. FLEET

GEOLOGICAL SOCIETY SPECIAL PUBLICATION NO. 118

Tectonic, Magmatic, Hydrothermal and Biological Segmentation of Mid-Ocean Ridges

EDITED BY

C. J. MacLEOD
University of Wales, Cardiff, UK

P. A. TYLER
University of Southampton, UK

and

C. L. WALKER
University of Leeds, UK

1996
Published by
The Geological Society
London

THE GEOLOGICAL SOCIETY

The Society was founded in 1807 as the Geological Society of London and is the oldest geological society in the world. It received its Royal Charter in 1825 for the purpose of 'investigating the mineral structure of the Earth'. The Society is Britain's national society for geology with a membership of around 7500. It has countrywide coverage and approximately 1000 members reside overseas. The Society is responsible for all aspects of the geological sciences including professional matters. The Society has its own publishing house, which produces the Society's international journals, books and maps, and which acts as the European distributor for publications of the American Association of Petroleum Geologists, SEPM and the Geological Society of America.

Fellowship is open to those holding a recognized honours degree in geology or cognate subject and who have at least two years' relevant postgraduate experience, or who have not less than six years' relevant experience in geology or a cognate subject. A Fellow who has not less than five years' relevant postgraduate experience in the practice of geology may apply for validation and, subject to approval, may be able to use the designatory letters C Geol (Chartered Geologist).

Further information about the Society is available from the Membership Manager, The Geological Society, Burlington House, Piccadilly, London W1V 0JU, UK. The Society is a Registered Charity, No. 210161.

Published by The Geological Society from:
The Geological Society Publishing House
Unit 7, Brassmill Enterprise Centre
Brassmill Lane
Bath BA1 3JN
UK
(*Orders*: Tel. 01225 445046
Fax 01225 442836)

First published 1996

British Library Cataloguing in Publication Data

A catalogue record for this book is available from the British Library

ISBN 1–897799–72–1

Typeset by Type Study, Scarborough, UK.

Printed by The Alden Press, Osney Mead, Oxford, UK.

Distributors

USA
AAPG Bookstore
PO Box 979
Tulsa
OK 74101–0979
USA
(*Orders*: Tel. (918) 584–2555
Fax (918) 560–2652)

Australia
Australian Mineral Foundation
63 Conyngham Street
Glenside
South Australia 5065
Australia
(*Orders*: Tel. (08) 379–0444
Fax (08) 379–4634)

India
Affiliated East-West Press PVT Ltd
G-1/16 Ansari Road
New Delhi 110 002
India
(*Orders*: Tel. (11) 327–9113
Fax (11) 326–0538)

Japan
Kanda Book Trading Co.
Tanikawa Building
3–2 Kanda Surugadai
Chiyoda-Ku
Tokyo 101
Japan
(*Orders*: Tel. (03) 3255–3497
Fax (03) 3255–3495)

Contents

Preface

Up until the early 1980s it was confidently believed that mid-ocean ridges were simple, linear volcanic features interrupted only by transform faults spaced hundreds or thousands of kilometres apart. It was only when ridge axes were first surveyed with the new, high-resolution multibeam echosounding systems that we discovered that the zone of active volcanism is further sub-divided into discrete spreading segments on a variable, but much smaller, scale. Small lateral offsets or misalignments of these spreading cells occur, and a whole new class of ridge axial discontinuities was recognized: 'overlapping spreading centres', 'non-transform offsets', 'deviations from axial linearity' etc. These findings have given rise to a completely new view of the global mid-ocean ridge system, as a much more complex, dynamic environment, and in turn spawned intensive study of how and why ridges are segmented in this way. We now know that along-axis variations in the morphology of the ridge crest reflect significant differences in the structure of the sub-seafloor magma reservoirs and the extent and nature of upwelling of partially molten mantle beneath the axis. Segments ends are starved of magma relative to segment centres, and plate separation, especially at slow-spreading ridges, may be accommodated by variable degrees of tectonic stretching as well as by magmatic accretion. Seawater penetration into the crust and eventual outflow at black smoker vents depends heavily upon the location of the magmatic heat source and the nature and distribution of permeability in the crust, leading to a relationship between the locus of hydrothermal discharge and the tectonomagmatic segmentation. In turn, the mechanisms of faunal colonization of vent sites and the evolutionary history of the vent organisms are also affected by segmentation of the ridges, depending strongly upon the spatial distribution of vents and the hydrodynamics of dispersal of the vent fluids.

Segmentation, therefore, plays a vital role in all physical, chemical and biological processes at the mid-ocean ridge system. The continuum of processes clearly pays no need to traditional discipline boundaries, yet in order to comprehend any one part of the system it is vital that we understand how they inter-relate. The purpose of this book is to explore the causes and consequences of this ridge segmentation from the tectonic, magmatic, hydrothermal and biological viewpoint. It comprises 15 chapters of recent findings that span and attempt to link the entire spectrum of mid-ocean research. Much of the work presented herein was carried out under the aegis of the British mid-ocean ridge 'BRIDGE' programme, a Thematic Initiative supported by the UK Natural Environment Research Council.

Papers with a predominantly tectonic theme span a variety of scales. **Sempéré et al.** examine the segmentation characteristics of the intermediate-spreading rate Southeast Indian Ocean ridge adjacent to the Australian–Antarctic Discordance. Within the Discordance, which is a region of deep seafloor thought to be underlain by anomalously cold mantle, the spreading segments have morphologies akin to those of slow-spreading ridges; to the east of it, the ridge and ridge axial discontinuities are similar in appearance to those at fast-spreading ridges, with rifts propagating towards the Discordance. This shows that mantle temperature rather than spreading rate is the underlying control on segment characteristics. **Allerton et al.** and **McAllister & Cann** examine faulting processes and the formation of valley-wall faults at the slow-spreading Mid-Atlantic Ridge between the Kane and Atlantis fracture zones. Both studies use the Southampton Oceanography Centre's deep-towed sidescan sonar system, TOBI. TOBI data are of such high resolution when compared to conventional multibeam bathymetery data that these authors are able to erect detailed models for the processes of initiation, growth and degradation of faults within and between segments. **Blondel** introduces a technique for the automated classification of TOBI data, applying it to a section of the Mid-Atlantic Ridge south of the Azores in order to identify variations in the relationship between volcanism and tectonism along axis.

Lawson et al. examine the same TOBI images (from two segments immediately north of the Kane fracture zone) as Allerton et al., but with the aim of understanding the volcanic geology on a small scale, and the effects of segmentation on magmatism. The TOBI data they ground-truth using near-bottom photographic traverses and dredges that have been located precisely relative to seabed volcanic features. From this extremely detailed study they obtain new insights into the plumbing system of the neovolcanic zone and, on a larger scale, assess the effects of the transform fault and non-transform discontinuity on magma transport and evolution. **Robinson et al.** discuss the geochemistry of dredge samples collected from the ultra-slow-spreading Southwest Indian Ridge. They show that melting is restricted in comparison to other, faster-spreading ridges, consistent with the suggestion that conductive heat loss is a significant factor in these very slow-spreading environments. **Batiza** presents a comprehensive overview of the link between magmatic and tectonic segmentation, based upon a compilation of geochemical studies of lava suites dredged from around the world's mid-ocean ridge system, and ranging from the global scale – from mantle source domains and implications for upwelling and melting processes – to the smallest scale of segmentation yet resolvable: individual volcanic edifices and the products of individual eruptions. **Edwards et al.** examine the effects of segmentation-related variations in melting beneath ridges by the complementary approach of studying the mantle peridotite residue from which the lavas were extracted. They compare the petrologies and geochemistries of peridotites recovered by dredging and/or Ocean Drilling Program drilling in the tectonic windows of Hess Deep and Garrett Deep. Both areas expose lithospheric mantle from beneath the East Pacific Rise, but the former is believed to be derived from beneath the centre of a segment, and the latter from the vicinity of a major segment end: the Garrett transform fault. Lower degrees of melting at Garrett Deep are related by the authors to a colder thermal regime adjacent to the axial discontinuity.

Despite the completely separate approaches adopted in those papers in this volume concerned with hydrothermal processes, a concensus is emerging as to the relative influence of both magmatism and tectonism upon hydrothermal circulation. **Haymon** updates conceptual models for ridge-crest hydrothermal systems in the light of new observations from the East Pacific Rise made from submersibles and remotely-operated vehicles. Venting appears to be associated either with 'magma-rich' or 'magma-poor' settings: in the former case discharge is magmatically controlled, associated spatially and temporally with a volcanic eruption or dyking episode; in the latter, the heat source is at greater depth, and venting is related to the spatial distribution of tectonically-induced faulting and fissuring. **MacLeod & Manning** investigate the latter, documenting the cooling history and development of permeability in high-level gabbros from the East Pacific Rise drilled in Hess Deep. They demonstrate that tectonically-induced fracturing at segment ends is an important means of allowing seawater deep into the lower crust at or near the ridge axis. **German et al.** come to a similar conclusion, but based upon evidence from seafloor and water-column data for abundant hydrothermal venting in spatial association with segment terminations on the Mid-Atlantic Ridge south of the Azores hotspot. This tectonic setting contrasts markedly with those of the other known vent sites elsewhere in the Atlantic, for example Broken Spur, which are situated in segment centres and fall in Haymon's 'magma-rich' category. With its abundant venting the Azores area also differs from the Reykjanes Ridge, which occupies a similar setting south of the Iceland hotspot, but along which hydrothermal activity is near-absent. On a slightly different theme, **Zaykov et al.** describe the setting and occurrence of fossil hydrothermal deposits in the Urals palaeo-ocean basin/island arc complex. Mineralization was apparently located close to the centres of the original oceanic segments, and in the more central parts of the palaeo-ocean.

Although the distribution of vents and their fauna is driven by physical and chemical processes, the fauna at vents is also a function of the regional plate tectonic history and segmentation pattern. **Tunnicliffe et al.** examine the distribution of present-day faunas with respect to the global configuration of the plates and, in particular, the spatial location of ridge axes. We often think of the mid-ocean ridge as a single, long, interconnected chain, but there are a number of places such as the Juan de Fuca Ridge where the spreading axis has been isolated long enough for a fauna to evolve that is related to, but distinctly different from, that of the East Pacific Rise and Galapagos Rift. Understanding the differences between the faunas of the mid-ocean ridge systems of the Pacific and Atlantic represents a future challenge. **Southward et al.** address the possible mechanisms for speciation at separate vent sites. Initially it was believed that two species of *Ridgeia* exist on the Juan de Fuca Ridge; however, the authors show, using molecular methods, that they are one, and therefore that the transform faults represent a minimal obstacle to genetic exchange. The last paper in the volume, by **Nisbet & Fowler**, presents some thought-provoking ideas on the origin of life at vents and particularly how bacteria have been central to this concept.

We would like to thank Lindsay Parson, who convened the original, highly successful meeting of the same name at Burlington House, London; his encouragement has been much appreciated. We are indebted, too, to Angharad Hills of the Geological Society Publishing House, for her calmness, patience and promptness during the preparation of this volume, especially at those times when we showed none of these qualities. Finally, we would like to thank the following individuals for refereeing the manuscripts submitted for publication in this volume; without their help we could not have maintained the high scientific standards for which we have striven: S. Allerton, R. Batiza, A. Briais, P. Browning, J. Cann, D. Christie, J. Cope, P. Cowie, G. Foulger, M. Fowler,

P. Gente, L. Hawkins, P. Kempton, E. Klein, M. Kleinrock, C. Manning, D. Mason, N. Mitchell, B. Murton, R. Nesbitt, I. Parkinson, R. Pockalny, H. Prichard, A. Saunders, R. Searle, M. Somers, R. Taylor, V. Tunnicliffe and E. Valsami-Jones.

The cover image, kindly provided by Roger Searle, shows the bathymetry of a spreading segment on the Mid-Atlantic Ridge between 28°40′ and 29°30′N. Warmer colours represent shallower regions. The blue region striking NNE through the centre of the image is the axial valley, containing the active plate boundary and site of most recent volcanism. The axial valley deepens towards the ends of the segment (as shown by the purple tones), and the plate boundary is offset right-laterally by approximately 20 km at each end. The image uses Sea Beam data collected by G. M. Purdy *et al.* (*Marine Geophysical Researches*, **12**, 247–252, 1990), illuminated for the NW and gridded at approximately 170 m intervals.

C. J. MacLeod
P. A. Tyler
C. L. Walker

The Southeast Indian Ridge between 127° and 132°40′E: contrasts in segmentation characteristics and implications for crustal accretion

JEAN-CHRISTOPHE SEMPÉRÉ[1], BRIAN P. WEST[1], & LOUIS GÉLI[2]

[1]*School of Oceanography, Box 357940, University of Washington, Seattle, WA 98195-7940, USA*

[2]*Département de Géosciences Marines, IFREMER, B.P. 70 29280 Plouzané, France*

Abstract: The Australian–Antarctic Discordance is an anomalously deep section of the Southeast Indian Ridge which overlies a colder than normal region of the upper mantle. The Southeast Indian Ridge exhibits large contrasts in its geophysical and geochemical characteristics across the eastern boundary of the Discordance. We present new geophysical data collected along the Southeast Indian Ridge between 127°00′E and 132°40′E which define the segmentation characteristics of this portion of the Indo-Australian–Antarctic plate boundary. The Southeast Indian Ridge within our survey area can be broken into three first-order segments bounded by one transform fault and two propagating rifts. The transform fault, located at the west end of the study area, forms the eastern boundary of the Australian–Antarctic Discordance.

The morphology of the spreading axis is that of an axial high akin to that found along the East Pacific Rise. The non-transform discontinuities found along the axis of accretion are also similar to those encountered at fast-spreading centres. These characteristics are in marked contrast with those of the spreading axis within the Australian–Antarctic Discordance, which resembles a slow-spreading centre despite its intermediate spreading rate ($c.$ 74 mm a^{-1}). We suggest that the geophysical and geochemical transitions that occur along the Southeast Indian Ridge near the eastern boundary of the Australian–Antarctic Discordance may be ascribed to varying mantle temperature and melt production rate along the spreading centre at constant spreading rate.

The two propagating rifts, located near 127°45′E and 131°E, are actively migrating toward the Discordance down the regional, along-axis gradient in bathymetry at the rates of 43 and 46 mm a^{-1}, respectively. Propagating rifts play a dominant role in the reorganization of the spreading centre east of the Australian–Antarctic Discordance. Satellite gravity data reveal a number of V-shaped, west-pointing structures on the flanks of the Southeast Indian Ridge. These structures appear to have been spawned near the George V ridge–transform intersection at 139°E. Four of these features display a clear association with active propagating rifts. Some of the remaining features may be due to the migration of melting anomalies entrained by asthenospheric flow along the axis of the Southeast Indian Ridge.

The Southeast Indian Ridge (SEIR) extends from the Macquarie Triple Junction to the Rodriguez Triple Junction. This intermediate-spreading centre separates the Indo-Australian and Antarctic plates. Between 120°E and 128°E, the SEIR traverses the Australian–Antarctic Discordance (AAD), one of the most enigmatic features of the ocean basins. The AAD is a region of anomalously deep seafloor character-ized by rough topography and closely-spaced fracture zones (Hayes & Conolly 1972; Weissel & Hayes 1974; Sempéré *et al.* 1991). Seismic studies have shown that the AAD is underlain by a mantle which has faster than normal shear wave velocities and therefore which may be unusually cold (Forsyth *et al.* 1987; Kuo 1993). The residual depth anomaly which characterizes the AAD extends NNE and SSE across the

Australian and Antarctic plates, respectively (Hayes 1988; Marks *et al.* 1990); hence, it appears to have migrated westward with time with respect to the SEIR (Marks *et al.* 1990).

The SEIR within the AAD is associated with a deep rift valley whereas the SEIR east of the AAD corresponds to an axial high (Weissel & Hayes 1974; Anderson *et al.* 1980; Sempéré *et al.* 1991). The existence of a sharp change in axial morphology at the eastern boundary of the AAD is confirmed by satellite-derived gravity data (Sandwell & Smith 1992). The geophysical contrast across this single transform coincides with distinct changes in major and trace element geochemistry (Anderson *et al.* 1975; Christie *et al.* 1990; Klein *et al.* 1991). The existence of such geophysical and geochemical contrasts at con-stant spreading rate provides an opportunity to

From: MacLeod, C. J., Tyler, P. A. & Walker, C. L. (eds) 1996, *Tectonic, Magmatic, Hydrothermal and Biological Segmentation of Mid-Ocean Ridges*, Geological Society Special Publication No. 118, pp. 1–15.

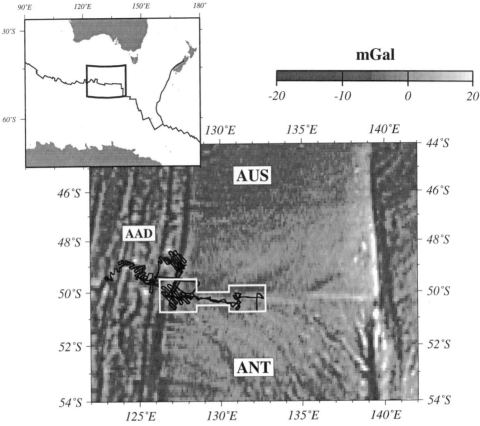

Fig. 1. The study area is located over the Southeast Indian Ridge between 127°E and 133°E east of the Australian–Antarctic Discordance (see inset). We have acquired SeaMARC II bathymetry and side-scan sonar data over the spreading centre. Ship tracks are overlain on satellite gravity data (Sandwell & Smith 1992). The transform fault located near 127°E at the west edge of the survey area is the eastern boundary of the Australian–Antarctic Discordance (AAD). AUS, Australian plate; ANT, Antarctic plate.

examine parameters other than spreading rate that control crustal accretion.

The AAD contains the boundary between the Pacific and Indian Ocean isotopic reservoirs (Klein *et al.* 1988). Pyle *et al.* (1992) have proposed that this boundary has migrated westward over the last 30 Ma, and that migration of Pacific mantle may have started when the South Tasman Rise separated from Antarctica about 30 Ma ago. In addition to its association with the migration of the isotopic boundary and residual depth anomaly, the AAD is a zone of convergence of multiple propagating rifts along the SEIR (Vogt *et al.* 1984). Satellite-derived gravity data confirm the existence of propagating rifts, but indicate the presence of other V-shaped structures pointing toward the Discordance east of the AAD (Fig. 1). Some of these

structures may be linked to the development of the AAD approximately 30 Ma ago, and to westward asthenospheric flow along the plate boundary.

In this paper, we present new bathymetric and side-scan sonar data obtained along the axis of the SEIR immediately east of the AAD between 127°E and 132°40′E (Fig. 1). We use these observations to establish the nature of the transition in axial topography at the eastern boundary of the AAD and to contrast the different segmentation characteristics observed at constant spreading rate (*c.* 74 mm a^{-1}) within this region. Finally, we use satellite gravity data to study the evolution of the segmentation and we speculate on the origin of the numerous V-shaped structures present on the flanks of the Southeast Indian Ridge east of the AAD.

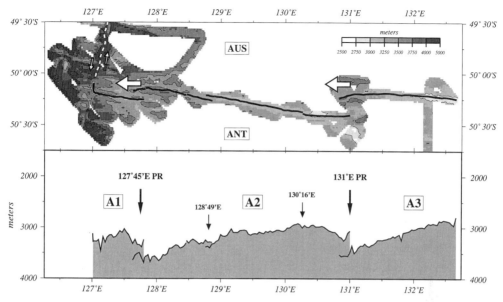

Fig. 2. (Top) Bathymetry of the Southeast Indian Ridge between 127°E and 132°40′E (contour interval = 500 m). The large arrows indicate the direction of migration of the two propagating rifts (PR) surveyed. The small arrows show the relative motion across the transform fault at the west end of our coverage. This transform fault marks the eastern boundary of the Australian–Antarctic Discordance. The bold line indicates the inferred location of the spreading axis. (Bottom) The along-axis depth profile shows the regional increase in depth to the west toward the Australian–Antarctic Discordance. The two propagating rifts, as well as minor discontinuities along the plate boundary, correspond to local depth maxima.

Data acquisition

The data presented in this paper were acquired using the SeaMARC II side-scan sonar system during leg MW8801 on the RV *Moana Wave* (Fig. 1). SeaMARC II was a 11–12 kHz, shallow-towed sonar system operated by the University of Hawaii and was able to gather bathymetry and backscatter information simultaneously (Blackington *et al.* 1983). Resolution of the SeaMARC II side-scan sonar is 5 m and 175 m in the across- and along-track directions, respectively (Tyce 1987). SeaMARC II bathymetry is sufficiently accurate to be contoured with a 100 m contour interval for our survey area. The side-scan sonar data were edited for bad pings and merged with corrected navigation. Contrast maps used to translate the 256 levels of recorded side-scan sonar information into 16 grey shades have been produced following corrections for changes in system gain. Survey lines were generally run at 45° to the strike of the seafloor fabric to optimize the use of the side-scan sonar. Navigation consisted of GPS fixes for about 10 hours per day, and a combination of Transit satellites and dead reckoning when GPS was not available.

The Southeast Indian Ridge between 127°E and 132°40′E

General characteristics

Between 127°E and 132°40′E, the Southeast Indian Ridge consists of three first-order segments, spreading at a rate of *c.* 74 mm a^{-1}, separated by a transform fault and two propagating rifts (Fig. 2). The left-slipping transform fault, which is located near 127°E and strikes 010°, forms the eastern boundary of the Australian–Antarctic Discordance. The age offset across this 175 km long transform is approximately 3.8 Ma (Vogt *et al.* 1984; Palmer *et al.* 1993). The morphology of the SEIR within the AAD west of this transform is discussed in Palmer *et al.* (1993). The other two major discontinuities along the plate boundary within our study area are propagating rifts located near 127°45′E and 131°E. Based on satellite gravity data (Sandwell & Smith 1992), the next major discontinuity along the SEIR east of our coverage is a third propagating rift located near 135°E. Within our survey area, the depth of the spreading centre progressively increases by

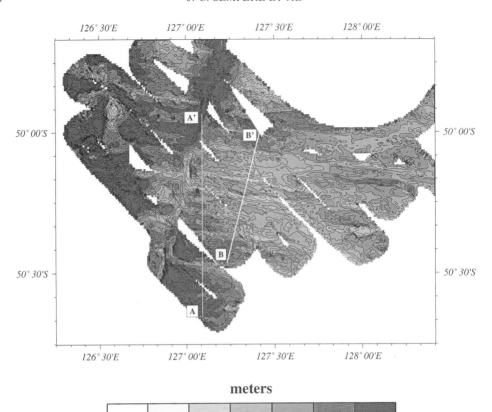

meters

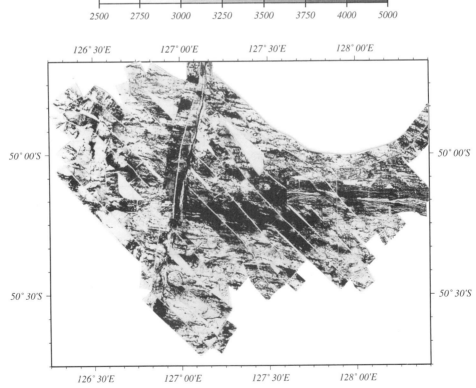

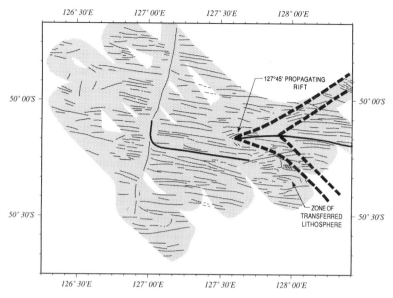

Fig. 3. (a) SeaMARC II bathymetry, (b) side-scan sonar imagery and (c) structural interpretation of the SEIR between 127°00′E and 128°25′E. Lineations observed on the side-scan sonar images are shown in the interpretive diagram. A bold line indicates the location of the spreading axis; dashed lines mark the inner and outer pseudofaults. Cross-sections of the spreading axis along profiles A–A′ and B–B′ are shown in Fig. 4.

c. 400 m from east to west toward the Australian–Antarctic Discordance. This regional increase in depth is not monotonic. Superimposed on this long-wavelength increase the axial depth profile reveals local maxima associated with the two propagating rifts mentioned above (Fig. 2).

Segmentation of the spreading centre

The first first-order segment (segment A1) starting from the west within the survey area extends from its intersection with the eastern boundary transform of the AAD to the 127°45′E propagating rift (Figs 2 and 3). Side-scan sonar and bathymetric data were acquired out to about 45 km off axis (*c.* 1.2 Ma) on each flank of the spreading centre. Westward migration of the 127°45′E propagator will eventually lead to the disappearance of this 55 km long segment. The minimum depth within segment A1 is found at *c.* 3050 m near 127°30′E in the eastern half of the segment close to its mid-point. Segment A1 is associated along its entire length with an axial ridge. The spreading axis corresponds to a region of high acoustic reflectivity as observed in our side-scan sonar mosaic (Fig. 3b). The ridge–transform intersection is located near 127°01′E, 50°11′S. The axial high curves sharply clockwise into the valley of the 127°E transform

and forms a 300 m high ridge which extends for *c.* 13 km within the transform domain. The crest of this ridge lies at a minimum depth of 3125 m. Away from the ridge transform intersection, the spreading axis is located on top of a 200–300 m high, 3–4 km wide, neovolcanic ridge trending *c.* 098° (Fig. 3c). The morphology of the axis is shown in cross-section in profiles A–A′ and B–B′ in Fig. 4. The location of the spreading axis is confirmed by examination of a bathymetric profile extracted along one of the aeromagnetic lines of Vogt *et al.* (1984) (A–A′ in Fig. 4). Within our coverage away from the axis in segment A1 we observe several linear abyssal hills which appear to be fault-bounded on their inward-facing flank on each side of the axis. Abyssal hills on the south flank of the axis mimic the geometry of the spreading axis by sharply curving north as they approach the transform fault.

Between the 127°45′E and 131°E propagators, the SEIR consists of a 240 km long, first-order segment (segment A2; Fig. 5). The west end of this segment is actively migrating west whereas its eastern end constitutes the doomed rift of the 131°E propagating system. Two small second-order discontinuities further partition this segment into distinct accretionary cells. These discontinuities are located near 128°48′E and 130°16′E and have offsets of 4.5 and 1.2 km,

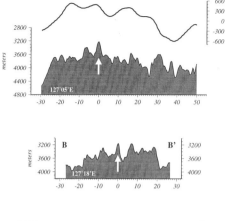

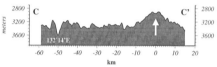

Fig. 4. Bathymetric profiles extracted from SeaMARC II bathymetric grids near 127°05′E, 127°18′E and 132°14′E. The axis is indicated by a white arrow in each profile. Cross-section A–A′ is coincident with one of the aeromagnetic profiles of Vogt *et al.* (1984). The magnetic data are shown above the bathymetric cross-section. Note that the ridge indicated by the arrow is located approximately in the middle of the central magnetic anomaly. The location of profiles A–A′ and B–B′ is shown in Fig. 3a, that of profile C–C′ in Fig. 6a. The longitude of the ridge axis crossings is indicated.

respectively. The 128°48′E left-stepping discontinuity, an overlapping spreading centre system, and the 130°16′E offset, a small non-overlapping offset, are akin to non-transform discontinuities observed along the fast-spreading East Pacific Rise (i.e., Macdonald *et al.* 1988). The lack of off-axis coverage in this area prevents us from determining the migration direction and rate of these minor offsets, as well as the morphology of abyssal hills. The shallowest point along the axis of the spreading centre between the two propagating rifts occurs near 130°12′E at a depth of *c.* 2925 m (Fig. 2). At the tip of the 127°45′E propagator, the axis lies at a depth of *c.* 3700 m. The depth of the spreading centre at the eastern end of this segment is *c.* 3200 m. The shallowest portions of the spreading centre usually correspond to the widest part of the axial ridge. The two small offsets of the axis of accretion near 128°48′E and 130°16′E correspond to local depth maxima. The spreading centre is associated with a prominent axial high along most of the length of this segment. East of 128°10′E, the spreading

axis corresponds to a 200–300 m high and 3–4 km wide ridge superposed on a broader swell. Near 130°12′E, where the spreading centre is shallowest, the spreading axis lies on a ridge about 400 m high and 7–8 km wide. The axis corresponds to a distinctive textural character indicative of fresh volcanic flows cut by fissures (Fig. 5b). The spreading axis is flanked by inward-facing faults which develop within a few kilometres (Fig. 5b and c). The axial high evolves into a shallow fault-bounded depression west of 128°10′E as the 127°45′E propagating rift is approached (Fig. 3). The morphology of the propagator tip is discussed below.

The easternmost segment (A3) within our survey area is bounded to the west by the propagating limb of the 131°E migrating offset (Fig. 6). The eastern end of this segment was not surveyed during our field program. We have obtained coverage over *c.* 150 km along the axis of this segment. The shallowest axial depth surveyed here is 2800 m near 132°40′E at the eastern end of our survey area; however, since we have only obtained partial bathymetric coverage of this segment, this depth minimum is unlikely to correspond to the shallowest point in the segment. The spreading centre increases in depth with a gradient of *c.* 5.6 m km^{-1} from the eastern end of our coverage to the tip of the 131°E propagator, which lies at a depth of *c.* 3450 m. Away from the 131°E propagating rift tip, the spreading centre corresponds to an axial high on which is sometimes superposed a shallow (50–100 m) and narrow (2 km) depression. Our crossing of the spreading axis near 132°14′E shows that the spreading centre is located on top of a 500 m high, 15 km wide ridge (profile C–C′ in Fig. 4). The small depression at the crest of the ridge may be similar to the axial summit caldera present along the East Pacific Rise (Haymon *et al.* 1991). Side-scan sonar data along and across axis show that inward-facing, normal faults dominate the fabric of the flanks of the spreading axis (Fig. 6b and c). West of 131°48′E, the axial high of the SEIR evolves into a shallow, fault-bounded depression (100–300 m deep) as the tip of the 131°E propagator is approached.

The 127°45′E propagating rift

We use the nomenclature defined by Kleinrock & Hey (1989*a*) in our discussion of the Sea-MARC II data collected over the two propagating rifts present in our study area. The 127°45′E propagator is a left-stepping offset of *c.* 13 km (Fig. 3). Based on backscatter data, we estimate that the two limbs of the propagating system may

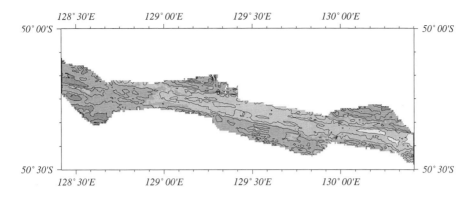

meters

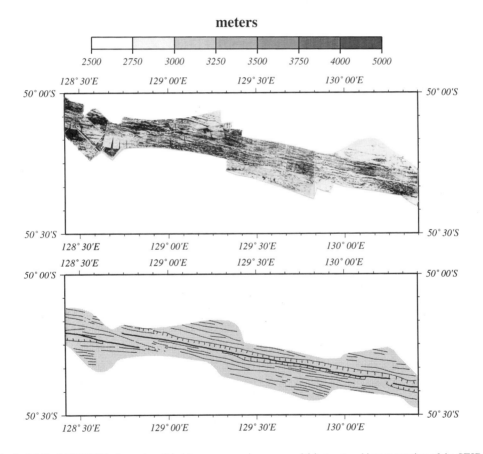

Fig. 5. (**a**) SeaMARC II bathymetry, (**b**) side-scan sonar imagery and (**c**) structural interpretation of the SEIR between 128°25′E and 130°25′E. Symbols are the same as in Fig. 3. The location of the axial summit caldera is shown by hash marks.

overlap by as much as 15 km. The propagating limb curves by *c.* 15° as the two axes overlap. This curvature is likely to be in response to the shear arising from the interaction between the two overlapping axes. The propagating ridge axis is located within a fault-bounded valley as

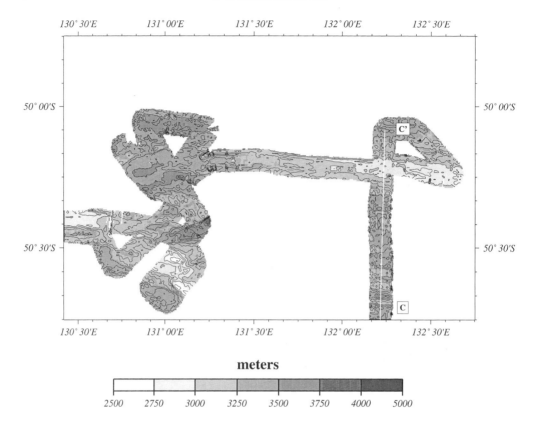

meters

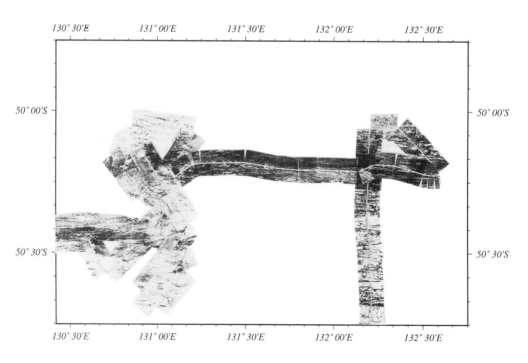

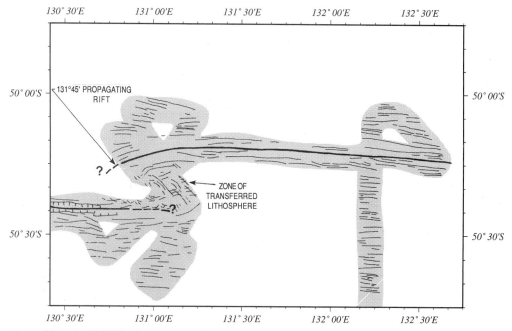

Fig. 6. (a) SeaMARC II bathymetry, (b) side-scan sonar imagery and (c) structural interpretation of the SEIR between 130°25′E and 132°40′E. A cross-section of the spreading axis along profile C–C′ is shown in Fig. 4. Symbols are the same as in Fig. 5.

mentioned above. The walls of this valley have a relief of *c.* 200 m. We interpret the tip of the propagator to be located within a wedge-shaped depression approximately 300 m deep. The maximum depth in this basin is >3800 m. The highly reflective character of the terrain near the tip of the propagating rift indicates relatively recent volcanic activity (Fig. 3b). This region forms a smooth plateau near the triangular depression noted previously (Fig. 3a). The failed spreading centre forms a lineament consisting of a series of right-stepping basins and ridges extending southeast from the failing ridge axis. The ridges present within the wake created by the failed spreading centre are at *c.* 3200 m depth. Their orientation and fine-scale characteristics are not constrained adequately by our existing coverage.

The two offset spreading axes are separated by a broad shear zone similar to that observed at the 95.5°W Galapagos propagating rift (Kleinrock & Hey 1989*b*). This broad shear zone is marked in the side-scan sonar mosaic as a region of intersecting lineations oblique to the otherwise prevailing seafloor fabric (Fig. 3b and c). This zone is observed southeast of the tip of the propagator. Following Kleinrock & Hey (1989*b*), we interpret this as the zone of transferred lithosphere. We have identified the

inner and outer pseudofault zones associated with the propagating system on the basis of bathymetry and structural information provided by our side-scan sonar data (Fig. 3c). The outer and inner pseudofault zones have an orientation of *c.* 40° away from the propagator tip. Because our off-axis coverage has a limited extent, the orientation of the pseudofaults may be in error by a few degrees. The orientation of these features is similar to the regional trends of the same propagating rift wakes identified by Phipps Morgan & Sandwell (1994), and obtained later in this study, based on satellite gravity data. The inner wake discussed in Phipps Morgan & Sandwell (1994) corresponds to the trace of the failed spreading centre rather than the inner pseudofault as defined by Kleinrock & Hey (1989*a*). The trace of the failed axis (Fig. 3a) determined in the present study merges on the flanks of the SEIR with the inner wake identified by Phipps Morgan & Sandwell (1994). We calculate a propagation rate of 43 mm a^{-1}, similar to that of Phipps Morgan & Sandwell (1994) (40 mm a^{-1}).

The 131°E propagating rift

Like the 127°45′E offset, the 131° propagator is a left-stepping discontinuity which is migrating toward the west (Fig. 6). The propagating and failing rift are separated by *c.* 20 km and appear to overlap by *c.* 18 km. The propagating limb of the system is located within a shallow rift valley with a relief of <300 m. The exact tip of the propagator is difficult to locate from our data alone, but it may lie within a small basin whose floor lies at a depth of 3550 m near 130°50′E. West of 131°07′E, the propagating axis curves south by 18°, presumably due to the shear stresses resulting from the interaction of the offset spreading centres.

The shear zone separating the two offset spreading centres is characterized by the juxtaposition of two sets of lineaments with distinct trends (ENE and SE; Fig. 6b and c). Both these trends are oblique to the orientation of the spreading centre and flanking abyssal hills away from the propagator tip. Some of these acoustic reflectors are cross-cutting indicating a complex tectonic interaction as a result of the migration of the instantaneous transform zone that accommodates the propagating offset. Our coverage off axis is too limited to trace the extent of this broad shear zone on the south flank of the axis, and to identify the location and orientation of the inner and outer pseudofault zones. As noted by Phipps Morgan & Sandwell (1994), a prominent lineation can be seen in satellite gravity data on the south flank of the 131° propagator (Fig. 1). This lineation lies in direct continuation with the location of the failed rift axis based on our bathymetric coverage. From the orientation of this lineation with the doomed ridge axis, we obtain a migration rate of 46 mm a^{-1}. This value is consistent with the estimate of Phipps Morgan & Sandwell (1994) (49 mm a^{-1}) for this propagating system.

Discussion

Geological and geochemical contrasts along the SEIR near the AAD

Slow-spreading centres (0–40 mm a^{-1}) are typically associated with a rift valley whereas fast-spreading centres (>100 mm a^{-1}) display an axial high (e.g., Macdonald 1982). The intermediate-spreading Southeast Indian Ridge between 127°00′E and 132°40′E is, in general, associated with an axial high. The morphology of this high is similar to that found along the East Pacific Rise (Macdonald *et al.* 1984) despite the intermediate spreading rate of the Southeast Indian Ridge. It is only within several kilometers of the tip of the two propagating rifts located within our study area that the axial high of the spreading axis evolves into a shallow rift valley.

In contrast, Palmer *et al.* (1993) have shown that the dominant axial morphology within the AAD is a highly segmented rift valley similar to that found along the Mid-Atlantic Ridge. For instance, in the first AAD segment west of our study area, the rift valley of the SEIR evolves into a 15 km wide and 1500 m deep valley (Palmer *et al.* 1993). Also in contrast with observations within our study area, the morphology of non-transform discontinuities within the AAD is similar to those found along slow-spreading centres. Thus, the eastern boundary of the AAD coincides with a significant transition in axial morphology and segmentation characteristics, as first proposed by Weissel & Hayes (1974). Because spreading rate is constant (*c.* 74 mm a^{-1}) throughout this region, the morphological transition cannot be ascribed to variations in spreading rate such as those observed between the East Pacific Rise and the Mid-Atlantic Ridge, and other controlling parameters must be considered.

Analysis of satellite gravity data suggests that the upper mantle beneath the AAD is *c.* 100°C cooler than more 'normal' surrounding mantle (West *et al.* 1994). These results are consistent with the conclusions of Forsyth *et al.* (1987), based upon analysis of surface waves. Such variations may account for the change in apparent segmentation characteristics from a crenulated geometry in the AAD to a more linear configuration east of the AAD (e.g., Rabinowicz *et al.* 1993). Assuming that the residual gravity anomalies result only from variations in crustal thickness, thereby ignoring all possible variations in mantle density, West *et al.* (1994) further showed that the average crustal thickness is significantly higher (by 3 km) beneath the SEIR within our survey area than within the AAD (Fig. 7). These results are also consistent with the preliminary results of the recent seismic refraction study of Tolstoy *et al.* (1995). The presence of thicker crust beneath the spreading centre east of the AAD may explain the differences in axial morphology and segmentation characteristics between the SEIR within and east of the AAD. Chen & Morgan (1990*a, b*) proposed that the axial topography of mid-ocean ridges is the result of the relative width of a decoupling zone beneath the ridge, and of the zone where stresses in the brittle plate due to mantle flow exceed the yield strength. The decoupling zone may be due to the higher ductility of crustal rocks over mantle rocks at high temperatures. Variations in upper mantle temperature and crustal thickness along a spreading centre, at constant spreading rate, may therefore cause transitions in axial topogra-

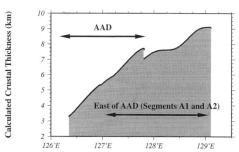

Fig. 7. Calculated crustal thickness for wavelengths greater than 60 km plotted along the axis of the Southeast Indian Ridge within and east of the Australian–Antarctic Discordance (from West *et al.* 1994). These values were obtained by downward continuation of satellite, residual gravity anomalies to Moho depth.

phy from an axial high to a rift valley, as discussed by West *et al.* (1994) for this region. Such variations may also allow the ductile lower crust to flow along axis and smooth out potential crustal thickness variations resulting from three-dimensional mantle upwelling, thus explaining the differences in the geometry of segmentation between the SEIR in and east of the AAD (e.g., Bell & Buck 1992; Phipps Morgan & Chen 1993*a*, *b*).

In terms of basalt geochemistry, the AAD defines one end-member of a global array of mid-ocean ridge basalt compositions (e.g., Klein & Langmuir 1987, 1989). Lavas from the AAD constitute a very different geochemical population from those east of the AAD. East of the AAD, lavas have a broader range of MgO contents, and other fractionation-dependent chemical variables, and an apparently narrow range of parental magma compositions (Christie *et al.* 1990). The chemical variability of lavas from the axial valleys of the AAD appears to be controlled predominantly by deep-seated processes in the mantle, including mean depth and extent of fractional melting, and, possibly, high-pressure crystallization (Christie *et al.* 1990). In contrast, the variability of lavas from the axial ridges east of the AAD is controlled by shallow crystal fractionation processes. These distinct populations have been referred to as 'source-dominant' and 'fractionation-dominant' populations, respectively (Christie *et al.* 1990). Such distinct populations correspond to the differences in geochemical variability observed between fast- and slow-spreading centres, but they are observed here at constant spreading rate. The fractionation-dominant basalts recovered along the axial high of the SEIR east of the AAD are likely to be the product of a high

magma supply associated with a taller melting column in a warmer mantle and more uniform melting process (Christie *et al.* 1990). In contrast, the major and trace element characteristics of the AAD samples indicate that they are derived from low pressures of melting and low average degrees of melting of a more fertile, clinopyroxene-rich source mantle relative to the SEIR to the east (Pyle 1994).

This study of the Southeast Indian Ridge between 127°E and 132°40′E, coupled with the study of segmentation within the AAD of Palmer *et al.* (1993), demonstrates that the range of variability in axial morphology and segmentation characteristics normally ascribed to spreading rate can be reproduced at constant spreading rate. Similar observations can also be made elsewhere along the mid-ocean ridge system (e.g., Reykjanes Ridge, Pacific–Antarctic Ridge). Both the geophysical and geochemical contrasts observed between the Southeast Indian Ridge within and east of the AAD may arise from differences in magma supply and from the presence of long-lived magma bodies at shallow crustal levels. These differences, which occur at constant spreading rate, may ultimately be due to variations in mantle temperature along the spreading centre (West *et al.* 1994).

V-shaped structures on the flanks of the Southeast Indian Ridge: constraints on the evolution of the segmentation

We have up to now focused on the present-day segmentation characteristics and ridge propagation along the SEIR based on SeaMARC II data. Ridge propagation has been a long-lived process along the Southeast Indian Ridge and continues to play an important role in the segmentation of this plate boundary. Temporal and spatial information on the long-term segmentation history of the SEIR may be obtained by studying satellite gravity data.

Structures possibly related to along-axis migration either of the mantle or of lithospheric features such as spreading centre discontinuities were first identified by Vogt *et al.* (1984) along the SEIR east of the AAD on the basis of aeromagnetic data. The satellite gravity map of Sandwell & Smith (1992) confirms the aeromagnetic work and reveals the presence of a series of west-pointing, V-shaped structures between 127°E and 140°E (Fig. 8). Some of the V-shaped structures coincide with the pseudofaults of the propagating ridge segments present along the SEIR near 127°45′E, 131°E and 135°E (Phipps Morgan & Sandwell 1994, and this study).

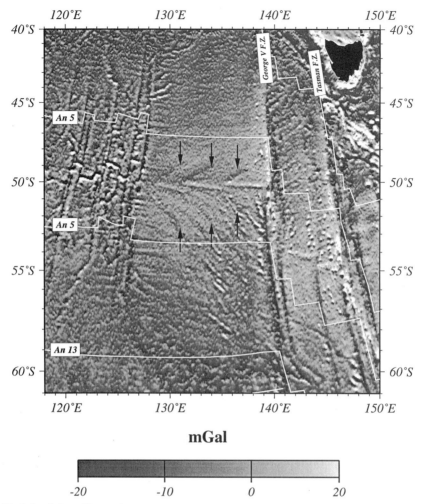

mGal

Fig. 8. Shaded-relief satellite gravity map of the Southeast Indian Ridge east of the Australian–Antarctic Discordance (Sandwell & Smith 1992). Isochrons corresponding to magnetic anomalies 5 and 13 are from Müller *et al.* (1993). Arrows indicate the V-shaped structures that characterize the wakes of the propagating ridge segments present along the Southeast Indian Ridge near 127°45′E, 131°E and 135°E. Note the presence of oblique alignments of small gravity highs on the south flank of the spreading axis. The alignments located between 51°30′S and 52°30′S and between 135°E and 137°E do not point towards the tip of any of the three propagating rifts present along the spreading centre.

On the south flank of the SEIR, the wakes of the three propagating rifts can easily be traced from the ridge axis to the George V Fracture Zone near 139°E, where they originated at different times (Phipps Morgan & Sandwell 1994). The trace of the 127°45′E propagating rift appears to cross the fracture zone, extending as far south and east as 60°S, 145°E; however, in the absence of additional bathymetric data, it is difficult to confirm whether the traces observed on each side of the fracture zone are part of a single, continuous feature, or are two different features fortuitously aligned across the fracture zone. The traces associated with the 127°45′E and 131°E propagating rifts on the south flank are *c.* 40° oblique to the ridge axis, and the trace of the 135°E propagating system is oblique by *c.* 45° to the ridge axis. On the north flank, and for crustal ages less than about 4 Ma, the outer pseudofaults of the three propagating rifts form linear troughs, 30–35° degrees oblique to the axis of the SEIR; for crustal ages greater than about 4 Ma, the angle of obliquity is greater than 40°. The irregular and sometimes subtle signature of the off-axis wakes of the propagating rifts make it difficult to determine how steady the migration rate has been over time.

Oblique alignments of small gravity highs appear on the south flank of the SEIR (Fig. 8). With the exception of an isolated high found near 53°50′S, 133°E, most of these structures are located east of 135°E, and aligned in a direction 40–50° oblique to the SEIR axis. The age of the crust on which these features appear varies from 33 Ma near 59°20′S, 139°E, to 23 Ma near 59°30′S, 135°E, and 4 Ma near 52°30′S, 136°E. It is interesting to note that the alignments located between 51°30′S and 52°30′S and between 135°E and 137°E do not point towards the tip of any of the three propagating rifts present along the SEIR. This suggests that they may have a different origin than the wakes of the propagating rifts. They may be associated with past propagating rifts that have become inactive with time, or they may reflect the westward migration along the axis of the spreading centre of melting anomalies, perhaps associated with along-axis flow from the region of warm mantle identified beneath the Tasman Zone by Rouland et al. (1985) and Roult et al. (1994) in their surface wave studies.

Although the satellite gravity data lack the necessary resolution required to resolve fine-scale, tectonic details, they provide constraints on the large-scale evolution of the propagating rifts and segmentation of the plate boundary. Satellite gravity data indicate that ridge propagation has been the dominant mode of segmentation evolution along the plate boundary. They also suggest that along-axis asthenospheric flow may be an important aspect of lithospheric accretion and upper mantle dynamics along the Southeast Indian Ridge as indicated by geochemical and geophysical studies (e.g., Vogt & Johnson 1973; Weissel & Hayes 1974; Forsyth et al. 1987; Marks et al. 1990, 1991; Pyle et al. 1992).

Conclusions

The analysis of SeaMARC II data collected along the Southeast Indian Ridge east of the Australian–Antarctic Discordance and of satellite gravity data on the flanks of the spreading centre allows us to draw the following conclusions.

(1) The Southeast Indian Ridge east of the Australian–Antarctic Discordance is associated with an axial high. Between 127°00′E and 132°40′E, spreading segments along the plate boundary are offset by one transform fault and several non-transform offsets, the morphology of which is similar to that of discontinuities encountered along fast-spreading centres. These characteristics are in marked contrast with those of the Southeast Indian Ridge within the Australian–Antarctic Discordance.

(2) Our study and that of Palmer et al. (1993) show that the global variability in axial morphology and segmentation characteristics normally attributed to spreading rate can be observed at constant spreading rate. Mantle temperature, and associated variations in melt production rate and crustal thickness, may be responsible for the observed geophysical and geochemical transitions across the eastern AAD transform.

(3) Ridge propagation has been the dominant process in the evolution of the segmentation of the Southeast Indian Ridge east of the AAD. Westward migration of propagating rifts and possibly of melt anomalies in the sub-ridge mantle initiated near the intersection of the Southeast Indian Ridge with the George V Transform appear to have occurred frequently over the last 30 Ma. Thus, along-axis asthenospheric flow toward the AAD may also play an important role in the segmentation of the Southeast Indian Ridge.

We are very grateful to S. Shor, the SeaMARC II group at the University of Hawaii, and the captain and crew of the RV Moana Wave for their help at sea. We also wish to thank S. Miller, J. Palmer, and D. Pyle for their assistance and for fruitful discussions. We used the GMT software package of Wessel & Smith (1991) and Smith & Wessel (1990) throughout this study. This work was supported by National Science Foundation grants OCE-8722980 and OCE-9420649 and an Office of Naval Research Graduate Fellowship award to B. P. West. Correspondence to B. P. West (e-mail: bpw@ocean.washington.edu).

References

ANDERSON, R. N., SARIOSU, D. J., WEISSEL, J. K. & HAYES, D. E. 1980. The interrelation between variations in magnetic anomaly amplitudes and basalt magnetization and chemistry along the southeast Indian ridge. *Journal of Geophysical Research*, **85**, 3883–3898.

BELL, R. E. & BUCK, W. R. 1992. Crustal control of ridge segmentation inferred from observations of the Reykjanes Ridge. *Nature*, **357**, 583–357.

BLACKINGTON, J. G., HUSSONG, D. M. & KOSALOS, J. 1983. First results from a combination of side-scan sonar and seafloor mapping system (SeaMARC II), In: *Offshore Technology Conference*, OTC 4478, pages 307–311.

CHEN, Y. & MORGAN, W. J. 1990a. Rift valley/no rift valley transition at mid-ocean ridges. *Journal of Geophysical Research*, **95**, 17571–17581.

—— & —— 1990b. A non-linear rheology model for mid-ocean ridge axis topography. *Journal of Geophysical Research*, **95**, 17583–17604.

CHRISTIE, D. M., PYLE, D., SEMPÉRÉ, J.-C. & PALMER,

J. 1990. Petrologic diversity, axial morphology and magma supply at the Australian–Antarctic Discordance (AAD). *Eos Transactions of the AGU*, **71**, 1388.

FORSYTH, D. W., EHRENBARD, R. L. & CHAPIN, S. 1987. Anomalous upper mantle beneath the Australian–Antarctic Discordance. *Earth and Planetary Science Letters*, **84**, 471–478.

HAYES, D. E. 1988. Age-depth relationships and depth anomalies in the southeast Indian Ocean and South Atlantic Ocean. *Journal of Geophysical Research*, **93**, 2937–2954.

—— & CONOLLY, J. R. 1972. Morphology of the Southeast Indian Ocean. *In*: HAYES, D. E. (ed.) *Antarctic Oceanology II: the Australian-New Zealand Sector*. AGU Antarctic Research Series, **19**, 225–245.

HAYMON, R. M., FORNARI, D. J., EDWARDS, M., CARBOTTE, S. M., WRIGHT, D. & MACDONALD, K. C. 1991. Hydrothermal vent distribution along the East Pacific Rise (9°09′–54′N) and its relationship to magmatic and tectonic processes on fast-spreading mid-ocean ridges. *Earth and Planetary Science Letters*, **104**, 513–534.

KLEIN, E. M. & LANGMUIR, C. H. 1987. Global correlations of ocean ridge basalt chemistry with axial depth and crustal thickness. *Journal of Geophysical Research*, **92**, 8089–8115.

—— & —— 1989. Local vs. global correlations in ocean ridge basalt composition: A reply. *Journal of Geophysical Research*, **94**, 4241–4252.

——, ——, ZINDLER, A., STAUDIGEL, H. & HAMELIN, B. 1988. Isotope evidence of a mantle convection boundary at the Australian–Antarctic Discordance. *Nature*, **333**, 623–629.

——, —— & STAUDIGEL, H. 1991. Geochemistry of basalts from the SEIR, 115°-138°E. *Journal of Geophysical Research*, **96**, 2089–2108.

KLEINROCK, M. C. & HEY, R. N. 1989a. Detailed tectonics near the tip of the Galapagos 95.5°W propagator: How the lithosphere tears and a spreading centre develops. *Journal of Geophysical Research*, **94**, 13801–13838.

—— & —— 1989b. Migrating transform zone and lithospheric transfer at the Galapagos 95.5°W propagator. *Journal of Geophysical Research*, **94**, 13859–13878.

KUO, B.-Y. 1993. Thermal anomalies beneath the Australian–Antarctic Discordance. *Earth and Planetary Science Letters*, **119**, 349–364.

MACDONALD, K. C. 1982. Mid-ocean ridges: Fine scale tectonic, volcanic, and hydrothermal processes within the plate boundary zone. *Annual Reviews of Earth and Planetary Science*, **10**, 155–190.

——, SEMPÉRÉ, J.-C. & FOX, P. J. 1984. The East Pacific Rise from the Siqueiros to the Orozco fracture zone: Along strike continuity of the neovolcanic zone and the structure and evolution of overlapping spreading centres. *Journal of Geophysical Research*, **89**, 6049–6069.

——, FOX, P. J., PERRAM, L. J., EISEN, M. F., HAYMON, R. M., MILLER, S. P., CARBOTTE, S. M., CORMIER, M.-H. & SHOR, A. N. 1988. A new view of the mid-ocean ridge from the behaviour of ridge axis-discontinuities. *Nature*, **335**, 217–225.

MARKS, K. M., VOGT, P. R. & HALL, S. A. 1990. Residual depth anomalies and the origin of the Australian–Antarctic Discordance zone. *Journal of Geophysical Research*, **95**, 17325–17338.

——, SANDWELL, D. T., VOGT, P. R. & HALL, S. A. 1991. Mantle downwelling beneath the Australian–Antarctic Discordance zone: Evidence from geoid height versus topography. *Earth and Planetary Science Letters*, **103**, 325–338.

MÜLLER, R. D., ROEST, W. R., ROYER, J.-Y., CAHAGAN, L. M. & SCLATER, J. G. 1993. *A digital map of the ocean floor, Scripps Institution of Oceanography*. University of California, San Diego.

PALMER, J., SEMPÉRÉ, J.-C., PHIPPS MORGAN, J. & CHRISTIE, D. M. 1993. Morphology and tectonics of the Australian–Antarctic Discordance. *Marine Geophysical Researches*, **15**, 121–152.

PHIPPS MORGAN, J. & CHEN, Y. J. 1993a. The genesis of oceanic crust: Magma injection, hydrothermal circulation and crustal flow. *Journal of Geophysical Research*, **98**, 6283–6297.

—— & —— 1993b. The dependence of ridge-axis morphology and geochemistry on spreading rate and crustal thickness. *Nature*, **364**, 706–708.

—— & SANDWELL, D. T. 1994. Systematics of ridge propagation south of 30°S. *Earth and Planetary Science Letters*, **121**, 245–258.

PYLE, D. G. 1994. *Geochemistry of mid-ocean ridge basalts within and surrounding the Australian–Antarctic Discordance*. PhD Thesis, Oregon State University.

——, CHRISTIE, D. M. & MAHONEY, J. J. 1992. Resolving an Isotope Boundary within the Australian–Antarctic Discordance. *Earth and Planetary Science Letters*, **112**, 161–178.

RABINOWICZ, M., ROUZO, S., SEMPÉRÉ, J.-C. & ROSEMBERG, C. 1993. Three-dimensional models of mantle flow beneath spreading centres. *Journal of Geophysical Research*, **98**, 7851–7869.

ROULAND, D., XU, S. H. & SCHINDELE, F. 1985. Upper mantle structure in the southeast Indian Ocean: surface wave investigation. *Tectonophysics*, **14**, 281–292.

ROULT, G., ROULAND, D. & MONTAGNER, J.-P. 1994. Antarctica II: Upper mantle structure from velocities and anisotropy. *Physics of the Earth and Planetary Interiors*, **84**, 33–57.

SANDWELL, D. T. & SMITH, W. H. F. 1992. Global marine gravity from ERS-1, GEOSAT and SeaSat reveals new tectonic fabric. *Eos Transactions of the AGU*, **73**, 43, 133.

SEMPÉRÉ, J.-C., PALMER, J., CHRISTIE, D. M., PHIPPS MORGAN, J. & SHOR, A. 1991. Australian–Antarctic discordance. *Geology*, **19**, 429–432.

SMITH, W. H. F. & WESSEL, P. 1990. Gridding with continuous curvature splines in tension. *Geophysics*, **55**, 293–305.

TOLSTOY, M., HARDING, A. J., ORCUTT, J. A. & PHIPPS MORGAN, J. 1995. A seismic refraction investigation of the Australian–Antarctic Discordance and neighboring Southeast Indian Ridges: Preliminary results. *Eos Transactions of the AGU*, **76**, 275.

TYCE, R. C. 1987. Deep seafloor mapping systems – a

review. *Marine Technology Society Journal*, **20**, 4–16.

VOGT, P. R. & JOHNSON, G. L. 1973. A longitudinal seismic reflection profile of the Reykjanes Ridge: Part II – Implications for the mantle hotspot hypothesis. *Earth and Planetary Science Letters*, **18**, 49–58.

——, CHERKIS, N. Z., MORGAN, G. A. 1984. Project Investigator-l: Evolution of the Australia–Antarctic Discordance from a detailed aeromagnetic study. *In*: OLIVER, R. L., JAMES, P. R. & JAGO, J. B. (eds) *Antarctic Earth Science*. Canberra.

WESSEL, P. & SMITH, W. H. F. 1991. Free software helps map and display data. *Eos Transactions of AGU*, **72**, 411, 445–446.

WEISSEL, J. K. & HAYES, D. E. 1974. The Australian–Antarctic Discordance: New results and implications. *Journal of Geophysical Research*, **197**, 2579–2587.

WEST, B. P., SEMPÉRÉ, J.-C., PYLE, D. G., PHIPPS MORGAN, J. & CHRISTIE, D. M. 1994. The importance of mantle temperature in crustal accretion: Evidence from the Australian–Antarctic Discordance and numerical modeling of mid-ocean ridge dynamics. *Earth and Planetary Science Letters*, **128**, 135–153.

Segmentation of the Mid-Atlantic Ridge south of the Azores, based on acoustic classification of TOBI data

PHILIPPE BLONDEL

BRIDGE Group, Southampton Oceanography Centre, Empress Dock, Southampton SO143ZH, UK

Abstract: The European MARFLUX/ATJ project has investigated volcanic, tectonic and hydrothermal processes along 500 km of the Mid-Atlantic Ridge near the Azores. During cruises CD89-CD90, multibeam bathymetry and high-resolution TOBI sidescan sonar data were acquired and added to the existing Geographical Information System. From the Azores Triple Junction, between 35° and 38°N, the ridge runs 227°, oblique to the 100° spreading direction. It is partitioned into several short ridge segments dominated by tectonism rather than volcanism, and is characterised by abundant hydrothermal activity. Textural analysis quantifies the second-order statistics of TOBI imagery, enabling the detection of details invisible to the human eye. The results of this analysis are used to quantify the relationship between tectonism and volcanism along-axis, and suggest declining neovolcanic activity away from the Triple Junction. They show that the discontinuities are heavily sedimented, whereas the segments are mostly volcanic and the segment ends are more densely faulted. Textural analysis quantifies the differences in style and activity of the different spreading segments, and the evolution of the discontinuities.

The Mid-Atlantic Ridge between 35°N and 38°N is a key section for the understanding of the smaller-scale processes of plate tectonics, not least because of its proximity to the Azores Triple Junction (Searle 1980; Detrick *et al.* 1995) (Fig. 1). Until recently, this portion of the MAR had been subject to only a few detailed studies (Searle 1980; Luis *et al.* 1994; Detrick *et al.* 1995), concentrated on segments FAMOUS (e.g. Laughton & Rusby 1975; Whitmarsh & Laughton 1976; Ballard & van Andel 1977; Macdonald & Luyendyk 1977; ARCYANA 1978; Gould & Karson 1985; Detrick *et al.* 1995) and Lucky Strike (e.g. Langmuir *et al.* 1993; Fouquet *et al.* 1994). In summer 1994, two cruises sponsored jointly by the European Community and the UK BRIDGE initiative mapped extensively the ridge axis with multi-beam bathymetry, TOBI sidescan sonar, and water column sensors (Parson *et al.* 1994; German *et al.* 1994). The volume of data generated by these sensors is huge (several Gigabytes for the acoustic imagery alone). This problem is not specific to TOBI data, and exists for all sensors of the new generation (DSL–120, EM–12, etc.) (Kleinrock 1992; Blondel & Parson 1994). To ensure efficient utilization of all data, new techniques must be devised for extracting various information and sup-plementing the interpreter with reliable quantitative material. The presentation of data sets as mere images is not enough, and these new techniques must support the interpretation with quantitative arguments going as far as possible beyond the human capabilities.

Textural analysis techniques describe the spatial organisation of grey levels within a neighbourhood, or texture. Textures can be intuitively described as smooth or rough, small-scale or large-scale, and are best quantified through stochastic methods. Grey-level co-occurrence matrices (GLCMs) have proved the most useful for remote sensing applications, and specifically for the analysis of sonar imagery (Reed & Hussong 1989; Blondel *et al.* 1993*a, b*). The present study presents its first application to TOBI imagery of the Mid-Atlantic Ridge. Comparison with earlier results from DSL–120 studies (Blondel *et al.* 1993*a*) will allow us to check and refine the methodology, widening the possible range of application to any kind of sonar image.

Results from GLCM textural analysis will be presented in detail over a small section of the ridge with sedimentary, tectonic and volcanic terrain, which are reflected in different acoustic backscatter and textural characteristics. Textural analysis is used to investigate the morpho-

From: MacLeod, C. J., Tyler, P. A. & Walker, C. L. (eds) 1996, *Tectonic, Magmatic, Hydrothermal and Biological Segmentation of Mid-Ocean Ridges*, Geological Society Special Publication No. 118, pp. 17–28.

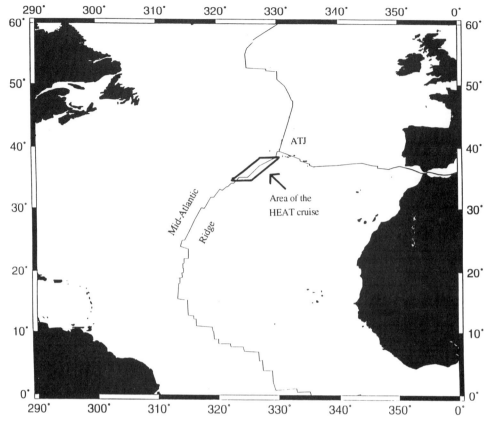

Fig. 1. Extended map of the area, showing the plate boundaries, and in particular the Azores Triple Junction (ATJ), the Mid-Atlantic Ridge, and the area of the HEAT cruise.

logical properties of the different elements of the ridge, and quantify the relationship between tectonism and volcanism along-axis. These results are compared with the conventional interpretation performed by German *et al.* (1996).

Textural analysis

The geological units comprising a remotely-sensed image can be mapped on the basis of their tonal or textural properties. Tonal information is directly related to the amount of acoustic energy backscattered and is expressed as grey levels, usually coded between 0 (black, no backscatter) and 255 (white, highest backscatter). Different statistical indices (extrema, contrast, median values) can be used to quantify local information. However, a natural image cannot be described on the basis of its grey levels alone (e.g. Haralick & Shapiro 1985; Shokr 1991). Local textures can be qualified as rough or smooth, local or regional, repetitive or random. Quantitative textural measurements

can be extracted from the image with various techniques, the most efficient ones being stochastic (e.g. Haralick *et al.* 1973). Experiments on human vision (Julesz 1973) demonstrated the eye could not distinguish between textures with different second-order statistics, proving the advantage of these methods. Grey-level co-occurrence matrices (GLCMs) are one of them and have been used successfully in various domains of remote sensing: sea ice monitoring (e.g. Shokr 1991), meteorology (e.g. Welch *et al.* 1988), agriculture (e.g. Rignot & Chellappa 1992). Reed & Hussong (1989) applied this method to SeaMARC-II sonar images, but their results were limited by the lack of ground-truth and the low resolution of the sensor. Results from the textural analysis of high-resolution sonar data with extensive ground-truthing were first performed by Blondel *et al.* (1993*a*) on DSL–120 imagery from the Juan de Fuca Ridge.

Grey-level co-occurrence matrices address the average spatial relationships between pixels of a small region (Haralick *et al.* 1973). The textural information is described by a set of matrices $\{P_D(i,j)\}$. Each element $P_D(i,j)$ expresses the relative frequency of occurrence of two points, with respective grey-levels i and j, at a distance $D(d,\theta)$ from one another. If the image is quantified with NG grey levels, the GLCMs will be NG × NG arrays. They are computed on finite windows, of dimension WDSZ × WDSZ. In homogeneous regions, differences between grey levels will be low, and elements close to the diagonal of the GLCMs will therefore have higher values. Less homogeneous regions will yield higher differences between neigbouring grey levels, and resulting GLCMs will therefore have higher values further away from the diagonal. More detailed discussions of the variations of co-occurrence matrices with the images are presented in Haralick et al. (1973) and Shokr (1991). One important point to note is the sensitivity of GLCMs to linear combinations of the grey levels. If an offset is added to all grey levels inside the computation window, entries in the GLCM will be displaced along the diagonal. However, if all grey levels inside the computation window are multiplied by a constant value, entries in the GLCM will move away from the diagonal. Co-occurrence matrices are not easy to interpret directly, and they are more effectively described by statistical measures called indices.

More than 25 textural indices are available from the current literature (Haralick et al. 1973; Haralick 1979; Pace & Dyer 1979; Reed & Hussong 1989; Shokr 1991). Usefulness of these indices to the specific case of sonar images in mid-oceanic ridge environments has been assessed (Blondel et al. 1993a) and only eight have been retained: inertia (INR), uniformity (UNF), correlation (CORR), entropy (ENT), local homogeneity (HOMOG), mutual information (INFO and INFOC), maximum probability (MAXPROB).

$$INR = \sum_{i=1}^{NG} \sum_{j=1}^{NG} \frac{(i-j)^2}{K^2} \times P_D(i,j) \qquad (1)$$

$$UNF = \sum_{i=1}^{NG} \sum_{j=1}^{NG} P_D^2(i,j) \qquad (2)$$

$$CORR = \frac{1}{\sigma_x \sigma_y} \left(\sum_{i=1}^{NG} \sum_{j=1}^{NG} \frac{(ij)}{L} \times P_D(i,j) - \mu_x \mu_y \right) \qquad (3)$$

$$ENT = -\sum_{i=1}^{NG} \sum_{j=1}^{NG} P_D(i,j) \times \log_{10} \cdot P_D(i,j) \qquad (4)$$

$$HOMOG = \sum_{i=1}^{NG} \sum_{j=1}^{NG} \left(\frac{P_D(i,j)}{1 + \left(\frac{(i-j)}{K}\right)^2} \right) \qquad (5)$$

$$INFO = \frac{(ENT - H_x - H_y)}{\max(H_x, H_y)} \qquad (6)$$

$$INFOC = \sqrt{1 - e^{-2(H_x + H_y - ENT)}} \qquad (7)$$

$$MAXPROB = \max\{P_D(i,j)\} \\ i = 1, NG; j = 1, NG. \qquad (8)$$

Factors K and L have been introduced to ensure invariance during linear grey level transformations (Shokr 1991; Blondel et al. 1993a):

$$K = \frac{\sum_{i=1}^{NG} \sum_{j=1}^{NG} |i - j| \times P_D(i,j)}{\sum_{i=1}^{NG} \sum_{j=1}^{NG} P_D(i,j)} \qquad (9)$$

$$L = \frac{\sum_{i=1}^{NG} \sum_{j=1}^{NG} ij \times P_D(i,j)}{\sum_{i=1}^{NG} \sum_{j=1}^{NG} P_D(i,j)} \qquad (10)$$

Some of these indices (Equations (3), (6), (7), (8)) use the marginal distributions P_x and P_y defined as (Haralick et al. 1973; Welch et al. 1988):

$$P_x(i) = \sum_{j=1}^{NG} P_D(i,j) \qquad (11)$$

$$P_y(j) = \sum_{i=1}^{NG} P_D(i,j) \qquad (12)$$

μ_x, μ_y, σ_x, and σ_y are the mean values and standard deviations, H_x and H_y the marginal entropies associated to each distribution P_x and P_y.

Inertia (INR, also called second-difference moment by Welch et al. 1988) is indicative of the contrast of the GLCM. Because of the $(i-j)^2$ term, inertia is very sensitive to the large differences inside the co-occurrence matrix. Highly contrasted regions will have a high inertia, whereas more homogeneous regions will have a low inertia. This textural index is therefore indicative of lithologic boundaries, fissures and faults, but will not yield information about the lava flows and their morphologies.

Uniformity (UNF) is sometimes referred to as the second-angular moment (Welch et al. 1988) or energy of the GLCM (Haralick 1979). The lowest value of uniformity is attained when all the $P_D(i,j)$ are equal, and there are no dominant grey levels. All grey levels, or most, are equally probable. This is characteristic of a rough texture (pillow lavas, talus, or speckle). Conversely, the highest values of uniformity show the presence in the co-occurrence matrix of high values. Only a few grey levels are dominant. The region inside the computation window is more homogeneous, or exhibits some regular structures. This would be the case of regular structures such as faults and fissures.

Correlation (CORR) quantifies the dependence of grey levels from one another for pixels separated by the distance **D**. A low correlation means that the grey levels are generally independent from one another, i.e. there is no regular structure in the image. However, if correlation is high, there is a high probability that one or several patterns repeat themselves inside the computation window. These could be talus, structures draped in sediments, or half-buried objects.

The entropy (ENT) measures the lack of spatial organisation inside the computation window. Entropy is high when all $P_D(i,j)$ are equal, which corresponds to a rough texture, and low when the texture is more homogeneous or smoother. According to their morphology (pillow, lobate, etc.), lava flows will exhibit different entropies, varying from intermediate to strong. All geological features that might be encountered in sonar images should be characterised by specific entropy levels, varying with their degree of roughness.

Local homogeneity (HOMOG) is also referred to as inverse-difference moment (Welch et al. 1988), and quantifies the amount of local similarities inside the computation window. Because it is inversely proportional to $(i-j)^2$, local homogeneity will be larger for GLCMs with elements concentrated near the diagonal. These GLCMs correspond to textures of organized and poorly contrasted features, with only a few grey levels at the same distance **D** from one another. Lower values of homogeneity will correspond to lower values of the $P_D(i,j)$, further away from the diagonal of the matrix, that is many similar grey levels at the same distance **D**. The differences in lava flow morphologies should be noticeable with local homogeneity. This parameter should also distinguish homogeneous regions from tectonized areas.

The next two parameters (INFO and INFOC) measure the degree of correlation inside the local texture. They quantify the difference between the entropy of the joint probability distribution $P_D(i,j)$, and the entropies of the marginal distributions P_x and P_y. INFO and INFOC are normalized measurements of mutual information. They are higher if the entropy of the joint distribution is closer to the entropies of the marginal distribution, i.e. if the textural information can be resolved along two orthogonal directions. Textures associated with interference fringes (for example in fine sediment covers: e.g. Huggett et al. 1992) or perpendicular cross-cutting tectonics would fit this description. Other geological features, which cannot be resolved in sub-perpendicular elements, would present lower values of correlation.

The maximum probability (MAXPROB) plays a role similar to uniformity, but is less sensitive to organised structures. The high values of MAXPROB are usually associated with homogeneous regions, and the lower values with heterogeneous regions. Because of its lack of sophistication, this parameter is rarely used in detailed studies. It may be used for rapid separation between smooth textures (lava lakes, shadows, etc.) and rough ones (pillow lavas, talus, etc.).

Classification of TOBI imagery

In the course of cruises CD89-CD90, acoustic backscatter data were collected along 350 km of the Mid-Atlantic Ridge, covering more than 3000 km² between 35°N and 38°N (Fig. 2). These sidescan sonar data were acquired by TOBI, the deep-towed multi-sensor sonar system developed by IOSDL (Flewellen et al. 1993). The system comprises a two-sided 30 kHz sidescan sonar, with a total swath of 6 km and a seabed footprint of 6 × 6 m after processing (Le Bas et al. 1995). During most of the survey, it was flown at altitudes of 200–350 m above the seafloor. TOBI data is processed with software adapted from the MIPS/WHIPS package (Le Bas et al. 1995). TOBI data are radiometrically and geometrically corrected, producing an image with square pixels. The different TOBI images are then geo-referenced, mosaicked and integrated into a Geographical Information System with other co-located datasets (Critchley et al. 1994; Blondel et al. in press).

The initial step in the classification of sidescan sonar imagery requires the definition of the optimal values of computation parameters and the useful textural indices. For this purpose, representative 'training regions' have been selected as follows: sedimented area (Fig. 3a), tectonized area (Fig. 3b), volcanic area (Fig. 3c). The distance $D(d,\theta)$ between pixels is very sensitive to the orientation θ. The insonification angle varies across-track, but also along-track (roll, pitch, yaw). Two identical structures may therefore be imaged with distinct look-angles, and their textural signatures seen as different.

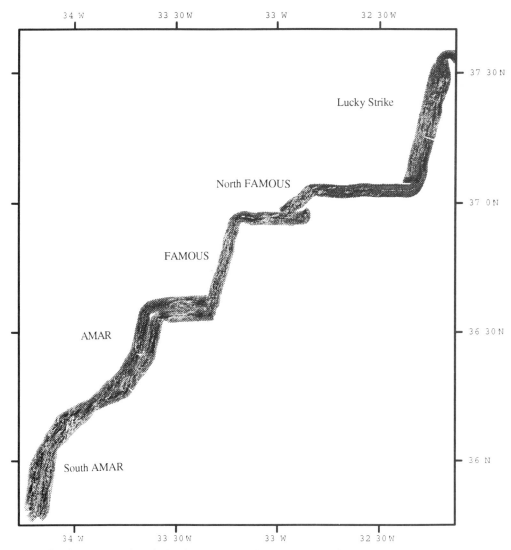

Fig. 2. TOBI imagery acquired during the MARFLUX/ATJ programme (courtesy of C. German and L. Parson, IOSDL). The imagery covers the central portion of the Mid-Atlantic Ridge, with a ground resolution of 6 × 6 m after processing. Swath width is 6 km (some areas have been imaged with two separate swaths).

To ensure that the textural indices of any non-isotropic texture are not significantly influenced by the angle at which it is insonified, I average the co-occurrence matrices for angles 0°, 45°, 90°, and 135°, following Reed & Hussong (1989), Shokr (1991), and Blondel *et al.* (1993*a*). GLCMs are computed in overlapping windows, so that the transition between textures can be fully observed. The optimal separation between the three classes occurs for a window size (WDSZ) of 20 × 20 pixels and an interpixel displacement (**D**) of 10 pixels inside the window. The number of grey levels NG does not influence

significantly the separation of the regions, but has a direct effect on the computation time. An optimal value of NG = 32 grey levels has been selected. These optimal parameters are the exact same ones as for the studies of DSL–120 images (Blondel *et al.* 1993*a*). They are related to the size in pixels of the geological features observed (i.e. ground resolution) but also depend on the imaging frequency. The relatively small size of WDSZ ensures the local angles of incidence stay close. Computation of the eight textural indices has been performed for this combination of parameters on the whole TOBI

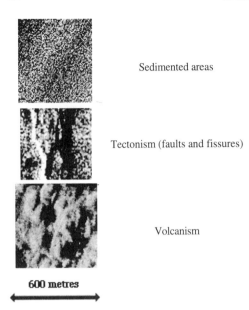

Sedimented areas

Tectonism (faults and fissures)

Volcanism

600 metres

Fig. 3. Examples of the training zones used to optimise the textural analysis parameters: (**a**) sediments ; (**b**) tectonized area; (**c**) volcanism.

dataset. It was found that several indices were highly correlated, and that entropy and local homogeneity were the least correlated, as found in earlier studies of DSL–120 images (Blondel *et al*. 1993).

The process of classification relates textural indices to the corresponding geological interpretations, and organises in 'classes' the multidimensional space of the indices. There is no theory of image classification *per se*, as emphasized in the exhaustive review written by Haralick & Shapiro (1985). Classification techniques are basically *ad hoc* and judged on their performance. Measurement-space guided (MSG) clustering is a technique particularly adapted to natural images (Gonzalez & Wintz 1977; Haralick & Shapiro 1985), especially in complex and noisy environments such as sonar images of mid-oceanic ridges. Tests have shown the MSG method is the most suited to this type of study, in particular because of its tolerance to localised heterogeneities inside homogeneously-textured regions (Blondel *et al*. 1993). The MSG technique partitions the measurement space and each pixel in the image can be assigned the label of the cell in the partition ('labeling' process). The regions in the image are then defined as the connected components of the pixels having the same label.

Entropies and local homogeneities computed on TOBI imagery were resampled into 256 intervals, along a logarithmic scale in the case of homogeneity. The corresponding 2D histogram (Fig. 4a) has been partitioned by locating the main geomorphological units in the feature space. For the sake of simplification, only the generic types of 'sediments', 'tectonics', and 'volcanism' have been represented (Fig. 4b). Entropies and local homogeneities computed for the training regions produce separate clusters of points in the measurement space diagram. They are used as 'seed points', around which region growing is performed with a conventional neighbourhood clustering algorithm. Entropies and local homogeneities of the training region labeled 'sediments' correspond to the upper left portion of the measurement space. Other regions of TOBI imagery interpreted as sedimentary also fall in the same portion of the diagram. They exhibit lower entropies and higher local homogeneities. This is due to the nature of sediment covers, smooth and with quite homogeneous textures. Sub-regions are visible in the 'sediment' polygon, due to different types of sediments and their relative thicknesses. Similarly, the points corresponding to the tectonised training region are situated in the middle parts of the diagram. The interval of local homogeneities is larger, and entropies are higher. The differences in homogeneities are related to the spatial organisation of local textures, tectonized areas with preferential directions having higher homogeneities than heavily tectonized zones where no pattern is discernible. The higher entropies are interpreted to be due to the presence of faults and fissures. Points corresponding to the 'volcanic' training region are located in the rightmost portion of the diagram. The other volcanic regions in the imagery cluster around these points. The polygon thus defined exhibits smaller homogeneities (on a larger range than sediments) and higher entropies than the other polygons. This is related to the nature of volcanic features, rougher and with varying degrees of organisation. Several clusters of points are visible in the 'volcanic' polygon, defining sub-regions that might be related to the different types of volcanism (lava flows, hummocks, etc.), but more systematic interpretation and ground-truthing are needed before concluding. Outside the three defined polygons lies a large region that represents only a minority of points in the backscatter images. It includes shadows, points at the limits of swaths or imaged at grazing angles, and some mixed uninterpreted textures.

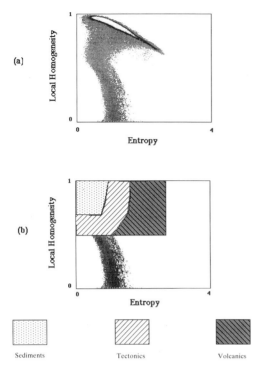

Fig. 4. (**a**) Bi-dimensional histogram of entropy v. local homogeneity for all TOBI data. The white background corresponds to empty intervals. Most of the points (>80%) are located in the upper parts of the diagram. (**b**) first-order interpretation of the regions visible in the histogram. These regions can be subdivided into several sub-regions with specific interpretations (types of sediments, density of tectonism, types of volcanism, etc.) The different polygons are slightly overlapping whenever the small windows in which textures are computed correspond to geological boundaries (for example interface between sediments and faults).

Segmentation of the MAR between 35°N and 38°N

Application to the 37°05′N discontinuity

This area is located between segments 'Lucky Strike' and FAMOUS, and is interpreted as a non-transform discontinuity (Parson *et al.* 1994; Detrick *et al.* 1995) (Fig. 5). The discontinuity is 46 km long, oriented 090°, and imaged with one TOBI swath only. This image illustrates two types of artefacts: horizontal white lines at the edges of the swath, which are explained by reflection from the sea surface, and fuzzy dark lines at the sonar's nadir. TOBI imagery shows significant sedimentation on the eastern part of the discontinuity, and fault scarps (oriented

roughly ENE) in the middle of the discontinuity. Some of them may be covered by a thin layer of sediments. Neovolcanic activity is mainly located at the ridge–transform intersection, increasing with proximity to the North FAMOUS segment (southeast corner of the image).

Entropy and local homogeneity have been computed with the parameters described above. Points in TOBI data whose textural indices fall within the defined polygons are represented in green (sedimented areas), red (volcanic areas) and blue (tectonized areas) (Fig. 6a). Points outside the polygons or ambiguous are left as grey levels. The results from textural analysis show the dominance of sediments, covering 60% of this image. Most of the areas classified as tectonics correspond to the scarps of en-echelon extensional faults. Regions classified as 'volcanic' are concentrated at the northern tip of the FAMOUS segment. Some regions inside the discontinuity are also classified as 'volcanic'. They are mainly located at the bottom of fault scarps, and the corresponding full-resolution backscatter imagery indeed shows highly-reflective lava flows. Comparison between TOBI imagery (Fig. 5) and textural analysis results (Fig. 6a) verify the accuracy of the classification.

The effect of 'mixed-windows' is not to be under-estimated. When a computation window covers two different regions, the texture measurement will be closer to the texture measurements of the region most represented in the window. For example, in a window covering both tectonic and volcanic terrains, the texture measurements will be closer to the 'Tectonic' polygon when tectonism is more important. As the window moves along and incorporates more volcanic terrains, the points will be closer to the 'Volcanic' polygon. The only problem is when proportions of both terrains are approximately equal, and the texture measurements are very distinct from the classes. These ambiguous points will be rejected by the classification algorithm, but, as computation windows are overlapping, the neighbouring texture measurements will be correctly classified, and thus the local textures will be correctly represented.

There are only two types of systematic misclassification, related to imagery artefacts. TOBI data at the sonar's nadir is either absent or erroneous. Some of the corresponding textures are classified as tectonic or volcanic, as can be seen along the sonar track (Fig. 6a). Areas misclassified as tectonic present acoustic fabrics close to the genuine tectonic areas, with elongated features, whereas areas misclassified as volcanic present rougher and brighter fabrics.

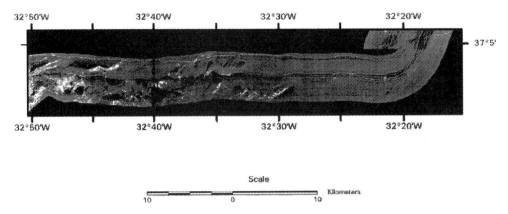

Fig. 5. TOBI sidescan imagery of the 37°05′N discontinuity. Ground resolution is 6 × 6 m after processing.

The classification was expected to be erroneous at the nadir, where the backscattering mechanisms differ from the rest of the image. The second type of misclassification occurs in some of the very dark regions interpreted as shadows. Histogram stretching of the corresponding TOBI imagery reveals that these shadows are not uniformly black, and that some lowly-contrasted patterns are visible. These minute variations from the darkest grey levels may be due to refraction from the sonar beam (Murton pers. comm.) on the obstacle creating the shadow, or by unprocessed noise from the sonar electronics at low levels. The surface reflection artefact is correctly recognised as noise by the classification algorithm.

Application to the whole TOBI dataset and interpretations

Because we are mainly interested in the segment-scale aspects, textures have been computed on the full-resolution TOBI dataset only every 60 m. This should smooth out the finer details, but preserve the integrity of the classification results and the subsequent interpretations. Entropy and local homogeneity have been computed with the same parameters, classified with the same partitions of the measurement space, and represented with the same colours (Fig. 6b).

The relative extents of the areas classified as sedimented, tectonized, and volcanic have been computed along the ridge, 1′ by 1′. The results are represented graphically in Fig. 7. The abscissae correspond to the latitudes, going from North to South, while ordinates correspond to the areal extents, normalised with respect to the local TOBI coverage and expressed in percentages. Entropy and local homogeneity were not

computed on the outer edges of the swaths, and points whose textural signatures are ambiguous (e.g. TOBI track, mixed lithologies) are not classified. Therefore, the normalized areal extents are conservative estimates of the actual extents. Nevertheless, there is an excellent agreement between the classification and the visual interpretation of TOBI imagery (e.g. Parson et al. 1994; German et al. 1996). For example, sediments mainly occur at non-transform discontinuities (Fig. 7a). Volcanic terrain (Fig. 7b) is anti-correlated with sedimentary processes, and the peaks in the diagram correspond to the different segments. Neovolcanic activity declines with distance from the Azores, which is consistent with previous observations (Parson et al. 1994; Detrick et al. 1995; German et al. 1996). The large volcanic construct in the centre of the 'Lucky Strike' segment (cf. Detrick et al. 1995) is clearly visible as an additional peak in the diagram. The differences in activity along different parts of individual segments are brought into evidence, for example in the FAMOUS segment (e.g. Detrick et al. 1994; Fouquet et al. 1994; Parson et al. 1994). Tectonic activity (Fig. 7c) is conversely increasing with distance away from the hot-spot, but is not restricted to the spreading segments. Tectonism also occurs in some of the southern non-transform discontinuities, as described by Parson et al. (1994). The trends in the profile enhance the distinction between the northern segments, more affected by volcanism, and the southern ones, where tectonic processes play a more important role. According to German et al. (1996), this boundary marks the limit of the influence from the Azores hot-spot.

Textural analysis proves a definite plus for the accurate classification of TOBI backscatter imagery and the quantification of the geological

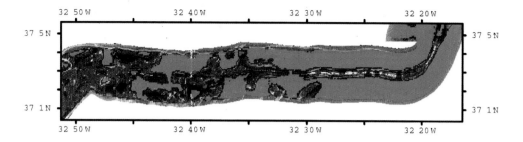

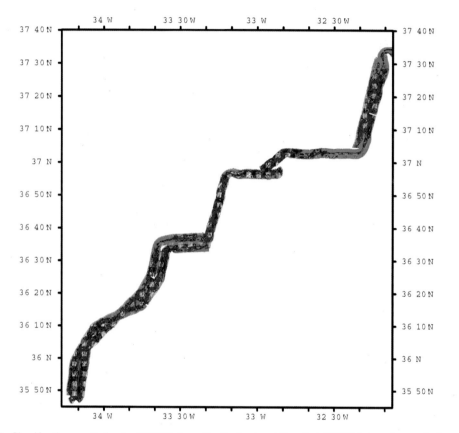

Fig. 6. Classification results for the 37°05'N discontinuity (top) and the whole TOBI dataset (bottom). Areas classified as sedimented are represented in green, tectonic areas in blue, volcanic areas in red.

processes at play. Although textures were computed in this last example on lower-resolution images (60 m pixel size), the results from textural classification compare well with interpretation (Parson *et al.* 1994; Detrick *et al.* 1995; German *et al.* 1996). Furthermore, they allow the quantification of the areas of the different terrain types.

Summary

The purpose of this study was twofold. The first goal was to demonstrate the potential of textural analysis with grey-level co-occurrence matrices for quantitative seafloor characterisation. The second was to apply this method to TOBI sidescan sonar data from the Mid-Atlantic Ridge

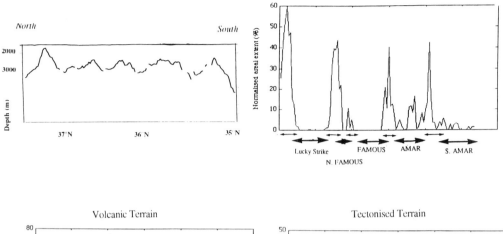

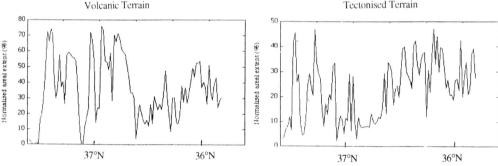

Fig. 7. Bathymetry (from Detrick *et al*. 1995) and normalized areal extents of the three different processes (sedimentary, tectonic, volcanic) along decreasing latitudes. The arrows indicate the extent of each morpho-tectonic unit. Thicker arrows are used for segments, thinner arrows for discontinuities.

(German *et al*. 1994; Parson *et al*. 1994; German *et al*. 1996).

GLCMs prove a useful tool for the classification of acoustic backscatter imagery. Only two indices, entropy and local homogeneity, are necessary to describe accurately the different textures. The computations performed on various sets of TOBI imagery have defined the optimal parameters such as window size and inter-pixel displacement. Supervised classification has been conducted with measurement-space guided clustering to recognise the basic geologic types of 'sediments', 'volcanic', and 'tectonic'. In the majority of cases, classification of the textural indices is accurate. The classified images present readily the different patterns of segmentation, in particular the interplay between tectonism and volcanism in the different segments, and the importance of sedimentation in the discontinuities. Precise location of the different geologies permits along-axis quantification of the areal extents, and, when coupled with bathymetry, the different volumes. Neo-volcanic activity is seen to decline away from the Azores Triple Junction and the Azores hot-spot. The textural maps confirm the conventional interpretation of these regions (e.g. Parson *et al*. 1994; Detrick *et al*. 1995, German *et al*. 1996) and supplement them quantitatively. The classification presented here result from a coarse subdivision between three types of geology (sediments, tectonics, volcanism). Many other types of geology can be distinguished in the TOBI images, and in the {entropy; homogeneity} space. Current studies aim at refining the classification with more detailed classes, validating the accuracy for the recognition of specific terrains (e.g. hummocks v. circular volcanoes, serpentinite v. pelagic sediment), and comparing with the results from conventional interpretations (German *et al*. 1996). Extension to other areas imaged with TOBI and other types of acoustic sensors is planned, and will allow extension of the methodology for applying grey-level co-occurrences matrices to geological interpretation.

Members of the HEAT cruise, as well as officers and crew of RRS Darwin, are thanked for their efforts in the collection of the data. L. Parson and C. R. German were co-PIs on this cruise, and I would like to thank them for making their data accessible to present the potential of textural analysis. TOBI imagery was processed by R. Cave. Constructive comments from N. C. Mitchell and two anonymous reviewers helped improve the text. Funding from the MAST-II EEC Project no. CT.93–0070 is gratefully acknowledged. Ph.B. is sponsored through a BRIDGE Fellowship.

References

ARCYANA 1978. *FAMOUS: photographic atlas of the Mid-Atlantic Ridge: rift and transform fault at 3000 meters depth*. Gauthiers-Villars & CNEXO, Paris.

BALLARD, R. D. & VAN ANDEL, T. H. 1977. Morphology and tectonics of the inner rift valley at latitude 36°50'N on the Mid-Atlantic Ridge. Geological Society of America Bulletin, **88**, 507–530.

BLONDEL, PH. & PARSON, L. M. 1994. *Sonar Processing in the UK*. BRIDGE Position Paper 1.

——, SEMPÉRÉ, J.-C. & ROBIGOU, V. 1993a. Textural Analysis and Structure-Tracking for geological mapping: Applications to sonar data from Endeavour Segment, Juan de Fuca Ridge. *Proceedings of OCEANS'93, IEEE-OES*, Victoria, B.C., 209–213.

——, ROBIGOU, V., SEMPÉRÉ, J.-C. & DELANEY, J. R. 1993b. Textural Analysis and Geological Mapping of High-Resolution Sonar Images: Applications to Endeavour Segment, Juan de Fuca Ridge. *EOS Transactions of the AGU*. **74**, 43, 573.

——, PARSON, L., CRITCHLEY, M., COLLER, D. & the MARFLUX/ATJ Scientific Party – Marine Geoscience Data in a GIS: Geophysical Observations of the Mid-Atlantic Ridge near the Azores. *Proceedings of the Environmental GIS and Remote Sensing*. Conference Remote Sensing Society, in press.

CRITCHLEY, M. F., COLLER, D. W., GERMAN, C. R., BLONDEL, PH., FLEWELLEN, C., PARSON, L., ROUSE, I., TEARE, D., BOUGAULT, H., NEEDHAM, D. & MIRANDA, M. 1994. Integration of Deep Tow Sidescan Sonar Imagery, Bathymetry and Other Data along the Mid-Atlantic Ridge. *EOS Transactions of the AGU*. **75**, 44, 579.

DETRICK, R. S., NEEDHAM, H. D. & RENARD, V. 1995. Gravity anomalies and crustal thickness variations along the Mid-Atlantic Ridge between 33°N and 40°N. *Journal of Geophysical Research*, **100**, B3, 3767–3787.

FLEWELLEN, C. G., MILLARD, N. W. & ROUSE, I. P. 1993. TOBI, a vehicle for deep ocean survey. *Electronic & Communications Engineering Journal*. **5**, 85–93.

FOUQUET, Y., CHARLOU, J.-L., COSTA, I., DONVAL, J.-P., RADFORD-KNOERY, J., PELLÉ, H., ONDRÉAS, H., LOURENÇO, N., SÉGONZAC, M & TIVEY, M. K. 1994. A detailed study of the Lucky Strike hydrothermal site and discovery of a new hydro-thermal site: Menez Gwen. *InterRidge News*, **3**, 2, 14–17.

GERMAN, C.R., PARSON, L. M. & THE HEAT SCIENTIFIC TEAM 1996. Hydrothermal exploration near the Azores Triple Junction: tectonic control of venting at slow-spreading ridges. *Earth and Planetary Science Letters*, **138**, 93–104.

——, PARSON, L. M. & THE HEAT SCIENCE PARTY. 1994. Hydrothermal exploration at the Azores Triple Junction. *EOS Transactions of the AGU*. **75**, 44, 308.

GONZALEZ, R. C. & WINTZ, P. 1977. *Digital image processing*. Addison-Wesley.

GOULD, M. R. & KARSON, J. A. 1985. Tectonics of short-offset, slow-slipping transform zones in the FAMOUS area, Mid-Atlantic Ridge. *Marine Geophysical Researches*, **7**, 4, 489–514.

HARALICK, R. M. 1979. Statistical and structural approaches to texture. *Proceedings of the IEEE*, **67**, 786–804.

—— & SHAPIRO, L. G. 1985. Image Segmentation Techniques. *Computer Vision, Graphics and Image Processing*, **29**, 100–132.

——, SHANMUGAM, K. & DINSTEIN, R. 1973. Textural features for image classification. *IEEE Transactions Systems, Man, and Cybernetics*, **SMC–3**, 610–621.

HUGGETT, Q. J., COOPER, A. K., SOMERS, M. L. & STUBBS, A. R. 1992. Interferences fringes on GLORIA side-scan sonar images from the Bering sea and their implications. *Marine Geophysical Researches*, **14**, 47–63.

JULESZ, B. 1973. Inability of humans to discriminate between visual textures that disagree in second-order statistics – Revisited. *Perception*, **2**, 391–405.

KLEINROCK, M. C. 1992. Capabilities of some systems used to survey the deep-sea floor. *In*: GEUER, R. A. (ed.) *CRC Handbook of Geophysical Exploration at Sea, Hard Minerals*, CRC Press, Boca Raton, 36–90.

LANGMUIR, C. H. & THE LUCKY STRIKE TEAM. 1993. Geological setting and characteristics of the Lucky Strike vent field at 37°17'N on the Mid-Atlantic Ridge. *EOS Transactions of the AGU Fall Meeting*, **74**, 43, 99.

LAUGHTON, A. S. & RUSBY, J. S. M. 1975. Long-range sonar and photographic studies of the median valley in the FAMOUS area of the Mid-Atlantic Ridge near 37°N. *Deep-Sea Research*, **22**, 279–298.

LE BAS, T. P., MASON, D. C. & MILLARD, N. C. 1995. TOBI Image Processing – The State of the Art. *IEEE Journal of Oceanic Engineering*, **20**, 85–93.

LUIS, J. F., MIRANDA, J. M., GALDEANO, A., PATRIAT, P., ROSSIGNOL, J. C. & VICTOR, L. A. M. 1994. The Azores Triple Junction evolution since 10 Ma from an aeromagnetic survey of the Mid-Atlantic Ridge. *Earth and Planetary Science Letters*, **125**, 439–459.

MACDONALD, K. C. & LUYENDYK, B. P. 1977. Deep-tow studies of the structure of the Mid-Atlantic Ridge crest near latitude 37°N. *Geological Society of America Bulletin*, **88**, 621–636.

PACE, N. G. & DYER, C. M. 1979. Machine classification of sedimentary sea bottoms. *IEEE Transactions Geosciences and Electronics*, **GE–17**, 52–56.

PARSON, L. M., COLLER, D. W., GERMAN, C. R., NEEDHAM, H. D. & THE HEAT SCIENTIFIC PARTY 1994. Interrelationship between volcanism and tectonism at the MAR, 38–36°N. *EOS Transactions of the AGU*, **75**, 44, 658.

REED, T. B. & HUSSONG, D. 1989. Digital Image Processing Techniques for Enhancement and Classification of SeaMARC II Side Scan Sonar Imagery. *Journal of Geophysical Research*, **94**, B6, 7469–7490.

RIGNOT, E. & CHELLAPPA, R. 1991. Segmentation of synthetic-aperture-radar complex data. *Journal of the Optical Society of America*, **8**, 000–000.

SEARLE, R. 1980. Tectonic pattern of the Azores spreading centre and Triple Junction. *Earth and Planetary Sciences Letters*, **51**, 415–434.

SHOKR, M. 1991. Evaluation of second-order texture parameters for sea ice classification from radar images. *Journal of Geophysical Research*, **96**, C6, 10 625–10 640.

WELCH, R. M., SENGUPTA, S. K. & CHEN, D.W. 1988. Cloud field classification based upon spatial resolution textural features. *Journal of Geophysical Research*, **93**, D10, 12 663–12 681.

WHITMARSH, R. B. & LAUGHTON, A. S. 1976. A long-range sonar study of the Mid-Atlantic Ridge near 37°N (FAMOUS area) and its tectonic implications. *Deep-Sea Research*, **23**, 1005–1023.

Initiation and evolution of boundary-wall faults along the Mid-Atlantic Ridge, 25–29°N

EDDIE McALLISTER & JOHNSON R. CANN

Department of Earth Sciences, University of Leeds, Leeds LS2 9JT, UK

Abstract: The major faults that bound the median valley on slow spreading ridges form a staircase that migrates upwards and outwards from the valley floor like a slow escalator. Each boundary wall fault is generated at the margin of the median valley floor at about 2 km from the spreading axis. Boundary wall faults that appear to be single scarps on the relatively low resolution Seabeam multibeam echosounder maps are shown by high resolution TOBI sidescan sonar images in some cases to be single faults, but in others to be complex anastomosing fault zones. The major faults in such zones are often scalloped in plan view. There is no systematic relation between fault zone geometry and segment type.

We conclude that during the growth of boundary wall faults the strain is first accommodated over a wide area, in which an array of small faults dissects the volcanic surface. However, with time, and with continued extension, the strain becomes organized into increasingly narrower zones of deformation until a major boundary wall fault develops. Faults grow by propagation and linkage with other nearby faults. The early stages of linkage include tilted ramps, but at later stages brittle faults cut across the ramps, and eventually become part of a scalloped fault trace. In some places, linkage is through zones of small-scale diffuse faulting, rather than through a single fault. Fault capture and linkage can occur over lateral distances (perpendicular to the mean azimuth of the faults) of up to 1.5 km.

This deformation takes place within a fault growth window fixed relative to the spreading axis, through which the lithosphere passes as seafloor spreading continues. Most of the growth of each boundary fault takes place before the next fault is initiated. We conclude that the location and size of the narrow fault growth window is controlled by the changing, mechanical properties of the ocean lithosphere as it spreads away from the axis. Its inner boundary represents the position at which a normal fault reaches the sea floor after originating at the base of the lithosphere at the spreading axis. Its outer boundary is interpreted as the point at which the strength of the boundary wall fault, increasing as the lithosphere ages and thickens, becomes greater than the strength of the lithosphere at the spreading axis, when a new fault is generated. The process of fault generation at mid-ocean ridges has relevance to fault generation in other extensional environments.

Slow-spreading ridges, such as the Mid-Atlantic Ridge, are with the exception of the Reykjanes Ridge (Searle & Laughton 1981) characterized by the presence of a major median valley, 20–30 km wide and up to 2 km deep (Heezen *et al.* 1959), within which the spreading axis lies. The floor of the median valley is 8–12 km wide; within it the spreading axis is usually marked by an axial volcanic ridge composed of a myriad of small volcanic constructions (Smith & Cann 1992). The volcanics of the median valley floor are dissected by numerous small impersistent faults but not by any major faults; the first major normal fault scarp, as shown by multibeam bathymetry, defines the margin of the median valley floor and occurs usually 2–6 km from the spreading axis. The walls of the median valley are made of a staircase of similar major normal faults, all throwing down towards the axis (Macdonald & Luyendyk 1977). As the lithosphere spreads apart, and the median valley floor grows wider, major normal faults are periodically nucleated within the valley floor. Each new fault isolates a slice of the median valley floor and adds it to the staircase. We term the major faults that form the staircase *boundary wall faults*. These faults are seen often to be made up of several fault strands when imaged at high resolution. This paper uses new high resolution surveys to investigate the origin and evolution of these boundary wall faults.

At a large scale, all mid-ocean ridges are divided along their length into individual tectono-magmatic spreading centres, tens of kilometres long, which reflect the three-dimensional nature of crustal accretion (Whitehead *et al.* 1984; Schouten *et al.* 1985; Fox *et al.* 1991; Grindlay *et al.* 1991; Sempéré *et al.* 1993). At slow spreading ridges this segmentation is especially marked. Figure 1 shows in outline the 18 segments between the Kane and Atlantis fracture zones in Mid-Atlantic Ridge. The axial

From: MacLeod, C. J., Tyler, P. A. & Walker, C. L. (eds) 1996, *Tectonic, Magmatic, Hydrothermal and Biological Segmentation of Mid-Ocean Ridges*, Geological Society Special Publication No. 118, pp. 29–48.

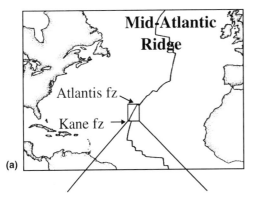

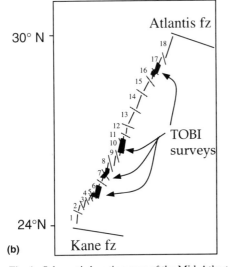

Fig. 1. Schematic location map of the Mid-Atlantic Ridge between the Kane and Atlantis Fracture zones showing the spreading segments and the location of the survey areas over which both multibeam bathymetry and TOBI sidescan sonar data is available.

cause shoaling of isotherms and hence thinning of the elastic lithosphere at segment centres (Kuo & Forsyth 1988; Lin *et al*. 1990; Blackman & Forsyth 1991; Morris & Detrick 1991).

Most previous studies of faulting at the Mid-Atlantic Ridge have used multibeam bathymetric surveys to examine fault patterns on the segment scale. Shaw (1992) and Shaw & Lin (1993) measured fault height and fault spacing from multibeam surveys. They showed that the scarp height of the boundary wall faults and the spacing between faults show a systematic variation with position within spreading segments. Boundary wall faults near to segment ends have in general higher scarps (1000 m) and are spaced further apart (9 km) than those at segment centres, where a mean scarp height and spacing of 300 m and 2.5 km is found. Near segment ends their faults include the large scarps that have been interpreted as low-angle detachment faults (e.g. Mutter & Karson 1988). This interpretation is controversial at present. Wolfe *et al*. (1995) show that earthquakes with steep nodal planes underly one such scarp at 29°N, and Cann *et al*. (1992) have suggested that these scarps are the surfaces of landslides generated from serpentinite diapirs rising along steep faults.

In this paper we investigate the faults near segment centres, omitting the controversial scarps at segment ends. We use high resolution TOBI sidescan sonar imagery as well as lower resolution Seabeam multibeam bathymetry to investigate the detailed anatomy of boundary wall faults in seven spreading segments from the Mid-Atlantic Ridge between 24° to 29°N (Fig. 1). Further constraints come from the relationship between distance from the spreading axis and crustal age which enables us to study the evolution of the boundary wall faults through time.

Data collection and study areas

In this paper we use two complementary sources of data, relatively low resolution multibeam bathymetric surveys and higher resolution side scan sonar images. Both are essential for the interpretation. The multibeam bathymetry comes from a Seabeam survey of the near axis region of the Mid-Atlantic Ridge between the Atlantis and Kane Fracture Zones which reaches out to approximately 30 km from the axis (Purdy *et al*. 1990). This bathymetry has been used to investigate the distribution and size of median valley volcanoes and faults (Shaw 1992; Smith & Cann 1992; Sempéré *et al*. 1993). The spatial resolution of the multibeam data is relatively coarse (the ocean floor footprint is

volcanic ridges and boundary wall faults are offset laterally at segment ends. The longitudinal topographic profile of a slow-spreading segment is in general strongly arched, with a bathymetric high at segment centres and bathymetric lows at segment ends. Similarly, segment centres are marked by a negative mantle Bouguer anomaly (e.g. Lin *et al*. 1990). This indicates that crust is thicker or mantle is lighter, or both, at segment centres than at segment ends. Both effects could result from a higher rate of magma supply to segment centres because of the focussed mantle upwelling (Kuo & Forsyth 1988; Phipps-Morgan & Forsyth 1988). A high rate of magma supply would be expected to

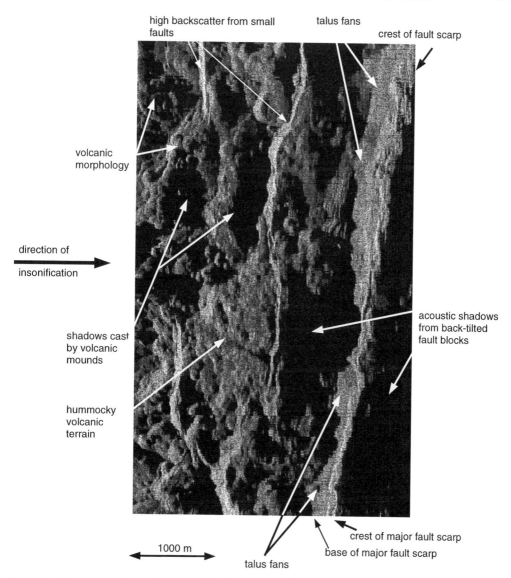

high backscatter from small faults

talus fans

crest of fault scarp

volcanic morphology

direction of insonification

shadows cast by volcanic mounds

hummocky volcanic terrain

acoustic shadows from back-tilted fault blocks

1000 m

crest of major fault scarp

base of major fault scarp

talus fans

Fig. 2. A sidescan sonar image of a group of fault scarps insonified from the left of the image. Continuous white traces define the fault scarps. The mid grey hummocky texture of most of the image is characteristic of constructional volcanics. Black areas are acoustic shadows. Careful examination of the major fault scarp on the right of the image will distinguish talus fans at the base of the scarp and craggy outcrops near the crest of the scarp.

about 150 m across), and the system resolves slopes only of less than 30°. However the survey gives precise measurements of ocean floor depth and this information is essential to estimate fault scarp height.

The deep-towed sidescan sonar data used in this paper were collected in 1992 aboard the RRS Charles Darwin using the Towed Ocean Bottom Instrument (TOBI). This was de- veloped by and is operated by the Institute of Oceanographic Sciences Deacon Laboratory (now Southampton Oceanography Centre; Flewellen *et al.* 1993). During a sidescan sonar survey the sea floor is insonified by a fan of sound on each side of a towed vehicle. Some areas of the seafloor scatter the sound back strongly, including bare rock outcrops, especially if they are tilted towards the vehicle. Sediment covered surfaces and sea floor tilted away from the vehicle scatter the sound back weakly. Higher areas, such as volcanoes or fault scarps, cast shadows behind them. The result is an image that highlights relief and surface material, and is analogous in many ways to an aerial photograph taken when the sun is low. The TOBI sidescan sonar operates at a frequency of 30–32kHz, and collects a 6 km wide image swath, 3 km on each side of the vehicle. During the cruise, TOBI was towed at 400–600 m above the seafloor. The spatial resolution of the sidescan images is a function of range from the vehicle, but on average the image pixel size is about 10 m. A fault scarp with a length of a hundred metres and a scarp height greater than a few metres is thus well resolved. Faults with smaller scarp heights give a clear signal if the intensity of backscatter from the fault is much higher than that of the surroundings.

Identification of a normal fault from sidescan images depends on the direction of throw of the fault relative to the direction of insonification from TOBI. Figure 2 is a TOBI image of a representative section of the median valley floor. Here a number of normal fault scarps dissect the volcanic constructional topography. Volcanic topography is commonly seen as areas of low to moderate backscatter with a hummocky, almost cauliflower like, surface texture (Smith & Cann 1993). Normal fault scarps which throw down towards TOBI can be identified by linear zones of high backscatter. The width of the zone of backscatter is related to the height of the fault scarp, but cannot be used as a precise measure of scarp height unless the slope of the scarp is known. On faults with a displacement greater than about 100 m, talus fans can be distinguished at the base of the fault scarps, crags can be seen along the upper sections of the scarps, and shadows cast from the crests of the scarps may conceal back tilted fault blocks.

A normal fault which throws down away from the TOBI can be identified, though with less certainty, by the linear shadow cast over the fault scarp, the width of which is a function of scarp height and also of the small scale topography on the hanging wall block. Details of fault morphology cannot be observed with this direction of insonification, but it does allow the nature of the surface of the uplifted footwall block to be seen. Most of the faults discussed in this paper were insonified from both directions, so as to draw on both types of evidence. All of the illustrations of faults shown in this paper use images in which faults throw down towards TOBI. Only features positively identified from TOBI as faults are included in this analysis; some steep slopes that might from the bathymetry have

been identified as faults were shown by TOBI images to be steeply tilted volcanic surfaces.

A number of fault attributes were measured by using a combination of the multibeam bathymetry and the TOBI sidescan. Fault length and fault strike are best determined from the sidescan images where the high spatial resolution shows well-defined fault traces. Since TOBI does not measure bathymetry, the fault scarp height, which is taken here to approximate the vertical displacement of the fault, is measured from the multibeam bathymetry with a precision of about 20 m. For a small fault, the height of the fault scarp relative to a nearby large fault can be estimated from the relative widths of the high backscatter zones that defines the fault scarps, though this assumes a constant dip of the fault surface. The TOBI images show that single bathymetric scarps may be made of many individual fault strands. Where we quote a single scarp height, this refers to the total bathymetric scarp height of what may be a complex fault zone. A fuller description of fault identification from TOBI can be found in McAllister et al. (1995).

Seven spreading segments were insonified by sidescan imagery during RRS *Charles Darwin* Cruise 65 in 1992 (Fig. 1). They exhibit a wide variation in median valley width and depth, and in the size of the axial volcanic ridge. A description of the volcanic and tectonic structures from the insonified segments can be found in Smith et al. (1995). In this paper we select TOBI sidescan images from all segments surveyed to discuss the evolution of median valley faults as a class.

Large-scale morphology of boundary-wall faults

In all of the segments in this study the first (youngest) boundary-wall fault, defined from the bathymetry, is represented on the sidescan sonar images by a zone of deformation between 2 and 6 km on either side of the axis. What appears to be a single fault when imaged by Seabeam bathymetry may turn out to be a complex fault zone when imaged by TOBI, consisting of one or more normal faults throwing down towards the axis. Because previous authors, using bathymetry, have discussed these as single faults, we will retain the term 'boundary-wall fault' even when discussing a complex fault zone. The variation in segment morphology in the Kane to Atlantis region described by Smith & Cann (1992) and Sempéré et al. (1993), is matched by an equally diverse range in the morphology of the boundary-wall fault system, though no systematic relationship between segment morphology and a particular style of faulting has been observed.

In the seven spreading segments used in this study the sidescan sonar morphology of the first boundary wall fault, as defined by bathymetry, differs between segments, and on opposite sides of the same segment. To demonstrate this variability, three sidescan and bathymetry images of the boundary wall faults which span the range observed for all segments are shown in Figs 3, 4 & 5. The length of each boundary wall fault zone in Figs 3, 4 and 5 is about 20 km, while the width over which strain is accommodated varies between boundary wall fault zones. The cumulative scarp height of each fault zone, summed over all faults in the fault zone, and measured from the bathmetry, lies in the range 400–600 m of vertical relief.

Figure 3 shows the first boundary wall fault on the western side of segment 17, also known as the Broken Spur segment. The multibeam map shows a narrow concentration of bathymetric contours over a 0.5 km width, with the magnitude of the scarp height decreasing from the centre of the fault to its tip (Fig. 3a). The fault trace imaged on the bathymetry map is not straight, but curvilinear in plan view. The sidescan image of the same fault (Fig. 3b) confirms the bathymetric observations in general, showing a narrow zone of deformation with a single fault trace accommodating all of the strain. The sidescan image confirms the curvilinear trace of the fault; a series of scallops defines the crest of the fault scarp (Fig. 3b & c). The sidescan brings out details of the morphology of the fault scarp and shows the texture of the scarp with a talus fan at its foot and crags at its top. Minor fault splays and smaller faults can be seen on the footwall.

Figure 4 shows the boundary wall fault on the eastern side of segment 17, directly opposite the previous example. The boundary wall fault is defined from the bathymetry as a slope dipping towards the median valley floor, with a difference in topography of about 500 m from the floor of the median valley to the crest of the slope lying 5 km to the east (Fig. 4a). Within the slope a number of closely spaced bathymetric contours suggest the presence of at least one fault. The sidescan sonar images reveal that the slope is not the product of a single fault, but of a fault zone 5 km wide (Fig. 4b). The fault zone contains a complex array of anastomosing fault traces, in which the diffusely distributed strain creates an irregular staircase rising out of the median valley. Most individual faults within the boundary wall fault zone in Fig. 4 have scarp heights below the resolution of the Seabeam bathymetry

(less than 20 m). The maximum measured scarp height of 200 m corresponds to the feature visible in the bathymetric contours of Fig. 4a.

The first two cases have shown extremes of morphology. An intermediate example is the western boundary-wall fault of segment 11 (Fig. 5). The bathymetric map shows a ridge with steep sides on either side, which decreases in height to the south (Fig. 5a). Again it is only with the sidescan sonar imagery that the true complexity of the fault zone is revealed. The outward-dipping surface of the ridge is made of back tilted volcanics, while the inward-dipping surface is a complex fault zone. From north to south the boundary-wall fault zone becomes narrower and the summed scarp height across the fault zone reduces in magnitude (Fig. 5b). To the north three major faults accommodate the strain; further south the middle fault disappears to leave two faults which merge together to form a single fault trace at the southern end of the image. The summed fault scarp height reaches a maximum of about 500 m at the northern end of the image, similar to that found in previous segments. The westernmost fault has well-developed talus fans abutting the scarp face and lying on the hanging wall. There is talus present, though it is less well defined, on the easternmost fault. The two main faults on this image are scalloped in plan, in a similar way to the fault of Fig. 3. The scallops are 2.5 km on the western fault, while on the eastern fault they are about 1 km long.

Small-scale morphology of boundary-wall fault zones

Detailed examination of the sidescan images of the morphology of the boundary-wall faults in all seven segments highlights further the discontinuous and complex nature of the fault zones. Numerous discrete fault segments are observed, similar to those documented in fault zones on land (e.g. Pollard et al. 1982), and these are linked to each other in different ways.

From the TOBI images we have identified five separate geometries of fault linkage which are similar to fault linkages reported for normal faults in continental regions. We discuss fault interaction geometries using the terminology of Walsh & Watterson (1991). Soft-linkage is used where mechanical and geometric continuity is maintained between the principal faults by ductile strain of the rock volume, or where the rock volume is deformed in a brittle manner by faults below the resolution of observation. Hard-linked faults are faults linked by a

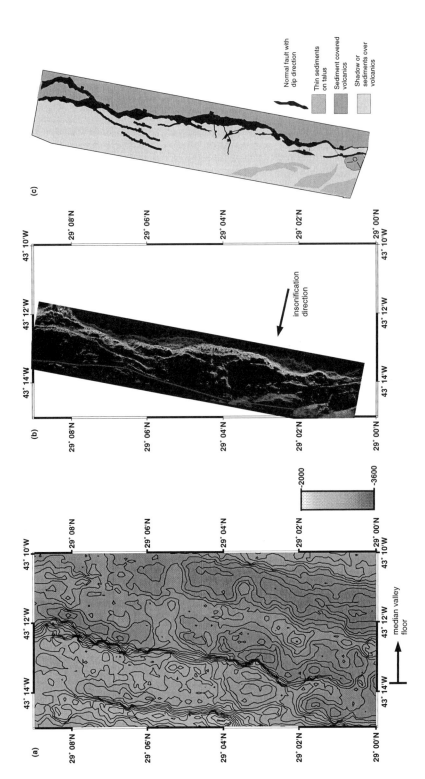

Fig. 3. Three images of the first boundary wall fault to the west of the axis in segment 17. (**a**) The gridded multibeam bathymetric map (contour interval 50 m) defines a fault scarp which throws down towards the median valley floor to the east. The maximum scarp height is 500 m, diminishing from the centre of the fault to the fault tips. (**b**) A TOBI sidescan sonar image of the same fault, insonified from the right of the image. The fault scarp can be seen to be composed of a single major fault trace, which has a scalloped plan. A number of smaller faults can be seen splaying from the major fault trace. (**c**) Geological interpretation of the TOBI image.

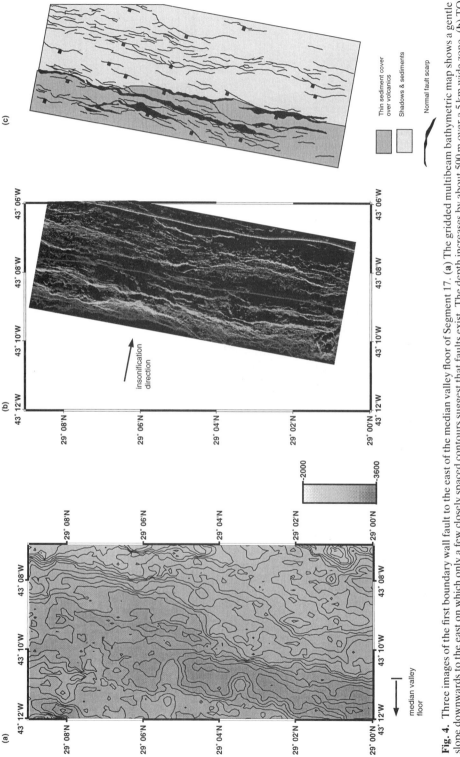

Fig. 4. Three images of the first boundary wall fault to the east of the median valley floor of Segment 17. (**a**) The gridded multibeam bathymetric map shows a gentle slope downwards to the east on which only a few closely spaced contours suggest that faults exist. The depth increases by about 500 m over a 5 km wide zone. (**b**) TOBI sidescan sonar image of the same fault, with insonification from the west. The slope can be seen to be composed of a complex staircase-like zone of anastomosing faults, all linking and overlapping to accommodate the strain over a broad zone. This boundary-wall fault zone has the same total scarp height as the previous fault, but the relief is distributed over a wider area. (**c**) A geological interpretation of same area.

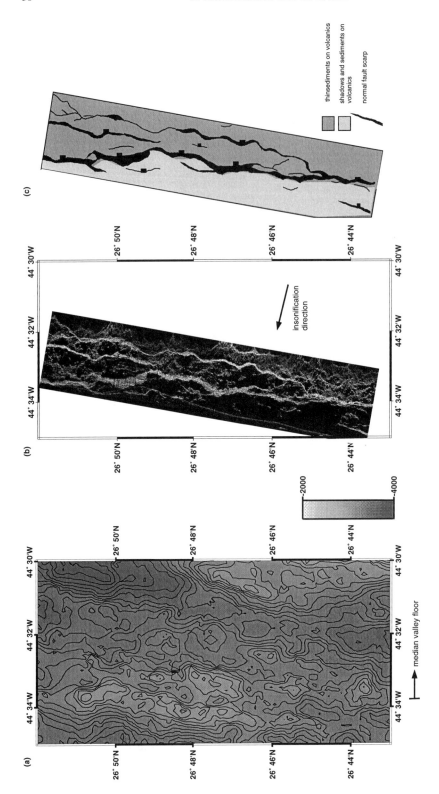

Fig. 5. Three images of the first boundary wall fault to the west of segment 11. (**a**) The gridded multibeam bathymetry map shows a number of closely spaced contours, suggesting the presence of faults throwing down towards the median valley floor to the east. (**b**) A TOBI sidescan sonar image of the same fault, insonified from the east. This fault zone is intermediate between the two previous examples. Three overlapping faults accommodate the strain at the north of the image, and these reduce to a single fault at the southern end. All faults show a scalloped trace in plan. (**c**) Geological interpretation of the same image.

coherent brittle structural feature, which, to be resolved on TOBI images, must have a minimum length of 100 m and a minimum displacement of a few metres. Commonly, linkage occurs between two overlapping, en-echelon principal faults. The region of overlap is termed a relay structure.

Nowhere in the survey area was a relay zone found in which all of deformation is soft-linked, but there is clear evidence that soft-linkage accommodates the early stages of linkage in many examples. In Fig. 6 a & b there is a 4 km overlap between two principal faults, with the backscatter width, and hence the displacement, decreasing for both faults in opposite directions through the region of overlap. Three hard linkage zones between the two overlapping faults can be seen to the left, centre and right of the image. The central linkage appears to represent a breach zone where the two faults have linked together across a narrow 500 m wide gap by a number of obliquely oriented linkage faults. Between the hard linkage zones are flat surfaces tilted from right to left, which we interpret as parts of an early soft-linked ramp, now dissected by the later brittle faults.

Hard-linked relay structures are the most common in the survey area, where a single fault is observed to form a hard link transfer fault between the overlapping faults. Transfer faults range in scale from a maximum length of 1.5 km (see Fig. 6e) down to the resolution limit of the sidescan. Two separate types of transfer fault have been found, a simple single fault trace, and an array of transfer faults. In the first case, a single oblique transfer fault links two non-overlapping, en echelon faults together close to their tips. The transfer faults make a 90° angle with the tips of the two principal faults. Such a case suggests that the transfer fault acts as a barrier to forward propagation (Fig. 6c & d), and the linked faults start to act as a single fault (Walsh & Watterson 1991).

Arrays of transfer faults are typically found when a substantial overlap exists between the en-echelon principal faults (Fig. 6e & f). In Fig. 6e, the scarp heights of the transfer faults decrease systematically from north to south, in the same direction as the scarp height of the eastern fault decreases. The change in scarp height of transfer faults may reflect the time since fault initiation, if all faults are currently active, or may indicate that only the most northerly transfer fault is active and that all strain is focussed into the principal faults and the northern transfer fault. All of the transfer faults are parallel to each other and make an angle of 60–70° with the main faults.

A rare but dramatic example of fault interaction, termed here dendritic tip deformation, occurs where a small fault (scarp height about 10 m) runs en echelon to a major fault (scarp height about 300 m) (Fig. 6g & h). Towards the tip of the small fault, the fault trace abruptly breaks down into a dendritic network of fault splays with an inter-splay angle of 30–50°. The area of intense deformation is 500 m by 500 m. The dendritic fault zone is made up of many similar sized (100 m) lozenges outlined by splays which implies that the fault splays are linking together. This dendritic pattern was only once observed in our survey. A similar fault geometry, though in a strike slip array, has been documented at the centimetre scale (McGrath & Davison 1995, fig. 9c).

The last example of fault interaction we describe is where two co-linear principal faults are linked by a complex zone of deformation (Fig. 6i & j). Here both principal faults have scarp heights of about 200 m. The zone of deformation between the two faults occupies an area of 2×2 km, through which no single major fault trace runs. The deformation zone is elliptical in plan, with fault splays leaving each parent fault at high angles. Within the deformation zone faults tend to curve around until they strike parallel to the two principal faults.

The fault geometries identified on the Mid-Atlantic Ridge are consistent with reported geometries observed in continental rift zones, and from normal and strike slip fault zones, where the following sequence of fault growth is proposed (Rosendahl et al. 1986; Morley et al. 1990; Peacock & Sanderson 1991; Gawthorpe & Hurst 1993; Trudgill & Cartwright 1994): a relay ramp develops when two propagating faults, with similar strikes but en echelon to each other, propagate close enough so that the continued growth of one fault starts to influence the growth of another. Since only a finite amount of strain can be accommodated by the fault system, either one fault must cease to grow or the faults must act together to accommodate the strain. The relay ramp develops in response to a period when both the faults are still growing, and the intervening crust deforms to accommodate the flexure of each fault. Once the yield strength of the intervening area is exceeded the ramp will be breached and a through-going transfer fault will form a bridging fault between the two original faults. Once a transfer fault has developed, the faults will continue to grow and to act as a single fault (Fig. 7). This evolutionary transition from relay ramp to transfer fault has been noted at the millimetre scale (e.g. Peacock & Sanderson 1992) and metre to kilometre scale (e.g. Morley

(a)

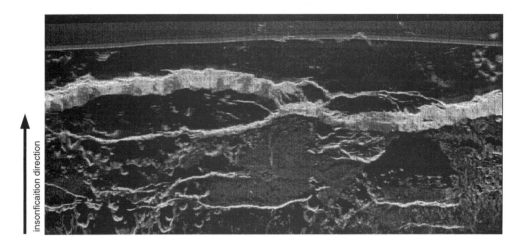

insonficaition direction

(b)

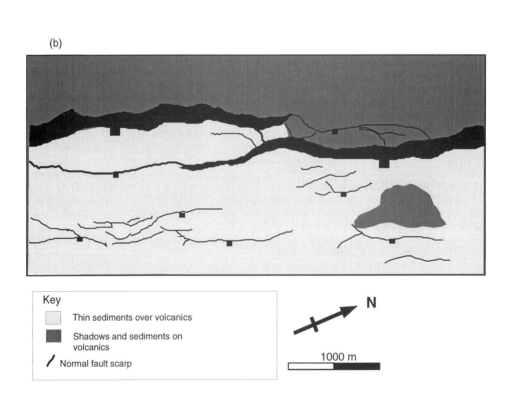

Key

Thin sediments over volcanics

Shadows and sediments on volcanics

Normal fault scarp

N

1000 m

(c)

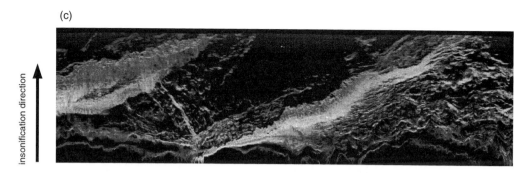

insonification direction

(d)

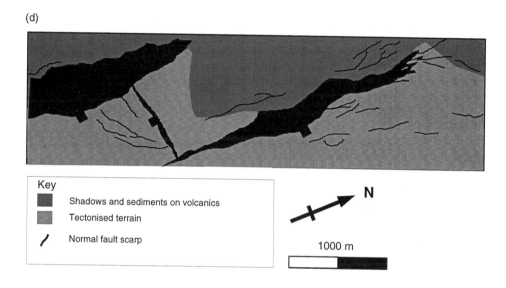

Key

■ Shadows and sediments on volcanics

■ Tectonised terrain

/ Normal fault scarp

N

1000 m

Fig. 6. Close-up views of relay structures and related features on five different fault zones. (**a**) & (**b**) TOBI sidescan image and geological map of a relay ramp insonified from the bottom of the image. Two large faults have overlapped and united. The relay zone was soft linked originally, as shown by the tilted blocks in the overlap zone, but the deformation has now been taken up by a brittle relay zone in the centre of the image. (**c**) & (**d**) TOBI sidescan image and geological map of a simple transfer fault. The fault zone is insonified from the bottom, with both the major en-echelon faults and the smaller transfer fault well imaged. There is only a small overlap between the two major faults. (**e**) & (**f**) TOBI sidescan image and geological map of a complex transfer zone, insonified from the bottom of the image. Here a large overlap exists between two major faults. Several oblique faults dissect the overlap region and at least one fault links the two major faults. (**g**) & (**h**) TOBI sidescan image and geological map of the tip of a small fault close to a large fault. Insonified from the bottom, the top of the image shows the talus ramp of a major fault, the scarp of which lies just outside the field of view. At the bottom, a small fault has propagated in from the right and has broken down into a dendritic network of faults. The size of the fault splays decreases down the network, away from the single small fault, to produce a complex zone of deformation over a 500 by 500 m area. (**i**) & (**j**) TOBI sidescan image and geological map of a broad zone of deformation between two collinear faults. The fault is insonified from the bottom of the image. The two major faults have scarp heights of about 200 m. As they approach one another, the scarps rapidly break down to produce a broad diffuse zone of faulting, which define an elliptical area of deformation linking the major faults.

(e)

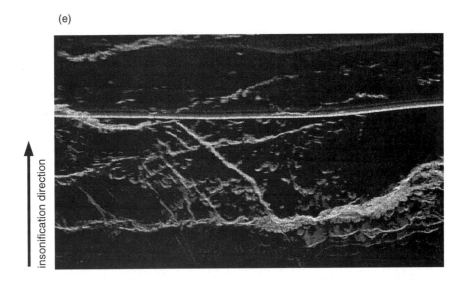

insonification direction

(f)

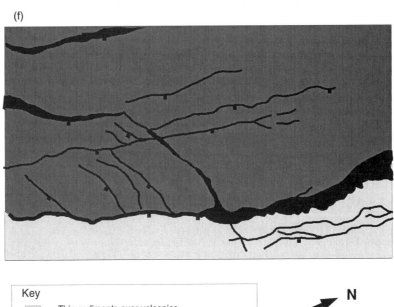

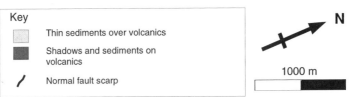

Key

Thin sediments over volcanics

Shadows and sediments on
volcanics

Normal fault scarp

N

1000 m

(g)

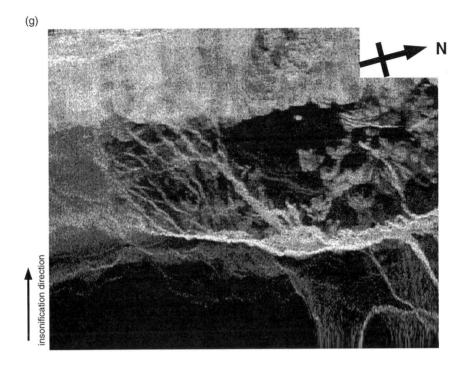

insonification direction

N

(h)

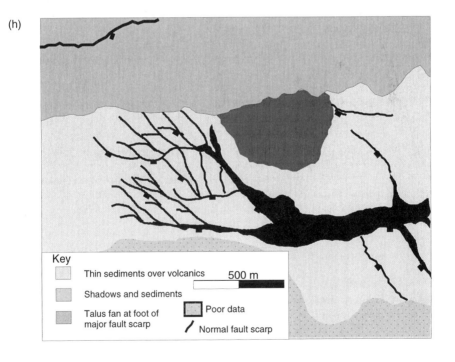

Key

Thin sediments over volcanics

Shadows and sediments

Talus fan at foot of major fault scarp

500 m

Poor data

Normal fault scarp

(i)

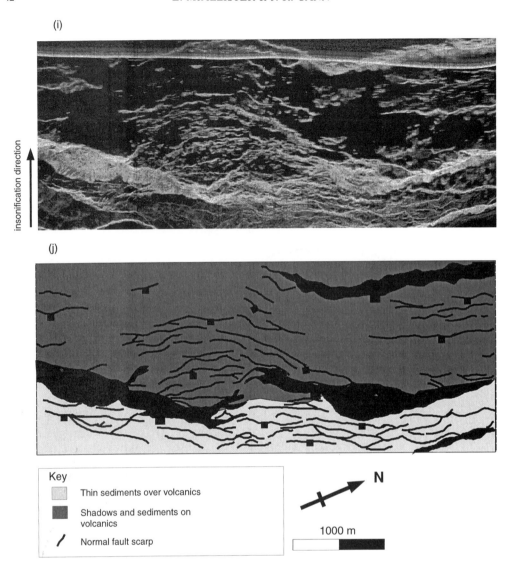

insonification direction

(j)

Key

Thin sediments over volcanics

Shadows and sediments on
volcanics

/ Normal fault scarp

N

1000 m

et al. 1990, Trudgill & Cartwright 1994). Our observations agree with this general picture, but it is clear that at times this simple evolution is complicated by the breaking down of single faults into dendritic fault zones or zones of more distributed deformation.

The fault growth window

The sidescan imagery has shown that the boundary-wall faults show a wide diversity in the style and linkage of faults. While the multibeam bathymetry locates major zones of deformation, the sidescan imagery shows that the boundary-wall faults may form a continuum between a narrow, single fault scarp and a broad zone of anastomosing and linking faults. The internal anatomy of a boundary-wall fault is similarly complex, with overlapping faults, relay structures and complex fault interactions all forming part of the fault architecture. Fully evolved boundary-wall faults are not straight but are defined by an irregular, at times scalloped trace in plan view.

Similar geometries have long been recognized in faults in continental areas. Fault segmentation on a scale of hundreds of metres to kilometres is well known, and jogs in the fault trace are believed to represent seismic boundaries (Schwartz & Sibson 1989). Field-based studies have indicated that the formation of major faults occurs by the linkage and propagation of subordinate faults, with linkage geometries forming over a range of scales (e.g. Pollard & Aydin 1984).

Plans of major faults show that faults are not

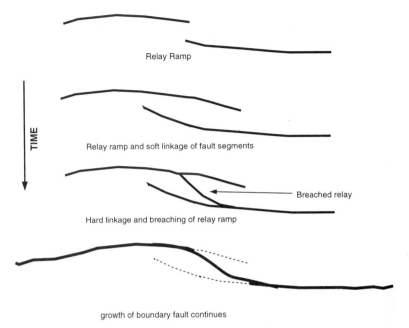

Fig. 7. Schematic illustration of propagation and linkage of faults (after Cartwright *et al.* 1995). The faults propagate towards one another, linking first through a tilted ramp, and later through a brittle breach in the ramp. Eventually the breach becomes part of the main fault trace, and its former position is shown by a scallop in the plan of the fault.

straight but curvilinear with splays frequently branching from the main trace (e.g Morley *et al.* 1990). Detailed mapping of normal faults by Trudgill & Cartwright (1994) has shown that the curvilinear nature of fault traces may reflect evolution of large faults by propagation and linkage of many smaller individual faults. Such an evolution has been demonstrated in plaster of Paris models of fault growth by C. Mansfield & J. Cartwright (pers. comm. 1994–5). In these models curvilinear fault traces result from the linkage of smaller faults. Significant changes in the strike of faults corresponding to relict relay zones.

Cowie *et al.* (1993*a*) use a statistical physics model to show that in a thin plate under constant strain rate, the fault pattern becomes progressively organized over time. An initially random distribution of faults will become organised into discrete, narrow zones of linked faults separated by areas within which little deformation occurs. The faults which develop have complex anastomosing structures. The spacing between the major faults has been attributed to elastic interaction effects (Cowie *et al.* 1995).

At the Mid-Atlantic Ridge the boundary wall faults evolve in a similar way. There are several well established constraints on the boundary wall faults: (a) they evolve in oceanic lithosphere as it spreads from the ridge axis, developing

between 2 and 8 km from the axis; (b) seismic evidence shows that they are probably inactive beyond 20 km from the axis (Bergman & Solomon 1992); (c) they have scarp heights of 400–600 m at the centre of segments; and (d) major faults are spaced about 5 km apart, with smaller faults in the gaps between them (Searle & Laughton 1981; Macdonald 1982). The regularity of spacing between boundary-wall faults and the consistent position of the inner boundary fault suggest to us a strong control on the location of the zone within which boundary wall faults grow. We will refer to this zone as the *fault growth window*. We envisage that almost all of the growth of the boundary-wall fault occurs within this window.

What are the dimensions and location of the fault growth window? We take the length of the window to be the length of a spreading segment, since the median valley floor of all slow-spreading segments are bordered by major boundary-wall faults.

Where is the inner limit of the fault growth window? Following the scenario for fault evolution outlined above, this is where the initial array of small faults nucleates and starts to grow. Bathymetric data shows that the inner boundary wall faults in all of the segments between the Kane and Atlantis transforms lie between 2 and 6 km from the spreading axis (Sempéré *et al.*

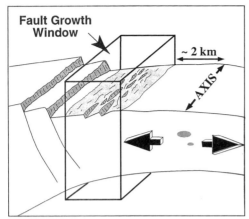

Fig. 8. Conceptual illustration of the fault growth window. The window is fixed in space relative to the spreading axis, and the lithosphere moves through the fault growth window with time. Almost all of the growth of the first boundary-wall fault occurs within the window. Note how the faults propagate downwards through the cooling and thickening lithosphere as they are spread outwards, and thus become stronger, while the lithosphere at the spreading axis maintains a nearly constant strength.

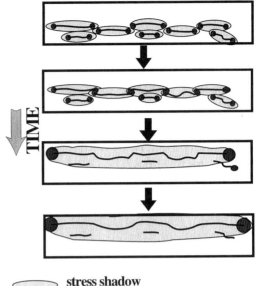

⬭ **stress shadow**

● **capture zone**

Fig. 9. Flowchart for self-organization of boundary-wall fault growth through propagation of faults, fault capture by capture zones at the tips of faults, and extinction of small faults within the stress shadows of larger faults. Modified from Cartwright *et al.* (1995).

1993). TOBI sidescan images confirm this conclusion. No major boundary wall faults have been found closer than 2 km from the axis (this study, Allerton *et al.* 1995). The lack of any major faults within 2 km of the axis may be the result of synvolcanic fault growth, if the effusion of lavas onto the median valley floor keeps pace with, or outstrips, the rate of fault growth. Small faults imaged by TOBI within the axial zone might be the surface expression of major synvolcanic faults. There are problems with such an interpretation, though. Volcanism at the Mid-Atlantic Ridge is considered to be episodic (Smith & Cann 1993). It is most unlikely that episodic volcanism would be so closely correlated with faulting that nowhere has a major fault outpaced the volcanic resurfacing and appeared at the surface within the axial zone. Further evidence comes from the obliquely spreading Reykjanes Ridge. There faults found within a few kilometres of the axis are oriented subparallel to the axial volcanic ridges and orthogonal to the spreading direction, while the boundary-wall faults strike parallel to the overall axis of the ridge and 20° from the orientation of the small axial faults (Searle & Laughton 1981; Murton & Parson 1993; McAllister *et al.* 1995). The near-axis faults identifiable from TOBI and submersible dives in the axial valley of the Mid-Atlantic Ridge (Ballard & van Andel 1978; Murton & Parson 1993; Allerton *et*

al. 1995; McAllister *et al.* 1995) are more likely to be related to the volcanic evolution of the segment, and to be superficial in their length and displacement. They do not seem to contribute towards the development of the boundary-wall fault zone. For these reasons we place the inner limit of the fault growth window at 2 km from the axis.

Where is the outer limit of the fault growth window? Earthquake epicentres within the crestal mountains (Bergman & Solomon 1988) show that some deformation continues for tens of kilometres from the spreading axis, but there is a variety of lines of evidence that most deformation is complete well before then. Using GLORIA sidescan images Searle & Laughton (1981) observed little change in the size of Mid-Atlantic Ridge boundary faults beyond the first fault. The Seabeam bathymetric maps of the area between the Kane and Atlantis fracture zones (Purdy *et al.* 1990) similarly show no systematic change in fault throw once the first boundary-wall fault is passed. Inspection of TOBI images from our survey areas shows little sign of reactivation of the talus slopes on faults more than a few kilometres from the axis. For comparison, Stein *et al.* (1991) have demonstrated from field observations from the Asal

Rift of the Afar region of East Africa that on this incipient slow spreading ridge system the zone of active faulting is less than 2 km wide. This suggests that most of the growth of the first boundary wall fault occurs within a fault growth window that is only a few kilometres wide, and that the growth of each boundary wall fault is essentially complete before the next starts to develop. Deformation clearly occurs further from the spreading axis, but probably contributes only a small fraction of the total strain.

We envisage that the position of the fault growth window remains fixed relative to the spreading axis, and that the oceanic lithosphere passes through the window (Fig. 8). Young oceanic lithosphere entering the window will contain minor faults and diffusely distributed fractures acquired on the valley floor (Ballard & van Andel 1977; Kong et al. 1988); while oceanic lithosphere leaving the fault growth window will have experienced almost all of the strain history active within the window.

Fault linkage

We consider that, within the fault growth window, fault nucleation and growth occurs in a stochastic manner (Davy et al. 1990; Sornette et al. 1993), and that numerous fractures and small faults nucleate near the inner margin of the window. Initially each fault will accumulate displacement and propagate as an isolated structure until it reaches a position where continued growth would result in the fault overlapping or joining another fault. At this stage the faults reorganize until either the faults link to create a larger fault or one of the faults continues to grow at the expense of the other. Linked faults will begin to act as a single fault, accumulating displacement over the whole length of the new fault zone. The linking of pairs of faults has two main consequences. First, as pairs of faults link, the length of the fault will increase rapidly, and, consequently, displacement will increase along the fault while the length remains constant (Cartwright et al. 1995). Second, the increased size of the new fault casts a wider stress shadow (the area over which the fault displacement relieves the elastic stress). A large fault, with a large associated stress shadow, can cause smaller faults within the shadow to switch off.

Fault linkage, fault growth and the changing size of stress shadows are critical factors in the self-organisation of a fault system. Fault linkage appears to be especially important for the geometry of the resulting fault zone. Segall & Pollard (1980, 1983) suggest that the lateral

distance over which a fault can link with another is a function of the size of the fault. The first boundary wall fault to the west of segment 17 has a strongly curvilinear trace (Fig. 3a); the fault trace wanders over a zone 1.5 km wide perpendicular to the mean azimuth of the fault, suggesting that 1.5 km is within the lateral linkage distance over which a relay structure can develop. Each active fault can be considered to have a capture zone at its tip (Fig. 9). As the fault grows, any fault which falls within the capture zone may be exploited as a fault pathway whether it is active or not, allowing a minimum amount of work to achieve a large increase in fault length. The curvilinear fault topology is in part a reflection of the history of fault linkage events, but also reflects the development of new faults and fault splays which develop in response to the changing stress field around the fault. Over time the fault may form a large wavelength curvilinear trace, with numerous branching and splaying smaller faults which act as accommodating structures.

Once a fault is larger than its neighbours it will grow preferentially, since it can accommodate strain over a wide region and represents a large zone of relative weakness. This leads to a feed-back mechanism in which continued fault growth leads to an increase in the area of stress relief, which in turn leads to continued growth of large faults and extinction of small faults. Eventually a state will be reached when a stable configuration of active faults is arrived at and growth will continue to proceed on these faults until the faults move outside the fault growth window.

How fast do boundary-wall faults grow?

The boundary-wall faults in the survey area have scarp heights of 400–600 m and spacings of about 5 km. We have argued above that the growth of each fault is nearly complete before the next fault starts to form. If the boundary faults on both sides of the median valley are active at the same time, and if fault growth is continuous, this suggests a rate of growth of 100 m for each 1 km of spreading, or a rate of $1.2 \, \text{mm a}^{-1}$. If, as is possible, only a single boundary-wall fault is active at any one time, the growth rate would be doubled to $2.4 \, \text{mm a}^{-1}$. Fault growth may not be continuous. The western boundary-wall fault of segment 17 is now 2.5 km from the axis. If it started to grow at 2 km from the axis, its growth rate would have been $12 \, \text{mm a}^{-1}$. Such a calculation depends critically on assumptions, and probably represents an upper limit. For comparison, Stein et al. (1991) calculated a mean growth rate of 1–2 mm a^{-1} from the Asal Rift in the Afar region, where the spreading rate is similar to that of the Mid-Atlantic Ridge. Cowie *et al.* (1993*b*) report fault growth rates of up to 8 mm a^{-1} for the East Pacific Rise, where the spreading rate is several times greater. These calculations are broadly consistent with each other, and suggest that the strain rate accommodated by faulting is about 10% of the total strain rate imposed by spreading, as calculated for the East Pacific Rise by Cowie *et al.* (1993*b*).

Model for the fault growth window

The concept of a fault growth window provides a framework within which a discussion of fault growth can take place. What physical factors might control its inner and outer limits, its size, and the evolution of faults within it? How can the more or less regular spacing of boundary wall faults on a transect away from the axis be explained? Why are boundary-wall faults nucleated at about 2 km from the axis and then cease to be active when they have been rafted 7 or 8 km away from the axis? There seem to be two critical factors involved, the changing strength of a fault as it spreads away from the axis, and the strength of the lithosphere at the axis.

A major fault is a zone of weakness in the lithosphere, as is shown by the focusing of strain onto major fault zones, and the development of stress shadows around them. Near the spreading axis, major faults probably penetrate through the entire thickness of the lithosphere, to judge from estimates that the elastic plate thickness is about 6 km there (Blackman & Forsyth 1991) and observations of earthquake epicentres at 7–10 km below the axial zone (Kong *et al.* 1992; Wolfe *et al.* 1995). A single normal fault that penetrates the entire lithospheric thickness to the magmatically active axial zone at the base of the lithosphere would reach the surface at about 2 km from the axis. This probably controls the inner limit of the fault growth window.

As the lithosphere is rafted away from the spreading axis it thickens by cooling through hydrothermal circulation, probably focused along the major faults. A continuously active boundary-wall fault must propagate downwards at the same rate to maintain its penetration of the lithosphere. As it does so, the normal stress across the fault will increase, since this is related to the mean lithostatic pressure on the fault plane, and increased normal stress will increase fault strength. Fault strength will be further increased if the fault rotates to a shallower dip during continued extension. The consequence is

that the boundary faults will become significantly stronger as they are rafted away from the spreading axis.

Lithospheric strength at the spreading axis will remain approximately constant through time, as new lithosphere is generated and new magma is intruded. At times of high magma input from below, the lithosphere will thin, and when magma input is low, it will thicken, but these fluctuations will be superimposed on a generally steady state.

As spreading continues, there will come a time when the growing strength of the boundary wall fault exceeds the strength of the lithosphere at the spreading axis. At that time, the lithosphere will rupture again at the axis, and the old boundary wall fault will become inactive. Because of the complexities of the variables involved, and especially because of the variation with time of the strength of the lithosphere at the axis, this will not happen at a precisely repeated interval, but with a mean interval that corresponds to a fault spacing of 5 km.

Such a mechanism would account for the pattern of faults seen at slow spreading ridges. Can the same ideas be applied in other extensional environments where spreading does not take place, but where tectonic thinning of the lithosphere coupled with partial melting of the mantle leads to evolution of the strength both of the lithosphere and of faults that cut through it in time and space? Certainly the relationships would be more complex in environments lacking the time dimension that spreading imposes on a mid-ocean ridge. But the problems of fault evolution are very similar in both cases, as is the range of fault geometries seen on both large and small scales. The Mid-Atlantic Ridge may hold important clues to the origin of extensional terrains in many other places.

We acknowledge the support of the Natural Environment Research Council in granting us time on *Charles Darwin*, the help of the Master of the Darwin and all of the officers and crew on Cruise 65, and the advice of the other scientists of the shipboard party. D. Smith, J. Lin, C. Ebinger and R. Knipe gave invaluable help with discussions of the ideas in this paper. We thank P. Cowie, S. Allerton and M. Kleinrock for perceptive and helpful reviews. Correspondence to E. McAllister (e-mail:McAllisterE@earth.leeds.ac. uk).

References

ALLERTON, S., MURTON, B. J., SEARLE, R. C. & JONES, M. 1995. Extensional faulting and segmentation of the Mid-Atlantic Ridge north of the Kane Fracture Zone (24°00′N–24°40′N). *Marine Geophysical Researches*, **17**, 37–61.

BALLARD, R. D. & VAN ANDEL, T. H. 1977. Morphology and tectonics of the inner rift valley at latitude 36°50'N on the Mid-Atlantic Ridge. *Geological Society of America Bulletin*, **88**, 507–530.

BERGMAN, E. A. & SOLOMON, S. C. 1988. Transform earthquakes in the North Atlantic: source mechanism and depth of faulting. *Journal of Geophysical Research*, **93**, 9027–9057.

BLACKMAN, D. K. & FORSYTH, D. W. 1991. Isostatic compensation of tectonic features of the Mid-Atlantic Ridge: 25–27°30'S. *Journal of Geophysical Research*, **96**, 11 741–11 758.

CANN, J. R., SMITH, D. K., DOUGHERTY, M. E., LIN, J., BROOK, B., SPENCER, S., MACLEOD, C. J., MCALLISTER, E., PASCOE, R. & KEETON, J. 1992. Major landslide zones in the MAR median valley: their role in crustal construction and plutonic exposure. *Eos, Transactions of the American Geophysical Union*, **73**, 43, 569.

CARTWRIGHT, J., TRUDGILL, B. & MANSFIELD, C. 1995. Fault growth by segment linkage – an explanation for scatter in maximum displacement and trace length data from the canyonlands grabens of SE Utah. *Journal of Structural Geology*, in press.

COWIE, P. A., VANNESTE, C. & SORNETTE, D. 1993a. Statistical physics model for the spatiotemporal evolution of faults. *Journal of Geophysical Research*, **98**, 21 851–21 821.

——, SCHOLTZ, C. H., EDWARDS, M. & MALINVERNO, A. 1993b. Fault strain and seismic coupling on mid-ocean ridges. *Journal of Geophysical Research*, **98**, 17 911–17 920.

——, SORNETTE, D. & VANNESTE, C. 1995. Multifractal scaling properties of a growing fault population. *Journal of Structural Geology*, **122**, 457–469.

DAVY, PH., SORNETTE, A. & SORNETTE, D. 1990. Some consequences for the proposed fractal nature of continental faults. *Nature*, **348**, 56–58.

FLEWELLEN, C., MILLARD, N. & ROUSE, I. 1993. TOBI, a vehicle for deep ocean survey. *Electronics and Communication Engineering Journal*, April 1993, 85–93.

FOX, P. J., GRINDLAY, N. R. & MACDONALD, K. C. 1991. The Mid-Atlantic Ridge (31°–34°S): Temporal and spatial variations of accretion processes. *Marine Geophysical Researches*, **13**, 1–20.

GAWTHORPE, R. L. & HURST, J. M. 1993. Transfer zones in extensional basins: their structural style and influence on drainage development and stratigraphy. *Journal of the Geological Society, London*, **150**, 1137–1152.

GRINDLAY, N. R., FOX, P. J. & MACDONALD, K. C. 1991. Second-order ridge axis discontinuities in the south Atlantic: Morphology, structure and evolution. *Marine Geophysical Researches*, **13**, 21–50.

HEEZEN, B. C., THARP, M. & EWING, M. 1959. *The Floors of the Oceans, 1 The North Atlantic*. Geological Society of America Special Paper, **65**.

KONG, L. S. L., DETRICK, R. S., FOX, P. J., MAYER, L. F. & RYAN, W. B. F. 1988. The morphology and tectonics of the MARK area from Sea Beam and Sea MARC 1 observations (Mid-Atlantic Ridge

23°N). *Marine Geophysical Researches*, **10**, 59–90.

——, SOLOMON, S. C. & PURDY, G. M. 1992. Microearthquake characteristics of a mid-ocean ridge along-axis high. *Journal of Geophysical Research*, **97**, 1659–1685.

KUO, B.-Y. & FORSYTH, D. W. 1988. Gravity anomalies of the ridge transform system in the South Atlantic between 31 and 34.5°S: upwelling centres and variations in crustal thickness. *Marine Geophysical Researches*, **10**, 205–232.

LIN, J., PURDY, G., SCHOUTEN, H., SEMPÉRÉ, J-C. & ZERVAS, C. 1990. Evidence from gravity data for focussed magmatic accretion along the Mid-Atlantic Ridge. *Nature*, **344**, 627–632.

MCALLISTER, E., CANN, J. R. & SPENCER, S. 1995. The evolution of crustal deformation in an oceanic extensional environment. *Journal of Structural Geology*, **17**, 183–199.

MACDONALD, K. C. 1982. Mid-ocean ridges: Fine scale tectonic, volcanic, and hydrothermal processes within the plate boundary zone. *Annual Reviews of Earth and Planetary Science*, **10**, 155–190.

—— & LUYENDYK, B. P. 1977. Deep-towed studies of the structure of the Mid-Atlantic Ridge crest near 37°N. *Geological Society of America Bulletin*, **88**, 621–636.

MCGRATH, A. G. & DAVISON, I. 1995. Damage zone geometry around fault tips. *Journal of Structural Geology*, **17**, 1111–1124.

MORLEY, C. K., NELSON, R. A., PATTON, T. L. & MUNN, S. G. 1990. Transfer zones in the East-African rift system and their relevance to hydrocarbon exploration in rifts. *Bulletin of the American Association of Petroleum Geologists*, **74**, 1234–1253.

MORRIS, E. & DETRICK, R. S. 1991. Three dimensional analysis of gravity anomalies in the MARK area, Mid-Atlantic Ridge 28°N. *Journal of Geophysical Research*, **96**, 4355–4366.

MURTON, B. J. & PARSON, L. M. 1993. Segmentation, volcanism and deformation of oblique spreading centres: a quantitative study of the Reykjanes Ridge. *Tectonophysics*, **222**, 237–257.

MUTTER, J. & KARSON, J. 1992. Structural processes at slow spreading ridges. *Science*, **257**, 627–634.

PEACOCK, D. C. P. & SANDERSON, D. J. 1991. Displacements, segment linkage and relay ramps in normal-fault zones. *Journal of Structural Geology*, **13**, 721.

—— & —— 1992. Effects of layering and anisotropy on fault geometry. *Journal of the Geological Society, London*, 149, 793–802.

PHIPPS-MORGAN, J. & FORSYTH, D. W. 1988. Three dimensional flow and temperature perturbations due to a transform offset-effects on oceanic crustal and upper mantle structure. *Journal of Geophysical Research*, **93**, 2955–2966.

POLLARD, D. D. & AYDIN, A. 1984. Propagation and linkage of oceanic ridge segments. *Journal of Geophysical Research*, **89**, 17–28.

——, SEGALL, P. & DELANEY, P. T. 1982. Formation and interpretation of dilatant en-echelon cracks. *Geological Society of America Bulletin*, **93**, 1291–1303.

PURDY, G. M., SEMPÉRÉ, J-C., SCHOUTEN, H., DUBOIS, D. L. & GOLDSMITH, R. 1990. Bathymetry of the Mid-Atlantic Ridge, 24°–31°N – A map series. *Marine Geophysical Researches*, **12**, 247.

ROSENDAHL, B. R., BURGESS, C. F., SANDER, S. & LAMBIASSE, J. 1986. Structural symptoms of continental rifting. *Bulletin of the American Association of Petroleum Geologists*, **70**, 641.

SCHOUTEN, H., KLITGORD, K. D. & WHITEHEAD, J. A. 1985. Segmentation of mid-ocean ridges. *Nature*, **317**, 225–229.

SCHWARTZ, D. P. & SIBSON, R. H. (eds) 1989. *Proceedings of workshop XLV: Fault segmentation and controls of rupture initiation and termination*. US Geological Survey Open File Reports, **89–315**.

SEARLE, R.C. & LAUGHTON, A. S. 1981. Fine-scale study of tectonics and volcanics on the Reykjanes Ridge. *Oceanologica Acta*: Proceedings 26th International Geological Congress, Geology of the Oceans Symposium, Paris, 7–17 July, 1980, 5–13.

SEGALL, P. & POLLARD, D. D. 1980. Mechanics of discontinuous faults. *Journal of Geophysical Research*, **85**, 4337–4350.

—— & —— 1983. Nucleation and growth of strike slip faults in granite. *Journal of Geophysical Research*, **88**, 555–568.

SEMPÉRÉ, J. C., LIN, J., BROWN, H. S., SCHOUTEN, H. & PURDY, G. M. 1993. Segmentation and morphotectonic variations along a slow spreading centre: the Mid-Atlantic Ridge (24°00′N–30°40′N). *Marine Geophysical Researches*, **15**, 153–200.

SHAW, P. R. 1992. Ridge segmentation, faulting, and crustal thickness in the Atlantic. *Nature*, **358**, 490–493.

—— & LIN, J. 1993. Causes and consequences of variations in faulting style at the Mid-Atlantic Ridge. *Journal of Geophysical Research*, **98**, 13 609–13 631.

SMITH, D. K. & CANN, J. R. 1992. The role of seamount volcanism in crustal construction at the Mid-Atlantic Ridge (24°–30°N). *Journal of Geophysical Research*, **97**, 1645–1658.

—— & —— 1993. Building the crust at the Mid-Atlantic Ridge, *Nature*, **365**, 707–715.

——, ——, DOUGHERTY, M., LIN, J., KEETON, J., MCALLISTER, E., MACLEOD, C. J. & SPENCER, S. 1995. Volcanism from deep-towed side-scan sonar images, Mid-Atlantic Ridge 25°–29°N. *Journal of Volcanology and Geothermal Research*, **67**, 233–262.

SORNETTE, A., DAVY, PH. & SORNETTE, D. 1993. Fault growth in brittle-ductile experiments and the mechanics of continental collision. *Journal of Geophysical Research*, **98**, 12 111–12 139.

STEIN, R. S., BRIOLE, P., RUEGG, J. C., TAPPONNIER, P. & GASSE, F. 1991. Contemporary, Holocene, and Quaternary deformation of the Asal Rift, Djibouti – Implications for the mechanics of slow spreading ridges. *Journal of Geophysical Research*, **96**, 21 789–21 806.

TRUDGILL, B. & CARTWRIGHT, J. 1994. Relay-ramp forms and normal-fault linkages, Canyonlands national-park, Utah. *Geological Society of America Bulletin*, **106**, 1143–1157.

WALSH, J. J. & WATTERSON, J. 1991. Geometric and kinematic coherence and scale effects in normal fault systems. *In*: ROBERTS, A. M., YIELDING, G. & FREEMAN, B. (eds) *The Geometry of Normal Faults*. Geological Society, London, Special Publications, **56**, 195–206.

WHITEHEAD, J., DICK, H. J. B. & SCHOUTEN, H. 1984. A mechanism for magmatic accretion under spreading centres. *Nature*, **312**, 146–148.

WOLFE, C. J., PURDY, G. M., TOOMEY, D. R. & SOLOMON, S. C. 1995. Microearthquake characteristics and crustal velocity structure at 29°N on the Mid-Atlantic Ridge: the architecture of a slow-spreading ridge. *Journal of Geophysical Research*, **100**, 24 449–24 472.

Bathymetric segmentation and faulting on the Mid-Atlantic Ridge, 24°00'N to 24°40'N.

SIMON ALLERTON[1], ROGER C. SEARLE[2] & BRAMLEY J. MURTON[3]

[1]*Department of Geology and Geophysics, Grant Institute, University of Edinburgh, West Mains Road, Edinburgh EH9 3JW, UK*

[2]*Department of Geological Sciences, University of Durham, South Road, Durham DH1 3LE, UK*

[3]*Southampton Oceanography Centre, Southampton, UK*

Abstract: This investigation of valley-wall faulting and its relationship to bathymetry focuses on the first two segments north of the Kane Fracture Zone. The valley-wall faults imaged by deep-towed sidescan sonar show aspects common to continental normal faults, and a similar erosional pattern, from a fresh fault surface, through eroded, talus covered scarps, to scarps draped with pelagic sediments. The first, southernmost segment is a symmetric graben at the segment centre, and an asymmetric half-graben at the northern end. The second, northern segment is an asymmetric half-graben for its entire length. The gross morphology is reflected in the faulting style, with multiple small faults at the symmetric graben, and one or two large faults on the eastern wall, defining the asymmetic half-graben. This difference in faulting style may reflect differences in depth to the brittle-ductile transition, with asymmetry occurring when the brittle-ductile transition is deep and brittle faults can cross the axial valley, locking faults on the other side. A gross estimate of horizontal strain accommodated by the valley-wall faults correlates well with valley-floor bathymetry, suggesting that bathymetric segmentation is partially due to tectonic thinning at segment ends.

Bathymetric segmentation (e.g. Schouten *et al.* 1985) correlates with variation in the gravity along the ridge axis, expressed as the mantle Bouguer anomaly. Mantle Bouguer anomaly lows at the segment centres may either reflect hot, low density crust or mantle, thicker crust, or some combination of both (Lin *et al.* 1990). Recent refraction studies in the southern Atlantic (Tolstoy *et al.* 1993) indicate that crust thins at segment ends. Magma chambers at slow spreading ridges have not been imaged on seismic reflection surveys (Detrick *et al.* 1990), and it is believed that they are relatively small, temporary features. The magma supply is considered to strongly affect local tectonics (Tapponier & Francheteau 1977; Harper 1985; Karson *et al.* 1987). Where the magma budget is high, spreading is accommodated mainly by accretion. Conversely, where the magma budget is low, the injection of new material is predicted to be episodic, punctuated by periods of intense faulting and fissuring in the upper crust, and development of high strain fabrics at depth. Faults at the segment ends are larger than those at the centres (Shaw 1992), so the crustal thinning observed at the ends of segments may be the result of either reduced magmatic accretion or increased extensional faulting.

In a recent paper Allerton *et al.* (1995) studied the relationships between valley-wall faulting and bathymetric and volcanic segmentation using Seabeam bathymetry data (Pockalny *et al.* 1988; Purdy *et al.* 1990) and deep-towed sidescan sonar data (TOBI, Flewellen *et al.* 1993) acquired during a cruise of the RRS *Charles Darwin* (CD57) from the Mid-Atlantic Ridge North of the Kane Fracture Zone (MARNOK) between 24°00'N and 24°40'N. They developed a model for a different styles of magmatic accretion associated with different segment morphologies.

In this paper we use the same data set from the MARNOK area to investigate the differences in structural style of valley-wall faulting between segment centres and ends, and the relationship between the strain on the valley-wall faults and the bathymetry of the valley floor.

The Mid-Atlantic Ridge between 24°00'N and 24°40'N

This study concerns the first two segments north of the Kane Fracture Zone between about

From: MacLeod, C. J., Tyler, P. A. & Walker, C. L. (eds) 1996, *Tectonic, Magmatic, Hydrothermal and Biological Segmentation of Mid-Ocean Ridges*, Geological Society Special Publication No. 118, pp. 49–60.

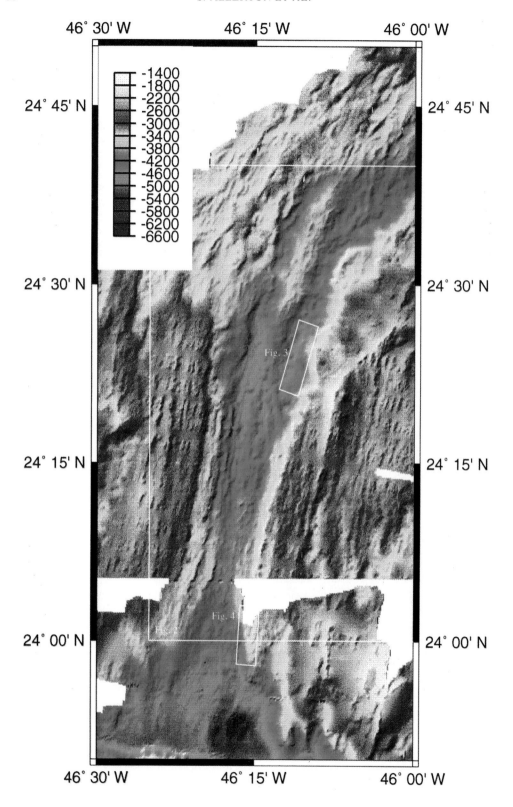

24°00′N and 24°40′N (Fig. 1). North of the second segment the ridge is offset by about 50 km east in a series of short en-echelon segments. There is no off-axis bathymetric expression of this discontinuity on the valley walls, and no offset of the Bruhnes–Matuyama boundary has been observed on surface magnetic data. The discontinuity at the northern end of the second segment leaves an off-axis wake, suggesting that this segment is propagating northwards into relatively old oceanic lithosphere (Sempéré *et al.* 1990).

First bathymetric segment

The first bathymetric segment extends from the Kane Fracture Zone at 23°50′N to 24°22′N. It rises from the nodal basin of the ridge-transform intersection of the Kane Fracture Zone and shallows and narrows to 3950 m, then deepens and widens to >4600 m at the segment end at 24°22′N. The bathymetry of the flanking walls mirrors that of the valley, particularly on the eastern side, where the shallowest points occur at the segment ends (the southernmost corresponding to the inside corner high of the ridge-transform intersection, which rises to minimum depth <1600 m. At the segment centre the minimum depth is >2700 m, and the valley walls are symmetric, rising in a series of 2 km wide terraces, each about 300 m above its innermost neighbour, up to about 13–15 km away from the centre of the valley. In contrast, the segment ends are highly asymmetric, with the eastern valley wall defined by a single fault scarp, and the western valley wall by a series of small fault bounded terraces.

The neovolcanic zone of this segment can readily be identified from TOBI images. This zone narrows at the segment centre, where deep-towed camera observations suggest that it is confined to a narrow (<0.5 km wide) graben at the centre of the valley. The bathymetric expression of the neovolcanic zone is relatively subdued, seldom rising more than 300 m above the adjacent valley floor. This interpretation differs from that offered in Allerton *et al.* (1995), who separated the neovolcanic zone into two separate segments. This separation is probably somewhat artificial.

The second bathymetric segment

The second segment defined by Sempéré *et al.* (1990) extends from the discontinuity at 24°22′N to 24°39′N. At its southern end it lies within the well defined axial valley walls, and trends approximately 010°. In the north, the valley opens up on the eastern side to form a major right lateral oblique zone which continues to about 25°N. The valley floor reaches a maximum depth of >4900 m at about 24°24′N on the west of the valley.

The neovolcanic zone is represented by a robust axial volcanic ridge which reaches heights of >500 m above the valley floor. This ridge terminates abruptly against a steep, ENE trending valley-wall fault at its northern tip. The contrast between the neovolcanic zone and the old, sedimented, tectonized terrain is clearer here than elsewhere along the axial volcanic ridge, suggesting that the boundary may represent the northward propagating tip of this axial volcanic ridge.

The eastern valley wall descends from 1600 m at 24°22′N to >3000 m north of 24°30′N, where the Mid-Atlantic Ridge becomes oblique to the spreading direction. The valley walls of this segment show a continued asymmetry with a single steep slope on the eastern side and regular terraces on the western side. In the northern part of the eastern valley wall, two trends of slopes can be identified. One, trending 010°, is parallel to the ridge axis, and the second, trending about 050° is approximately parallel to the oblique offset zone.

Axial valley-wall faulting

The valley walls are inward facing scarps resulting from the erosion of the original fault planes. The larger faults are imaged by Seabeam, but the vertical precision (*c.* 10 m) and the 'footprint' (*c.* 150 m × 150 m) means that the fine detail of these faults cannot be resolved. TOBI sidescan sonographs have a high resolution (*c.* 2 m at near range) and reveal intricate details of the fault geometries, including fault splays and fractures. We have combined the Seabeam bathymetry and TOBI sonographs to produce a tectonic map of the MARNOK area (Fig. 2). Where they dip towards the insonification

---

Fig. 1. Bathymetry (from Purdy *et al.* 1990 and Pockalny *et al.* 1988) of the Mid-Atlantic Ridge north of the Kane Fracture Zone (24°00′N to 24°40′N) also showing the locations of Figs 2, 3 and 4, and dredge CD57-22D (white triangle).

Mid Atlantic Ridge North of Kane
Structure

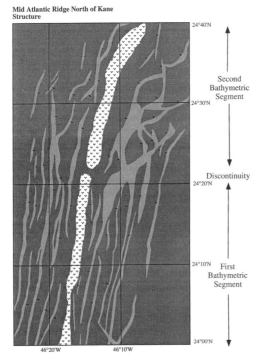

24°40'N

Second
Bathymetric
Segment

24°30'N

Discontinuity
24°20'N

24°10'N First
Bathymetric
Segment

24°00'N

46°20'W   46°10'W

Fig. 2. Structural map, showing normal faults (light shading) mapped from TOBI side-scan and Seabeam bathymetry. Also shown is the extent of the non-tectonized volcanic terrain, defined from the TOBI side-scan sonar (chevron shading).

direction, the valley-wall faults appear on TOBI sonographs as bright, relatively smooth (compared to neovolcanic morphologies) structures with typical normal fault geometries (Figs 3 and 4). These include clear examples of fault splays, where the major faults are linked by minor faults, and where minor faults branch off the tip of a major fault. Normal fault horses isolated between cross faults can be identified in a few examples.

The faults themselves show varying degrees of erosion and cover by talus fans and pelagic sediments (Fig. 5). A progression from relatively fresh faults to highly eroded and sedimented faults can be identified, which follow the morphological evolution described by Wallace (1977) for continental fault scarps. In the immature stage the scarp is close to the orientation of the original fault (Fig. 6a). A linear texture can be observed on the fault surface, and the base of the scarp is linear. These faults typically dip about 40°, determined from the Seabeam bathymetry. In the mature stage the fault scarps have been extensively eroded by

mass wasting (Fig. 6b). The tops of the slopes are marked by an oversteepened crest incised by debris chutes, and the lower parts of the slopes are extensively covered by talus fans which have a characteristic fine granular texture on the sonographs. The base of this scarp is irregular and scalloped. In the final, drape stage, pelagic sediments blanket the scarps (Fig. 3), although often a bright crest free-face appears on the TOBI images. In this phase pelagic sedimentation outpaces continued erosion and debris deposition.

In a few places major debris flows have left distinctive scours on the valley walls, and a large debris pile at the base of the scarps (Fig. 4). These debris piles have a distinctive coarse granular texture readily distinguishable from volcanic morphologies and pelagic sediments.

The fault scarps are typically planar, with dips between 30° and 40°. The upper surface behind the faults typically dip <10° away from the axial valley. Sedimentation does not seem to play a significant role in obscuring the effect of back-tilting as volcanic morphologies can often be seen peeping out through pelagic sediments on the fault terraces.

At the centre of the first bathymetric segment, between 24°00'N and 24°16'N, the axial valley-wall faults are regular in strike and throw. They are spaced across strike at about 2 km intervals, and have throws typically of only a few 100 m. They are sub-parallel to the axis and up to 40 km long. At this point the valley walls are symmetric (Figs 1, 2 & 7a). North of 24°16'N, to the discontinuity at 24°22'N, the throws on the faults increase to 0.5–2 km. Faulting becomes increasingly asymmetric northwards, with the inner axial valley-wall fault on the eastern side becoming dominant. A major fault, trending 050° on the eastern valley wall, has probably acted as a transfer fault, with a dominantly sinistral/normal sense of offset.

The second segment can be divided into two separate regions. Between 24°22'N and 24°30'N the valley is highly asymmetric, with large displacements (up to 5 km) on a single fault on the eastern valley wall (Fig. 7b). In the west the faults are continuous with those in the south, and there is no apparent change in structural style across the discontinuity.

The eastern valley-wall fault scarp is represented by a broad surface, dipping at about 30° towards the axis. There are some debris flows, but these are minor compared to the extent of the scarp surface. This margin corresponds to both the deepest point on the valley floor, and the shallowest point on the valley wall. The most pronounced back-tilting in the study area occurs

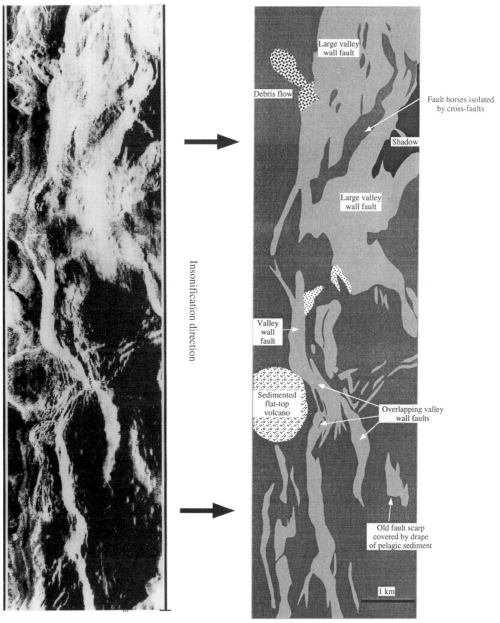

Fig. 3. Details of TOBI sonographs with interpretation. Bright surfaces face towards the instrument track. An overlapping pair of axial valley-wall normal faults, linked by a set of fine, small connecting splay faults, widen to form a large valley-wall fault at the discontinuity at 24°22′N.

at this point, with dips of about 18°. The vertical throw of this fault is at least 1.25 km, and is thus likely to expose sheeted dykes at the base of the scarp. Dolerite rocks, altered to greenschist facies conditions, were indeed recovered on Dredge CD57-22D.

North of 24°22′N, the valley broadens into a zone of oblique extension. There is an apparent reduction on the throw of the faults in this region, perhaps because faults to the east, in the region of oblique extension, continue to be active.

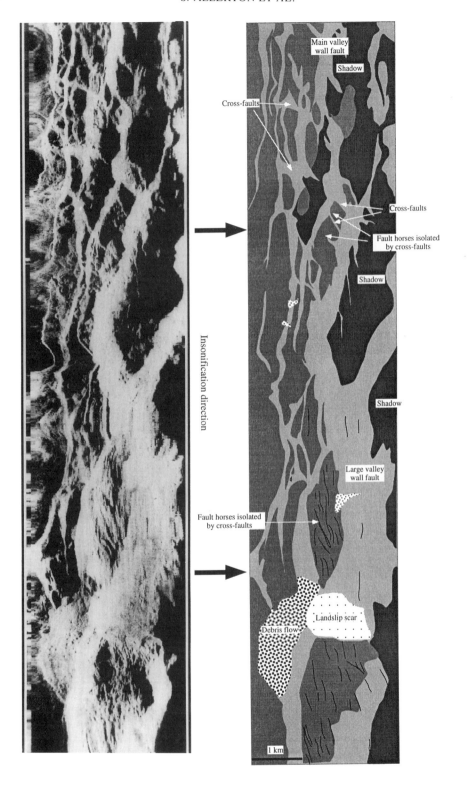

Insonification direction

Strain measurements

We have attempted to quantify the variation of the throw of the valley-wall faults along the axis by estimating the horizontal strain accommodated by the faults. Because the tilting of the faults is minimal, this can be estimated from the width of the faults observed on the Seabeam bathymetry, measured parallel to the plate separation vector (097°), over 20 km on each side of the axis. The width of the fault scarps is likely to be somewhat greater than the horizontal throw of the faults, because of erosion and the development of talus fans at the base of the faults, and because the technique does not account for back-tilting, so the measurement of strain will be something of an overestimate. It does, however, give a relative measurement of differences in faulting style along the axis.

At the segment centres the horizontal strains are approximately 20% (Fig. 8). The strains rise rapidly towards the segment ends, to values around 40%. The drop in strain to values <10% in the northern part of the second segment may reflect a broader region of deformation in the oblique offset zone, not included in our strain estimate. It may also reflect a shorter period of extension, consistent with the propagation of the second segment into old lithosphere.

The cumulative vertical component of fault displacement can be estimated from the height of the valley wall above the valley floor. This corresponds quite closely with the estimate of horizontal strain which we would expect if both are the result of normal faulting on faults with similar dips.

Discussion

The centre of the first segment is high, and the valley walls are symmetric, with multiple, relatively small-throw faults. Towards the ends of the segments the throws of the faults increase, and the asymmetry of the valley walls becomes more pronounced, with a single, major fault on the eastern side. This segment is essentially similar to the 'typical' MAR segment described by Shaw (1992), although he did not emphasis the asymmetry of the system.

In the second segment, the valley shallows slightly, to about 4000 m. Faulting remains dominantly asymmetric, with a large eastern valley-wall fault. The throw on the valley-wall fault decreases towards the segment centre. The geometry of the segment is quite different to the first segment, as it does not develop a symmetric graben at the centre, but remains asymmetric throughout. The valley floor is generally lower, and neovolcanic activity has produced the robust axial volcanic ridge which stands out above the valley floor.

There is a strong correlation between the bathymetry of the axial valley, the horizontal strain on the valley-wall faults and the height of the valley wall above the valley floor (Fig. 8). This suggests that the bathymetric variation may be at least partially the consequence of lithospheric stretching and thinning. The low strain encountered at the segment centres and the high strains at segment ends is consistent with the magma budget concept (e.g. Harper 1985). The axial valley faults at the segment ends have accrued about 10 km of extension, compared to about 4 km at the narrowgate. Presumably the difference between these cumulative extensions was taken up by greater accretion in the segment centre.

The differences in tectonic extension along the axis would produce crustal thinning at the segment end relative to the segment centre. Using the estimates of horizontal extension, we predict a difference of about 16%. This is significantly less than most estimates from other segments (e.g. Tolstoy *et al.* 1993), or from estimates assuming that the bathymetry of the valley floor is an isostatic response to crustal thickness variations. We therefore conclude that crustal thinning at the segment ends is the result of two associated processes; tectonic thinning and magmatic accretion. Further studies of extension and thinning associated with faulting, linked with studies of the crustal thickness are required to fully assess the relative contribution of these effects.

The large axial valley fault on the eastern side of the valley at the discontinuity between the segments corresponds to the deepest point of the valley and the shallowest part of the the eastern valley wall. Back-tilting also reaches a maximum of 18° at this point. This geometry suggests that

---

Fig. 4. Details of TOBI sonographs with interpretation. Bright surfaces face towards the instrument track. A set of normal splay faults is developing at the tip of a large axial valley-wall fault. These faults appear to cut a set of smaller faults, which may either be an early set of valley floor faults, or small connecting splays that developed at the same time as the main faults. This interconnected network of faults isolate individual fault blocks (or horses). A large axial valley-wall fault links two smaller faults. A slump occurs at the foot of the fault at this point, perhaps marking the collapse of an isolated fault horse.

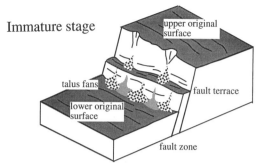

Immature stage

upper original surface

talus fans

lower original surface

fault terrace

fault zone

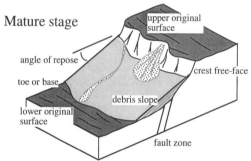

Mature stage

upper original surface

angle of repose

toe or base

crest free-face

debris slope

lower original surface

fault zone

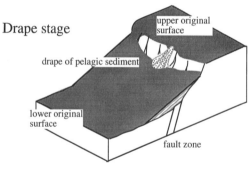

Drape stage

upper original surface

drape of pelagic sediment

lower original surface

fault zone

Fig. 5. Fault scarp evolution. (**a**) Immature stage. The original fault surface is exposed, with small talus fans developing at the base of the fault scarps and on terraces. The base of the fault is sharp and linear. Small erosive scars and debris chutes dissect the upper original surface. The measured angle of dip is close to that of the original fault. (**b**) Mature stage. The fault scarp is covered with a thick wedge of sediments produced by gravitational collapse of the unstable, steeper slope. The crest free-face (the sediment-free upper surface of the fault scarp) has eroded deeply into the upper original surface, and has a morphology typified by rock-buttresses and gullies. The debris slope will have a planar profile, controlled by the angle of repose of the sediment clasts, typically between 30° and 40°. The base of the slope has a cuspate aspect, marking the growth of individual fans. (**c**) Drape stage. The fault scarp is covered by a drape of pelagic sediments, which appear dark on TOBI sonograms. Occasionally the over-steepened crest free face can be observed at the top of the scarp as a bright scar on the sonograms. The boundaries of these surfaces are irregular and gradational. At this stage pelagic sedimentation is more rapid than continued erosion and debris deposition.

the normal faulting to form the half graben is also accompanied by flexural uplift and back-tilting of the footwall, such as that suggested by Buck (1988). The large valley-wall fault has a throw of at least 5 km, and the fault plane must continue at least this far beneath the axial valley. Such a fault would be expected to exhibit all the characteristics of a continental detachment fault (e.g. Lister & Davis 1989) and to juxtapose fault rocks of different grades: cataclastites and fault breccias in the hangingwall and mylonites in the footwall.

The switch from symmetric to asymmetric faulting does not appear to be controlled by the throw on the fault, as at the centre of the northern segment the valley walls are asymmetric, yet the strains recorded are around 20%,

similar to those observed in the centre of the first segment, where the valley walls are symmetric. The controlling factor may be the depth to the brittle–ductile transition, as has been suggested by Shaw & Lin (1993). If the brittle layer is thin, then valley-wall faults may be expected to propagate into the ductile layer beneath the axial valley. In this situation, the lithosphere would be weak, and faults would not accumulate large throws before new faults could develop. Alternatively, where the brittle–ductile transition is relatively deep, then, when one fault propagates more than half way across the axial valley, it may encounter faults from the opposite wall. It will either be cut by and lock up against, or cut and lock these faults. In the latter case the main fault can continue to be active, and will accommodate a large proportion of the fault movement. The relative strength of cold mantle relative to crust (e.g. Harper 1985) would favour the prolonged activity of a weakened pre-existing shear zone over fresh lithospheric fracturing.

This concept of the style of faulting being controlled by the depth to the brittle–ductile transition has implications for the thermal structure of the segments. We would predict that the northern segment is relatively cold, compared to the southern, narrowgate segment. This corresponds to observations which suggest that the northern segment may be relatively young, propagating northward into old, cold crust (Allerton *et al.* 1995). Also, the existence

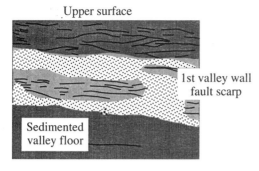

Upper surface

1st valley wall
fault scarp

Sedimented
valley floor

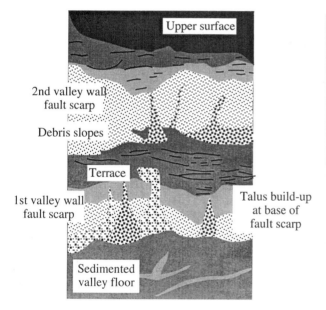

Upper surface

2nd valley wall
fault scarp

Debris slopes

Terrace

1st valley wall
fault scarp

Talus build-up
at base of
fault scarp

Sedimented
valley floor

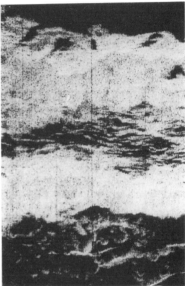

Fig. 6. Detail of a valley-wall faults imaged on high resolution TOBI sonograph, with interpretation. (**a**) Immature stage. A linear fabric is clearly distinguishable on the fault scarp. The base of the scarp is sharp and linear, suggesting that there are no large talus fans developed at the base. The scarp dips at a maximum of about 45°, shallowing to 30° on the lower slopes, as measured from the Seabeam data, which are smoother than the true bathymetry. (**b**) Mature stage. Small, individual fans can be identified on the lower slopes of the first valley wall scarp. Large debris slopes, marking a cuspate basal profile can be seen on the second valley wall scarp, suggesting a progressive increase in the amount of sediment on the scarps.

of the robust axial valley ridge, which stands up to 0.5 km above the valley floor, may argue for a deeper plate thickness than that seen at the narrowgate (e.g. Vogt 1974). Lawson *et al.* (this volume) demonstrates from deep-towed camera observations that the age of parts of the robust axial volcanic ridge may be as much as 40 ka, suggesting that the valley floor must have a sufficient elastic thickness to support this structure.

The large fault at the discontinuity may be expected to have been active for more than 0.25 Ma (assuming a plate separation half-rate of 12 mma^{-1}; McGregor *et al.* 1977; Allerton *et al.* 1995). This is longer than most estimates of recurrence-times of volcanic activity (e.g. 0.01 Ma, Bryan & Moore 1977; 0.04 Ma, Allerton *et al.* 1995), so it is likely that fault activity was accompanied by rising magma. In this

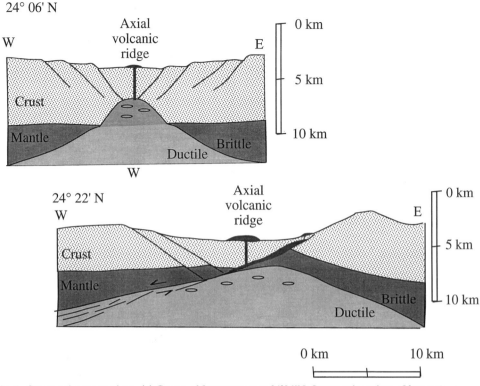

Fig. 7. Structural cross-sections. (**a**) Centre of first segment at 24°06′N: Symmetric graben. *Observations.*
Faults have throws of a few 100 m. Back-tilting is relatively minor, generally less than about 10°. Cumulative
strains are low, <25%. There is no clear axial volcanic ridge within the axial valley. *Interpretation.* High
thermal gradients due to the high magma budget place the brittle-ductile transition in the crust. Brittle normal
faults terminate into a high ductile strain region within the lower crust, indicated by strain ellipsoids. (**b**)
Discontinuity at 24°22′N: Asymmetric half-graben. *Observations.* Extension is mostly accommodated by
faulting on a single axial valley-wall fault to the east. Strains are high – approaching 50%. A distinct axial
volcanic ridge is observed within the axial valley. *Interpretation.* The geometry of the valley is that of a flexural
half-graben, with associated footwall uplift and back tilting, and shallowing of the main valley-wall fault. The
crust is thinner, because of tectonic stretching and limited accretion of new material. The fault will broaden into
a ductile shear zone at depth. The throw on this fault is large, and it has been active for a considerable period,
perhaps 0.25 Ma. This detachment may act as a migration path for rising magma. The asymmetry probably
initiates from a symmetric graben, when the depth to the brittle-ductile transition increases, so that individual
faults meet, and one locks. The other continues to remain active, accommodating large displacements.

situation the fault may have acted as a migration route for the magma. Flat-topped
seamounts (e.g. Lawson *et al.* this volume,
Allerton *et al.* 1995) commonly occur on the
valley floor next to large valley-wall faults, and
they are concentrated at the discontinuity
between 24°20′N and 24°25′N, adjacent to the
largest valley-wall fault. The extrusion of these
flat-topped seamounts may thus be controlled
by large crustal-scale detachments, with melt
being channelled to the surface along the
faults.

Conclusions

TOBI deep-towed sidescan sonar data produces
high resolution images of the valley-wall faults.
In detail, the fault geometries are typical of
normal faults, and include fine splays and
horses. Landslip scars and debris flows at the
bases of the faults can be identified in a few
places. An age progression can be identified
from fresh fault scarps, through highly eroded
scarps with extensive talus fans at their bases, to
scarps covered by a drape of pelagic sediments.

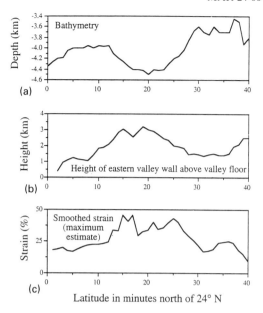

(a)

(b)

(c)

Latitude in minutes north of 24° N

Fig. 8. Variation along the axial valley (as minutes of latitude north of 24°00′N). (**a**) Bathymetry of the axial volcanic high, from Seabeam data (Purdy *et al.* 1990). (**b**) Height of the eastern valley wall above the valley floor. (**c**) Horizontal strain on axial valley-wall faults smoothed over a 3′ running window.

The cumulative strain on the valley-wall faults is about 25% at the segment centres and up to 40% at the discontinuity between the segments. This result is consistent with a relatively reduced magma budget at the segment ends. This can account for about 16% thinning of the crust at the discontinuity compared to the segment centres. This is significantly less than has been observed in other segments. The bathymetry of the axial valley correlates closely with the strain, suggesting that the bathymetric segmentation is at least partially a response to tectonic thinning. The height of the eastern wall above the valley floor, which roughly correlates with the vertical throw on the faults, correlates with the bathymetry and the horizontal strain.

The southern segment has a symmetric graben at the centre, with a series of small faults on each side of the valley. These faults increase in throw as the discontinuity is approached. North of 24°16′N the eastern valley wall becomes dominant, and the axial valley takes on a distinctly asymmetric aspect. The northern segment is asymmetric for its entire length. The style of faulting does not appear to be controlled by the extension they have accommodated, as the centre of the northern segment has strains similar to those at the centre of the southern segment. The style may rather be controlled by the depth to the brittle-ductile transition, with asymmetric faulting occuring where the transition is deep, and faults can propagate across the width of the axial valley, where they cut and lock faults from the opposite valley wall.

At the discontinuity, the valley floor reaches its deepest point, the eastern valley wall reaches its shallowest point, and back-tilting reaches a maximum of about 18°. Flexure is likely to be an important contributor to uplift and back-tilting. This fault has probably been relatively long-lived, and can have accommodated plate separation for about 0.25 Ma. It probably extends beneath the axial valley, locking up the faults on the western wall. This major crustal detachment may control the extrusion of flat-topped volcanoes.

References

ALLERTON, S., MURTON, B. J., SEARLE, R. C. & JONES, M. 1995. Extensional faulting and segmentation of the Mid Atlantic Ridge north of the Kane Fracture Zone (24°00′N to 24°40′N). *Marine Geophysical Researches*, **17**, 37–61.

BRYAN, W. B. & MOORE, J. G. 1977. Compositional variation in young basalts in the Mid-Atlantic rift valley near 36°49′N. *Geological Society of American Bulletin*, **88**, 556–570.

BUCK, W. R. 1988. Flexural rotation of normal faults. *Tectonics*, **7**, 959–973.

DETRICK, R. S., MUTTER, J., BUHL, P. & KIM, I. I. 1990. No evidence from multichannel reflection data for a crustal magma chamber in the MARK area on the Mid-Atlantic Ridge. *Nature*, **347**, 61–64.

FLEWELLEN, L., MILLARD, N. & ROUSE, I. 1993. TOBI: a vehicle for deep ocean survey. *Electronics and Communications Engineering Journal*, **5**, 85–93.

HARPER, G. D. 1985. Tectonics of slow spreading mid-ocean ridges and consequences of a variable depth to the brittle/ductile transition. *Tectonics*, **4**, 395–409.

KARSON, J. A., THOMPSON, G., HUMPHRIS, S. E., EDMOND, J. M., BRYAN, W. B., BROWN, J. R., WINTERS, A. T., POCKALNY, R. A., CASEY, J. F., CAMPBELL, A. C., KLINKHAMMER, G., PALMER, M. R., KINZLER, R. J. & SULANOWSKA, M. M. 1987. Along axis variation in seafloor spreading in the MARK area. *Nature*, **328**, 681–685.

LAWSON, K., SEARLE, R. C., PEARCE, J. A., BROWNING, P. & KEMPTON, P. 1996. Detailed volcanic geology of the MARNOK area, Mid-Atlantic Ridge, north of the Kane Transferal. *This volume.*

LIN, J., PURDY, G. M., SCHOUTEN, H., SEMPÉRÉ, J.-C. & ZERVAS, C. 1990. Evidence from gravity data for focused magmatic accretion along the Mid Atlantic Ridge. *Nature*, **344**, 627–632.

LISTER, G. S. & DAVIS, G. A. 1989. The origin of metamorphic core complexes and detachment faults formed during Tertiary continental extension in the northern Colorado River region, U.S.A. *Journal of Structural Geology*, **11**, 65–94.

McGREGOR, B. A., HARRISON, C. J. A., LAVELLE, J. W. & RONA, P. A. 1977. Magnetic anomaly patterns on Mid-Atlantic Ridge crest at 26°N, *Journal of Geophysical Research*, **82**, 231–238.

POCKALNY, R. A., DETRICK, R. S. & FOX, P. J. 1988. Morphology and tectonics of the Kane transform from Seabeam Bathymetry data. *Journal of Geophysical Research*, **93**, 3179–3193.

PURDY, G. M., SEMPÉRÉ, J.-C., SCHOUTEN, H., DuBois, D. L. & GOLDSMITH, R. L. 1990. Bathymetry of the Mid-Atlantic Ridge, 24–31°N: A map series. *Marine Geophysical Researches*, **12**, 247–252.

SCHOUTEN, H., KLITGORD, K. D. & WHITEHEAD, J. A. 1985. Segmentation of mid-ocean ridges. *Nature*, **317**, 225–229.

SEMPÉRÉ, J.-C., PURDEY, G. M. & SCHOUTEN, H. 1990. Segmentation of the Mid-Atlantic Ridge between 24°N and 30°40′N. *Nature*, **344**, 427–431.

SHAW, P. R. 1992. Ridge segmentation, faulting and crustal thickness in the Atlantic Ocean. *Nature*, **358**, 490–428.

—— & LIN, J. 1993. Causes and consequences of variations in faulting style at the Mid-Atlantic Ridge. *Journal of Geophysical Research*, **98**, 21 839–21 851.

TAPPONNIER, P. & FRANCHETEAU, J. 1978. Necking of the lithosphere and the mechanics of slowly accreting plate boundaries. *Journal of Geophysical Research*, **83**, 3955–3969.

TOLSTOY, M., HARDING, A. J. & ORCUTT, J. A. 1993. Crustal thickness on the Mid-Atlantic Ridge: Bull's eye gravity anomalies and focused accretion. *Science*, **262**, 725–729.

VOGT, P. R. 1974. Volcano height and plate thickness. *Earth and Planetary Science Letters*, **23**, 337–348.

WALLACE, R. E. 1977. Profiles and ages of young fault scarps, north – central Nevada. *Geological Society of American Bulletin*, **88**, 1267–1281.

Detailed volcanic geology of the MARNOK area, Mid-Atlantic Ridge north of Kane transform

KATE LAWSON[1,4], ROGER C. SEARLE[1], JULIAN A. PEARCE[1], PAUL
BROWNING[2] & PAMELA KEMPTON[3]

[1]*Department of Geological Sciences, University of Durham, DH1 3LE, UK*
[2]*Geology Department, University of Bristol, BS8 1TR, UK*
[3]*NERC Isotope Geosciences Laboratory, Keyworth, Nottinghamshire, NG12 5GG, UK*
[4]*Present address: Shell NAM, PO Box 28000, 9400HH, Assen Netherlands*

Abstract: We present the results of a detailed study of the volcanic geology of the median valley floor of the Mid-Atlantic Ridge in two adjacent but contrasting spreading segments immediately north of the Kane transform fault, based on a high-resolution deep-towed sidescan sonar (TOBI) survey, near-bottom photographic traverses, and geochemical analyses from 28 precisely sited sets of dredge samples. This has allowed us to assess the relative importance of small (intersegment) and large (transform) boundaries on magmatic processes at slow spreading ridges.

We find that the petrological effect of the transform (so-called 'transform fault effect') is minor compared to the manifestation of the second-order segmentation. The southern, 'narrowgate'-type segment has a poorly developed neovolcanic zone, which tapers towards its bathymetric centre where continuous faulting causes rapid dismemberment. Flat-topped seamounts are preferentially located at the ends of the segment, particularly near the non-transform offset. There is a striking variation in the degree of fractionation along the length of this segment: primitive lavas were sampled at the centre, while more fractionated basalts, showing a greater range of parental compositions, form discrete volcanic edifices at the ends. In contrast, the northern segment has a wider inner valley, and syn-magmatic faults extend up to 15 km along the crest of its robust axial volcanic ridge. The along-segment trend of increasing MgO towards the bathymetric crest of the axial volcanic ridge is similar to, but less well-defined than, that for the southern segment.

There is no variation in the bulk degree of melting along the segments as determined from the major element geochemistry; hence, melt migration is favoured over focused mantle upwelling as the main cause of the crustal thickness variations which define second-order segmentation. Radiogenic isotopes show only very small variations with no systematic pattern emerging within or between the ridge segments. The isotopic irregularities are not related to the variations in incompatible elements, implying that the latter result from dynamic melting processes rather than from long-lived source heterogeneity. A model is developed which relates the volcanology, tectonic style and the geochemical trends to the episodicity of magma supply and eruption.

Segmentation of Mid-Ocean Ridges

The full extent of the segmentation of the Mid-Atlantic Ridge (MAR) was first recognized during extensive surveys using multibeam swath mapping with SeaBeam (Purdy *et al.* 1990; Sempéré *et al.* 1990; Grindlay *et al.* 1992) combined with gravity and magnetics (Kuo & Forsyth 1988; Detrick *et al.* 1990; Lin & Phipps Morgan 1990; Deplus *et al.* 1992). Second-order (non-transform) segmentation is defined by negative mantle Bouguer anomalies (of about 40 mGals) which tend to coincide with bathymetric swells at intervals of 50–100 km along the ridge crest. Along-strike variations in crustal thickness (thinner crust towards the ends of the segment) are thought to predominate over mantle density variations in contributing to these gravity anomalies (Lin & Phipps Morgan 1992; Rabinowicz *et al.* 1993; Tolstoy *et al.* 1993). Off-axis magnetic anomalies, gravity and bathymetric traces left by non-transform discontinuities provide a record of the evolution of second-order segmentation, which is commonly characterized by along-axis migrations and a limited life span (Carbotte *et al.* 1990; Sempéré *et al.* 1993; Gente *et al.* 1995). Sempéré *et al.* (1993) have defined two types of segment on the basis of morphology and gravity signals: group 1 have narrow median valleys (so-called 'narrowgate') and large negative mantle Bouguer anomalies, whereas group 2 have broader valleys and smaller Bouguer anomalies.

In contrast, fracture zones have been traced for thousands of kilometres between the continental margins on either side of the Atlantic (Johnson & Vogt, 1973; Philips & Fleming 1978) demonstrating the longevity and stability of this first-order segmentation. At a ridge–transform intersection (RTI), the juxtaposition of a colder, thick edge of lithosphere against the truncated ridge axis is thought to have implications with regard to both the thermal regime and the 3D geometry of mantle upwelling (Fox *et al.* 1980; Fox & Gallo 1984). The axial depth typically increases by about 2 km over several tens of kilometres as large offset transforms such as the Kane are approached (Karson & Dick 1983; Fox & Gallo 1984; Fox & Gallo 1986; Pockalny *et al.* 1988). Variations in crustal thickness take place within 20–30 km of a transform, but very thin crust is found only in narrow strips which usually coincide with the fracture zone valley or wall (Louden *et al.* 1986; Abrams *et al.* 1988; Forsyth 1992). Petrological effects observed in lavas erupted near transforms include cooler magmatic temperatures, a wider range of compositions, higher pressure fractionation and slightly smaller extents of melting (Hekinian & Thompson 1976; Bender *et al.* 1984; Perfit *et al.* 1983; Langmuir & Bender 1984). These systematic geochemical and geophysical changes associated with the proximity of a transform are known as the 'transform fault effect' (TFE). Models explaining the TFE all predict a correlation between the magnitude of the edge effect and the age contrast of the plates juxtaposed at the RTI. Although a rough correlation exists between the maximum depth of a transform valley and the offset of the transform (Fox & Gallo 1986), the geochemical TFE does not appear to be similarly correlated (Langmuir & Bender 1984; Le Roex *et al.* 1989).

An as yet unresolved question is whether segmentation has a magmatic or lithospheric origin. Segmentation may arise from the three-dimensional nature of flow and melt migration in the underlying mantle or alternatively it may primarily be a consequence of lithospheric dynamics and tectonics. The decrease in crustal thickness towards the ends of both first- and second-order segments has been explained by three basic end-member mechanisms.

(1) *Focused mantle upwelling* under spreading centres causes corresponding variations in the extent of decompression melting and subsequent melt delivery. It is the mantle flow which transports the melt from the large region of

upwelling to the narrow region of crustal formation at the axis (e.g. Parmentier & Phipps Morgan 1990; Rabinowicz *et al.* 1993; Rouzo *et al.* 1995).

(2) *Focused melt migration* along the ridge axis causes the main variations in crustal structure. Melt is generated in a broad region below the ridge where the mantle itself upwells relatively uniformly along axis. Strong focusing of melt *delivery* occurs during segregation of the melt from the matrix or during transport in the upper mantle (e.g. Whitehead *et al.* 1984; Phipps Morgan *et al.* 1987; Sparks & Parmentier 1991).

(3) *Tectonic segmentation* is primarily a result of the way the lithosphere breaks under extensional stress. Along axis variations in mantle upwelling are then a consequence, not a cause, of the plate boundary configuration (e.g. Wilson 1965; Mutter & Karson 1992; Shen & Forsyth 1992).

Focused upwelling and focused melt delivery can be discriminated by variations in the degree of partial mantle melting observed along the segment. The latter produces a uniform melting signature along axis while the former should show a decrease in extent of melting towards the ends of the segment. As yet there is no methodology for discriminating whether tectonism controls magma delivery or vice-versa.

This study

We have carried out a combined geophysical and petrological study of ridge segmentation over two contrasting second-order segments of the Mid-Atlantic Ridge immediately north of the dextrally slipping Kane transform fault (Fig. 1). The study area, named MARNOK (Mid-Atlantic Ridge North of Kane), is situated on the western intersection of the MAR with the Kane Fracture Zone (KFZ) at about 24°00′N, 46°20′W. This area was chosen because of the large range in offset length (150 km across the transform as opposed to 1 km between the southern Segment 1 and the northern Segment 2), the contrasting morphology of the segments (Segment 1 appears to be of type1 and Segment 2 of type 2 in the classification of Sempéré *et al.* 1993), and its proximity to the well-studied MARK area immediately south of the transform. The two second-order segments are separated at 24°22′N by a third-order discontinuity marked by a 1 km right-stepping offset of the neovolcanic zone (NVZ; see Grindlay *et al.* 1991 for definitions of inter-segment discontinuities). The ridge here is far from the effects of the anomalous mantle temperatures and compositions that are associated with hotspots.

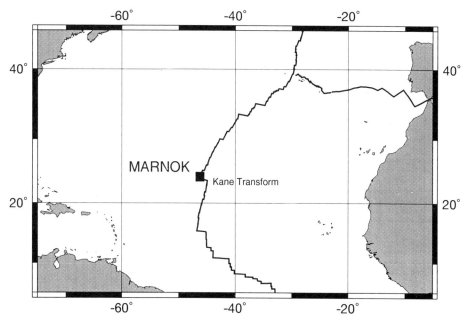

Fig. 1. Location map showing the present plate boundary configuration of the north Atlantic. The MARNOK study area is at the western intersection of the Kane Transform with the Mid-Altantic ridge.

It has a full spreading rate of about 30 mm a^{-1} in the direction 100°. The large variations in offset length allow the contribution of the TFE to be compared with the edge effects associated with small non-transform discontinuities. There is Sea Beam multibeam bathymetry coverage of both the transform and a 50 km wide band centred on the ridge crest to the north (Fig. 2; Pockalny *et al.* 1988; Purdy *et al.* 1990). In addition, submersible studies have been carried out by Karson & Dick (1983), Zonenshain *et al.* (1989) and Fujimoto *et al.* (1994).

Our main objective was to assess the volcanic, tectonic and petrological variations along the two contrasting segments, and in the light of these observations to assess the various mechanisms proposed for ridge segmentation: for example, focused melt migration models invoke gravitational Rayleigh–Taylor instabilities in a broad melt reservoir to predict extensive magma mixing within segments but isolation of the flow between segments (Whitehead *et al.* 1984; Crane 1985; Schouten *et al.* 1985). In addition we wished to assess processes occurring within the crust, including the length scales of petrological segmentation (Langmuir *et al.* 1986) as defined by the spatial distribution of neovolcanic zone lavas with common parental compositions. This should reflect the scales and organisation of

crustal plumbing and the possible role of lateral dyke injection.

Our major tools were: the deep-towed instrument package, TOBI (Towed Ocean Bottom Instrument; Huggett & Millard 1992) of the Institute of Oceanographic Sciences Deacon Laboratory (IOSDL, now renamed Southampton Oceanography Centre), comprising 30 kHz sidescan sonar, 7.5 kHz sediment profiler and magnetometer; IOSDL's towed camera system WASP (Wide-Angle Survey Photographer); and rock dredging for geochemical analysis. A major advantage of such an integrated study was that we were able to use the volcanology and structure revealed by the sidescan to determine the sampling sites as well as to aid interpretation of the geochemical data.

This paper concentrates on the fine detail of the inner valley floor and the geochemical systematics along axis. It is intended to complement Allerton *et al.* (this volume) who focus on tectonic aspects and the large scale morphological variation of the same two segments.

The MARNOK survey

The MARNOK survey was carried out on RRS *Charles Darwin* cruise 57 (CD57) in March–April 1991. It covered a 100 km length of the

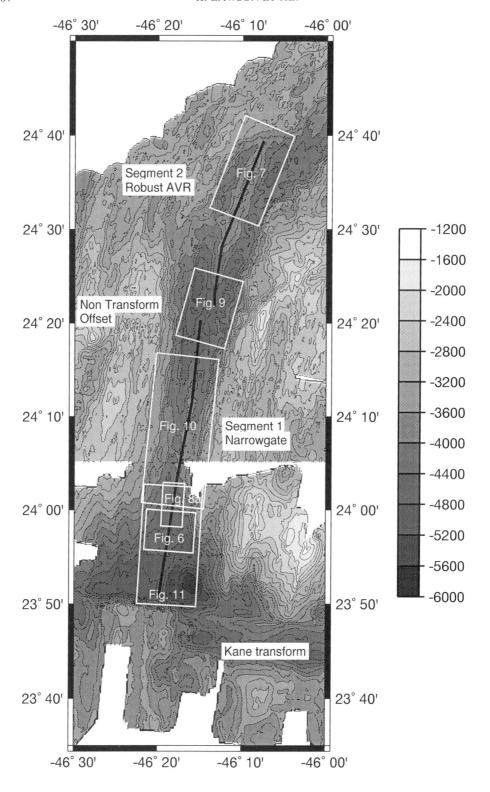

median valley floor with two ridge-parallel and three oblique sidescan swaths, the latter extending up to 25 km off axis. There were also four transform-parallel swaths covering the RTI area, of which two extend 50 km east along the transform valley (Fig. 3). Two traverses across the inner valley floor by the WASP towed camera system were designed to compare the tectonic and volcanic condition of the centres of the two segments and to groundtruth the sidescan sonar. 28 dredge stations covering a wide range of volcanic morphologies at a regular spacing along the median valley floor were selected with the aid of the sidescan images; all recovered relatively fresh basalt. Dredge stations are referred to below by numbers preceded by an initial 'D'.

As the relationship between the geochemistry and the volcanic morphology was an important aspect of this study, the spatial accuracy of the dredging and imagery was important. The ship was navigated by continuous GPS. TOBI, dredge and camera positions were located by a combination of estimated position relative to the ship based on instrument depth and tow-cable length, and comparison of TOBI images or measured topographic profiles with Sea Beam bathymetry. Although absolute positional errors in the Sea Beam maps can be up to 0.9 km (Purdy *et al.* 1990), we believe the relative locations of our samples with respect to seafloor features is considerably better than this, since we can confidently correlate features such as individual seamounts with their bathymetric expressions on the Sea Beam maps. In particular, we are confident of having accurately sampled selected targets such as small seamounts.

TOBI high-resolution sidescan sonar

TOBI was operated at altitudes of 400–500 m above the seafloor, producing a swath width of about 6 km. The along-track sonar resolution varies from 3.5 m to 43 m, depending on range, while cross-track resolution is potentially *c.* 2 m (Huggett & Millard 1992). The data shown in this paper are mosaics of data plots which had been subsampled to 2000 pixels per swath or an across-track display resolution of 3 m (less than

the size of the insonified footprint). The images are displayed with high amplitude returns as bright pixels and low amplitude as dark; acoustic shadows as black.

The amplitude of the sonar signal depends mainly on the angle of incidence, the acoustic impedance contrast between the seawater and seafloor, and the seafloor roughness. A high degree of roughness produces a greater return for the same acoustic impedance. Sediment causes a lower amplitude return because of acoustic penetration and attenuation in the subsurface. A 2–5 m thick cover may attenuate the backscatter by more than 20 dB (Mitchell 1993).

The sidescan images were initially interpreted by demarcating the boundaries between different textural terrains. Groundtruthing of these terrains was then carried out by a combination of deep-tow camera traverses, dredge sampling and sub-bottom profiler analysis. Additional high resolution studies, especially the submersible dives of Karson & Dick (1983), Zonenshain *et al.* (1989), and Fujimoto *et al.* (1994) helped to clarify the origin and geological significance of the terrains.

WASP deep-towed camera

For the MARNOK surveys, the WASP camera was loaded with black-and-white film, and towed along bottom for six to eight hours at a time. The vehicle's altitude was measured by a 35.5 kHz echo sounder with a range of 7–80 m and a resolution of 10 cm, and was used to guide the towing. At the optimum altitude of 10 m, frames are about 10 m across, and with a firing rate of one frame every every 15 s, we obtained almost continuously overlapping coverage. Camera depth was determined via a pressure sensor with a resolution and accuracy of ±1 m. By adding the instrument depth to its altitude we obtained very-high resolution bathymetric profiles of the sea floor, as shown in Figs 4 and 5, plotted against time. Since the tow speed was relatively constant, the time axis correlates approximately with distance. It has been scaled in the figures to produce no vertical exaggeration.

Fig. 2. Sea Beam bathymetry (Pockalny *et al.* 1988; Purdy *et al.* 1990) of the western Kane ridge-transform intersection and the two second-order segments to its north. The segments are separated at 24°22′ N by a third-order discontinuity corresponding to a right-stepping 1 km offset of the neovolcanic zone. On the basis of their morphology, the segment adjacent to the transform is a group 1 'narrowgate' type while Segment 2 to the north is a group 2 type with a wide U-shaped valley and a robust axial volcanic ridge (Sempéré *et al.* 1993). The northern half of Segment 2 lies within an oblique portion of the Mid-Atlantic Ridge; otherwise the axial valley is orthogonal to the spreading direction. Depth scale is in metres (contour interval is 200 m); thick black lines indicate segment axes.

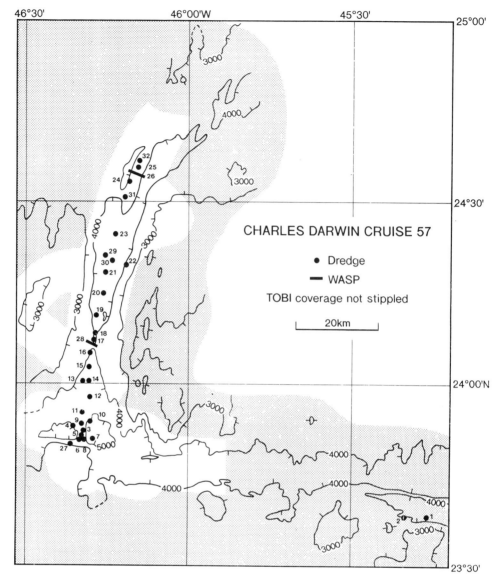

Fig. 3. Location of TOBI sidescan sonar coverage (unshaded areas), dredges (dots) and WASP deep-tow camera traverses (bars) from the RRS *Charles Darwin* cruise 57.

Gross morphology

Segment 1: the narrowgate

Segment 1 extends from immediately north of the fracture zone to 24°22′N. It is 57 km long, trends 010°, has a V-shaped cross-section, and an inner valley that narrows as its floor shoals towards a bathymetric saddle (3950 m deep) near 24°06′N (Fig. 2). Although there are no published gravity data in MARNOK, Segment 1's morphology places it in Sempéré *et al.*'s (1993) Group 1, or 'narrowgate'. The rift valley is symmetric at the saddle point, where the cumulative horizontal strain accommodated by faulting is low (*c.* 20%), but the valley walls become increasingly asymmetric to the north and south where the horizontal strain increases to about 40% (Allerton *et al.* this volume). As the valley deepens towards the segment ends, the relief on the eastern wall increases. At the RTI it achieves 3 km of elevation between the

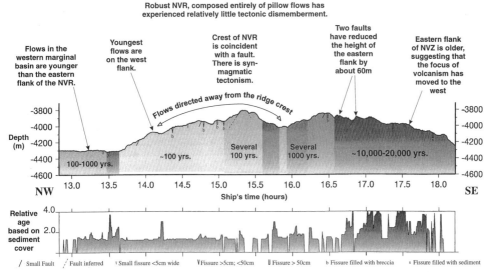

Fig. 4. High-resolution profile of camera traverse 26W over the neovolcanic ridge at the centre of Segment 2. Depth profile obtained from combination of WASP altimeter and vehicle depth, scaled to approximately no vertical exaggeration. The interpretative section is based on photoanalysis showing tectonic features and estimated crustal ages based on the relative age profile shown below. The relative age estimate uses the scale shown in Fig. 13.

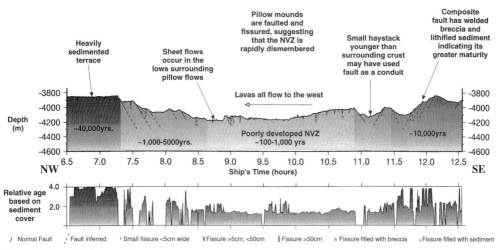

Fig. 5. High-resolution profile of camera traverse 28W across the narrowgate of Segment 1. Depth profile obtained from WASP altimeter and vehicle depth, scaled to approximately no vertical exaggeration. The interpretative section is based on photoanalysis showing tectonic features and estimated crustal ages based on the relative age profile shown below. The relative age estimate uses the scale shown in figure 13.

nodal basin and the inside corner high. In contrast, the western wall comprises a number of well-defined faulted terraces, each block being 1–2.5 km wide and having 200–400 m throw. To the north of the narrowgate the strike of the western faults is relatively constant at 012°, while to the south they deviate to 020°. The throw

gradually diminishes towards the transform, where the youngest faults are very small, leading to a wide inner valley floor and a 15 km wide basin on the non-transform side of the NVZ. The NVZ is bathymetrically poorly developed to the south of the saddle, consisting of a string of discontinuous edifices; in contrast, to the north

24°00'N —

46° 20'W

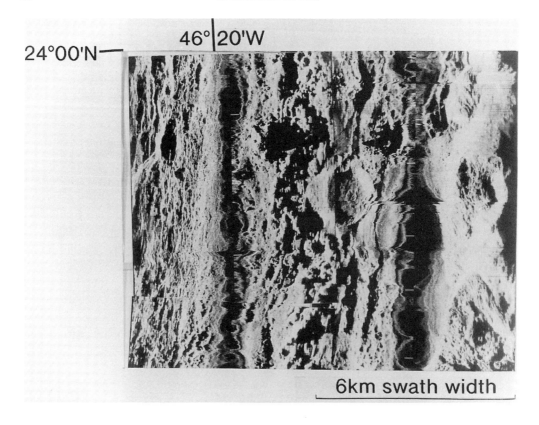

6km swath width

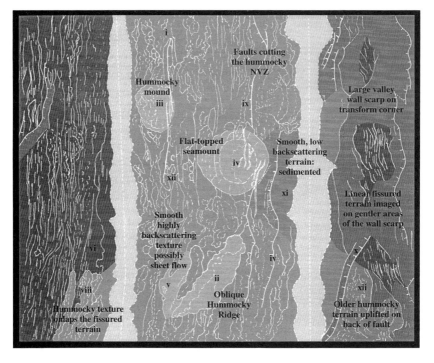

of the saddle there is a more continuous ridge with 50–150 m elevation.

Segment 2

Segment 2, defined by Sempéré et al. (1993), stretches from 24°22′N to 24°39′N. The morphology of its NVZ and rift valley contrast dramatically with Segment 1. Segment 2 is only 35 km long and its northern half lies within an oblique portion of the MAR which trends at 057° from 24°30′N to 25°00′N (45° to the average strike of the MAR between Kane and Atlantis transforms as defined by DeMets et al.1990). Two trends of valley wall faults, one ridge parallel and one oblique, mark the tectonic response to this oblique zone (see Allerton et al. this volume, fig. 2). Unlike Segment 1, the rift valley of Segment 2 remains asymmetric along its entire length with larger faults on the eastern side, although the magnitude of tectonic strain variations is similar (Allerton et al. 1995). The inner valley of Segment 2 is slightly wider than that of Segment 1 outside the RTI area.

The NVZ takes the form of a robust axial volcanic ridge (AVR) 400–600 m high, which is continuous for the entire length of the segment. The shape of the AVR is like an upturned boat, broadening towards the centre and tapering at the ends, with marginal basins separating the ridge from the inner walls of the median valley. The AVR reaches its shallowest depth of 3550 m towards the northern end of the segment near 24°37′N. The strike of the AVR crest is N–S at its southerly end and bends round to 020° in the north, where it terminates on the older crust of the oblique-trending median valley flanks (there being no well-developed wall fault at this point). Although the AVR is well developed compared to the NVZ of the narrowgate segment, the valley floor of Segment 2 is generally deeper than that of Segment 1, suggesting that the northern segment may in fact be the more magmatically starved of the two (Allerton et al. 1995).

Sidescan terrains

In the following three sections we describe the constructional (volcanic) and tectonic terrains recognised from the sidescan images, followed by a description of sedimentary and mass-wasting features.

Constructional terrains

(i) Hummocky neovolcanic texture. Figure 6 shows an area of sidescan mosaic from the northern edge of the RTI region in Segment 1, with an interpretation in terms of the terrains observed on the sidescan. The dominant constructional texture around the accretion axis is a fine-scale, hummocky volcanic terrain (Fig. 6i). It comprises a conglomeration of myriad, circular mounds, whose diameters range from a few tens of metres to about 500 m. The geometry of the shadows indicates that they are usually less than 50 m high, have rounded summits and no summit craters. TOBI data from the Reykjanes Ridge (Magde & Smith 1995; Murton & Parson 1993; Parson et al. 1993; Searle et al. 1994) and elsewhere on the Mid-Atlantic Ridge between 24° and 30°N (Smith et al. 1995a, b) show AVRs comprising a similar terrain. Our camera runs (Figs 4 & 5), which are described more fully below, showed that each hummock usually corresponds to an individual pillow flow, with the steep fronts and shallower tops of the flows causing the hummocky morphology. The hummocky terrain can vary in character from round, steep-edged mounds with strong shadows to dappled, lower relief areas where the contrast is much weaker. Such variations reflect the gross morphology of the flows, some of which are short with steep fronts while others are gentler, having undulations in the pillow flow surfaces which produce a dappled texture.

(ii) Hummocky ridges. Often the hummocks have a strong linear arrangement, as a narrow row of connected mounds or as a composite pile

---

Fig. 6. (**a**) TOBI sidescan sonar mosaic of the median valley inner floor near the northern edge of the ridge-transform intersection near 24°00′N, 46°15′W, showing a wide variety of sidescan terrains. Features i–xii are identified in (b). (**b**). Interpretation of (a) to illustrate various terrains observed on the sidescan sonar. Features i–xii are indicated on (a) and discussed in the text. (i) hummocky neovolcanic terrain; (ii) oblique hummocky ridge; (iii) hummocky mound; (iv) flat-topped seamount with summit depression and small faults cutting the summit plateau; (v) brightly backscattering smooth area with faintly dappled texture (possibly an area of sheet flow or low relief pillow lavas; (vi) fissured terrain with some small throw normal faults; (vii) gradational increase in number of faults and fissures cutting through neovolcanic zone; (viii) fissured terrain is onlapped by thin hummocky ridge forming a sharp contact; (ix) fault cutting the hummocky neovolcanic zone; (x) valley wall fault splays with the same trend as the oblique hummocky ridge; (xi) smooth low backscattering sedimented terrain; (xii) older hummocky terrain uplifted on back of fault.

Fig. 7. (a) Sidescan sonar mosaic of median valley floor at northern end of Segment 2, including axial volcanic ridge and flanking tectonized terrain. Features i–xi are identified in (b). (b) Interpretation of (a) to illustrate various terrains observed on the sidescan sonar. Positions of dredge sites 24D, 25D, and 32D and WASP traverse 26W are marked. Features i–xi are indicated on (a) and discussed in the text. (i) Fault gives strong linear character to the ridge crest; (ii) dappled hummocky texture of flow overlapping smooth talus indicates that faulting is syn-magmatic; (iii) talus fan has prograded out to form a brightly backscattering convex toe; (iv) large crestal flat-topped seamount; (v) southerly extent of 15 km long faulted hummocky ridge which extends under the centre of the crestal flat-topped seamount; (vi) sharp edge of faulted ridge that emerges from beneath the recent constructional activity; (vii) northern termination of faulted hummocky ridge bends to the east; (viii) bathymetric summit of the axial volcanic ridge is a high backscattering hummocky area to the NE of the crestal seamount; (ix) another sharp faulted ridge emerges from beneath the hummocky terrain and parallels the faulted crest to the west; (x) hummocky ridge bends to the east paralleling the faulted ridge to the west; (xi) poorly backscattering, sedimented crust of western marginal basin.

6km swath width

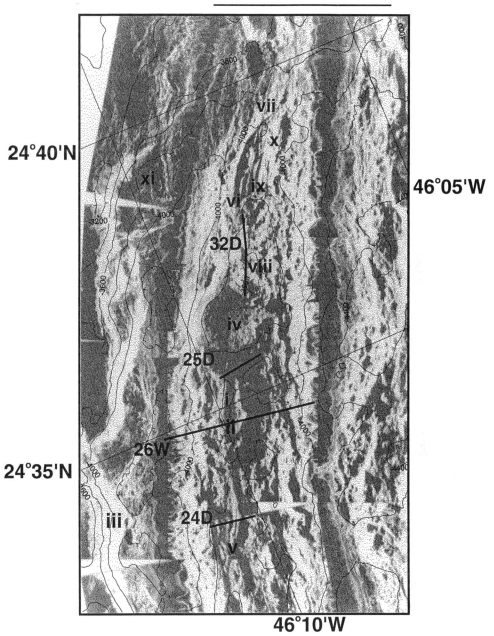

forming a constructional linear ridge (Fig. 6ii). Typically these hummocky ridges are several kilometres long and 100 m–2 km wide, often arranged in closely-spaced parallel rows, thus building up the NVZ as a composite pile of flows. In places the hummocky AVR has a gross alignment, but with no distinct, strongly linear hummocky ridges, suggesting that eruptions have occurred from a large number of conduits rather than from a dominant controlling structure. On the whole, these flows are directed away from the crest rather than along the strike of the feature (Figs 4 & 5). The alignment of the hummocky ridges is usually normal to the

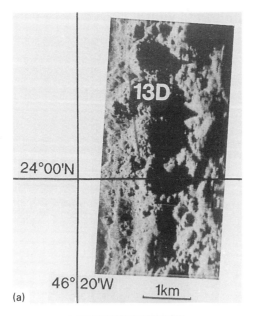

(a)

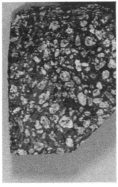

(b)

Fig. 8. (a) Sidescan image of composite seamount chain named Thatcher's Nose. The seamounts are topped by summit cones which cast spire-shaped shadows, from which the feature gains its name. (b) Rock sample dredged from Thatcher's Nose has about 40% modal plagioclase megacrysts.

spreading direction, suggesting that they are closely controlled by the stress regime. (The elongation of the ridges is also apparent in non ridge-parallel swaths, confirming that it is not an artifact of the directionality of the sidescan). In the RTI, the hummocky ridges adopt varying, oblique trends (Fig. 6ii), paralleling the edge of the nodal basin.

These hummocky ridges are interpreted as fissure eruptions similar to those commonly observed in subaerial basaltic volcanism. In Iceland and Hawaii eruptions which start with extrusion along the entire length of the fissure commonly focus into a number of discrete

points, usually spaced equally along the fissure (Thorarinsson 1969; Macdonald & Abbott 1970; Björnsson *et al.* 1979; Delany & Pollard 1982). In the extreme case, the flow of lava can become localized to a single vent (Richter *et al.* 1970; Thorarinsson *et al.* 1973). Smith & Cann (1993) comment that the ubiquitous nature of the hummocky ridges implies that for most eruptions the rate is not fast enough to sustain effusion along the entire fissure. There is some evidence that the ridges are fault- as well as fissure-related. Our camera runs revealed that when there is a sharp linear character to the ridge crest there is often a fault present; moreover, the brightly backscattering line of the fault scarp is sometimes imaged on the sidescan (Fig. 7i). Sometimes the smoother texture of fault talus is overlapped by the dappled hummocky texture of a flow, indicating that the faulting is syn-magmatic (Fig. 7ii).

A 4 km long chain of five distinctive composite seamounts named Thatcher's Nose (Fig. 8) seems to be an extreme end-member form of a hummocky ridge where almost complete focusing into regularly spaced centres has occurred over what is probably an eruptive fissure. Although each seamount is a discrete entity, the chain are clearly related by their similar morphologies, formed by an agglomeration of small rounded hummocks (<100 m diameter). The seamounts are up to 1 km across, about 100–200 m high, and are topped by summit cones that have slopes of 25°–30° and can provide up to another 100 m of relief.

Samples from Thatcher's Nose (dredge D13) are unusual in having approximately 2% olivine phenocrysts and 40% modal plagioclase megacrysts, about 20% more than from any of our other stations (Fig. 8b). An increase in lava viscosity is expected for such high crystal contents (McBirney & Murase 1984; Marsh 1989; Dragoni & Tallarico 1994) and may explain the unusually steep morphology. (Thatcher's nose is further discussed later.)

(iii) Hummocky domes. Hummocky cones or domes, reminiscent of cauliflower hearts (Fig. 6iii), are composite mounds up to 1.5 km across and 50–150 m high. They are relatively rare at MARNOK, though seen elsewhere on the MAR between 24°N and 30°N (Smith *et al.* 1995*a, b*) and common on the Reykjanes Ridge (Parson *et al.* 1993). The example in Fig. 6iii is surrounded by hummocky terrain and is faulted on its eastern side, but it is unclear whether it had an underlying linear constructional phase before the radial symmetry developed.

(iv) Flat-topped seamounts. Isolated seamounts with relatively smooth-textured, broad summit areas (referred to hereafter as 'flat-tops') form a distinctive volcanomorphology on the sidescan records (Fig. 6iv). They are rare in comparison with the hummocky volcanic terrain, occupying less than 5% of the NVZ. On the sidescan they appear to have flat tops, but in fact there is a very slight doming to the summit plateau. They are circular, 1–2 km in diameter and 50–200 m high, and have steep, strongly backscattering rims. There are often circular, 50–100 m diameter areas of shadow, probably small pit craters, in the centres of the circular summits. Figure 6iv shows a larger central collapse structure 600 m across. The radial symmetry and central depressions of these seamounts are strong indicators that they were fed from relatively profound central pipe conduits. Pillow lavas were the most common morphology recovered from dredges that targeted the flat-topped seamounts, although a small quantity of ropy pahoehoe implies that sheet flows may also be present.

These seamounts are similar in size to those seen elsewhere on the MAR (Batiza *et al.* 1989; Smith & Cann 1990, 1992; Smith *et al.* 1995*a*, *b*). They are similar in shape to seamounts near the axis of the East Pacific Rise (Searle 1983; Fornari *et al.* 1987), but fall in the lower end of that size spectrum. The texture seen on TOBI sidescan shows an evolution with age: compare the 'nested' seamounts (Fig. 9i) with a faulted seamount on older, poorly backscattering terrain (Fig. 9ii). Relatively young seamounts have a strongly backscattering summit plateau with a faintly dappled texture and slightly irregular margins. Their steep-sided flanks are very strongly backscattering but with a slightly sinuous base that tends to blend in with the surrounding hummocky terrain. Older seamounts have summit plateaux that become progressively more poorly backscattering as sediment accumulates, and are sometimes cut by faults. The edges of both the summit plateau and the base become smoother and more sharply defined as vigorous mass wasting degrades the steep constructional slope to produce a very uniform highly backscattering flank. The steep flanks of older seamounts contrast strongly with the summit plateau and the adjacent seafloor.

Studies from the Pacific have often located areas of satellite cones and lava fields around the base of seamounts (Batiza & Vanko 1983; Searle 1983). Satellite volcanism is thought to play an important part in the subsequent development of a seamount's summit plateau (Fornari *et al.* 1984). Eruptions with lower magmastatic heads

will preferentially erupt round the flanks and may drain off magma and precipitate summit collapse; summit eruptions will have higher magma pressure and more rapid extrusion rates. The off-axis flat-top shown in Fig. 9iii has a youthful texture and is surrounded by a strongly backscattering mantle of hummocky terrain overlying the older, highly sedimented crust to the north and south. We suggest this mantle was produced by satellite volcanism.

Figure 9i shows three flat-topped 'nested' or overlapping seamounts that have erupted with their centres 800 m apart with an included angle of 100°. This morphology suggests three discrete pulses of point-like rather than linear magma supply unrelated to a single fault or fissure. However, these seamounts occur at the junction of the two segments where the valley wall faulting suggests there is failure in two orientations, and where oblique structures are also present in the NVZ (Fig. 9iv & v). Thus their unique structure may be the result of two intersecting fracture systems providing some element of structural control on the magma conduit. Although most of the flat-topped seamounts are not obviously underlain by fractures, some of the larger ones do lie astride faults (Figs 6iv and 7ii), and Allerton *et al.* (this volume) have suggested that the extrusion of at least some of them is accomplished by the movement of magma up the valley-wall faults.

(v) Cratered cones. The cratered cone imaged near the nested seamounts in Fig. 9vi has neither the flat top of the other seamounts nor the elongation or composite flow morphology associated with the hummocky terrain. The well developed summit crater is about 600 m in diameter and suggests that the edifice is pipe fed. It is only about 60 m high, but has erupted onto an older flat-topped seamount that is already about 100 m tall and is cross cut by fractures (Fig. 9iv). This may be similar, though on a smaller scale, to the conical mounds sometimes built over the summit plateau in the final phase of Pacific seamount volcanism (Fornari *et al.* 1984).

(vi) Smoother neovolcanic terrains. Smoother, apparently flat, but highly backscattering areas are sometimes observed between the hummocks and seamounts that dominate the neovolcanic zone. The larger expanses of sheet flow identified during the camera survey coincide with completely smooth, flat and relatively strongly backscattering areas on the sidescan (Fig. 10i). Within these expanses, a number of different sheet flow morphologies were clearly identified during the photoanalysis but are not resolved on

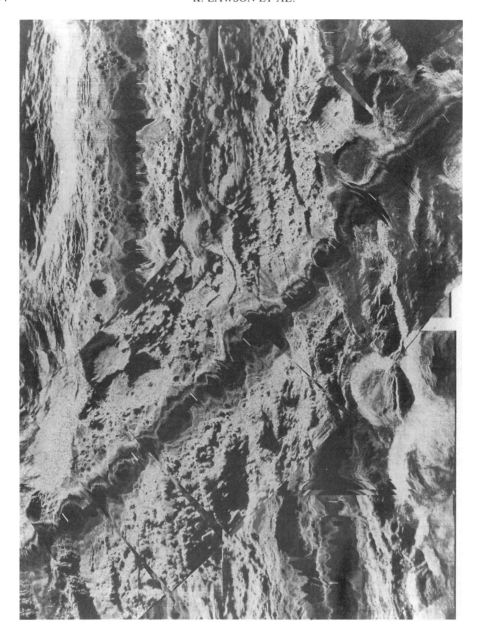

Fig. 9. (a) Sidescan mosaic of the area around the junction of segments 1 and 2. Features i–xi are identified in (b). (b) Interpretation of (a) to illustrate various terrains observed on the sidescan sonar. Positions of dredge sites 21D, 23D, 29D, 30D are marked. Features i–xi are indicated on (a) and discussed in the text. (i) Three nested flat-topped seamounts; (ii) older flat-topped seamount with faulted eastern flank; (iii) flat-topped seamount with mantle of brightly backscattering hummocky terrain adjacent to highly sedimented area at base of eastern median valley wall; (iv) oblique fractures cutting neovolcanic zone with same orientation as the oblique wall-fault to the east; (v) fault splays trending in same oblique orientation as fractures in the centre of the valley; (vi) cratered cone on summit of faulted seamount; (vii) landslide on western median valley wall; (viii) sediment slumping on eastern wall; (ix) sharply defined wall-fault emerges to the south of the area of sediment slumping; (x) dissected flat-topped seamount in centre of median valley; (xi) N–S fault cutting the hummocky terrain at end of axial volcanic ridge of Segment 1.

6km swath width

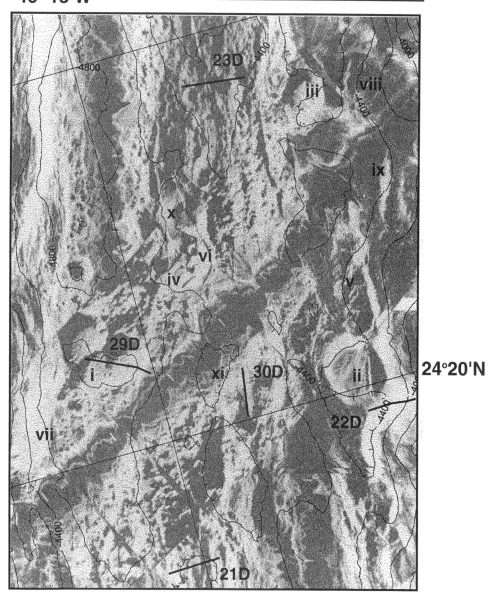

the sidescan images; nor were smaller sheet flows (<20 m across) or some with pillowed margins. Other low-relief areas have a faintly dappled texture on the sidescan images. One dredge station (D11) traversed one of these bright, low-relief areas (Fig. 6v) which is surrounded by the more typical hummocky terrain, and recovered only rounded pillow fragments. Either the indistinct hummocky texture is a low relief pillow field, or the dredge only sampled material from the surrounding higher relief hummocky terrain.

Tectonic terrains

(i) Tectonized terrain. Outside the NVZ, most of the median valley floor is characterised by a strongly lineated terrain (Fig. 6vi). We identify this with the zone of intense fissuring and small-throw normal faulting seen on submer-

Fig. 10. (**a**) Sidescan sonar mosaic of uniform hummocky ridge north of Segment 1. Features i–iv are identified in (b). (**b**) Interpretation of (a) to illustrate various terrains observed on the sidescan sonar. Positions of dredge sites 15D, 16D, 17D, 18D, 19D, 20D and WASP traverse 28W are marked. Features i–iv are indicated on (a) and discussed in the text. (i) strongly-backscattering, smooth-textured terrain identified as sheet flow by photoanalysis occurs where the neovolcanic zone narrows towards the narrowgate bathymetric saddle; (ii) very sharp contact between the hummocky terrain and the fissured terrain; (iii) smooth textured sedimented terrain on small faulted terrace on median valley floor just north of the saddle point; (iv) linear valley-wall fault breaks into complex duplex structure in the transform domain.

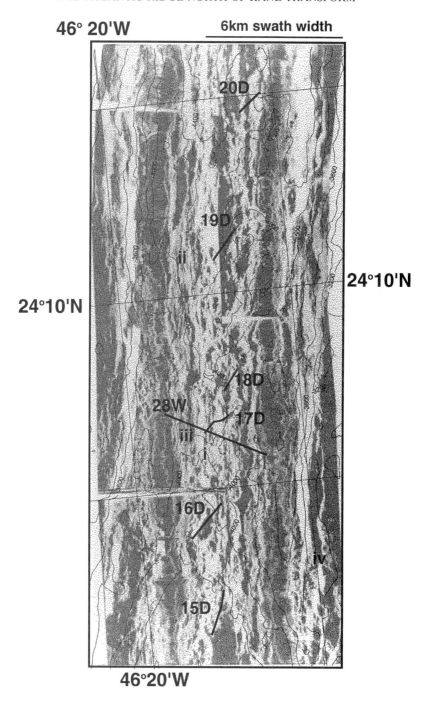

sible dives between the edges of the NVZ and the inner walls of the rift valley in the FAMOUS area (Ballard & Van Andel 1977). The contrast between this tectonized terrain and the hummocky texture usually allows the boundary of the neovolcanic zone to be easily delineated. This boundary is usually very sharp and clearly defined (Fig. 10ii), though occasionally it is more gradational, marked by a progressive increase in fissure density, or by fissuring between adjacent hummocky ridges (Fig. 9vii). Sometimes the contact is highly sinuous where the tectonized terrain is clearly onlapped by the younger neovolcanic terrain (Fig. 9viii). A wide gradational zone suggests that sometimes a fracture will tap magma and sometimes not, so either the depth or dip of the fractures is variable, or else the zone of magma is discontinuous in space and time. Where there is a sharp, well-defined limit to the NVZ the stages of volcanism and tectonism probably have a longer episodicity.

The tectonic fabric that we image has a high density of faults or fissures with a cross-strike spacing of less than 50 m. It includes fine scale features right down to the limit of resolution. Close examination reveals that many of the bright linear features lack a coupled shadow and so are actually small-throw fault scarps rather than true fissures. Individual faults can appear continuous for up to 6 km, although it is difficult to trace them along strike for more than several hundred metres at a time owing to shadowing from other features. Lineaments can be seen splaying and rejoining, sometimes with fluctuations in horizontal separation along their length, although they rarely cross as most are aligned subparallel to the ridge trend.

(ii) Faults and the median valley walls. Faults have sharply defined sub-parallel edges and a strong degree of linear continuity which cuts through the surrounding terrain (Fig. 6ix). They are usually strongly backscattering (Fig. 6x) because their steeply inclined surfaces present a high angle of incidence to the sonar beam and allow minimal accumulation of sediment.

On medium and long range sonars such as GLORIA (Laughton & Searle 1979; Roest *et al.* 1984) and Sea Beam (Purdy *et al.* 1990), which have lateral resolution of 100–200 m, the median valley walls generally appear as single scarps. The TOBI sidescan reveals a detailed structure which has only previously been observed on submersible dives (Ballard & Van Andel 1977; Macdonald & Luyendyk 1977) and on other high resolution sonars (Kong *et al.* 1988/89). Only where the bathymetry is steepest do the valley wall faults seem to form a continuous bright scarp. As the gradient of the wall decreases the complexity and composite nature of the scarp is revealed. A more detailed analysis of the faulting patterns recognised in TOBI and the development of the median valley wall faults is given in Allerton *et al.* (this volume) and McAllister & Cann (this volume).

Sedimentation and mass wasting

(i) Sedimentation of volcanic terrains. Within the limits of our study area, sedimentation and mass wasting become increasingly dominant with distance from the axis. Sedimented areas are identified by the lower amplitudes of their backscatter and the smoothed texture caused by ponding and draping. Our deep-towed photography showed that less than 1 m of sediment was required to cover horizontal pillow fields and sheet flows to form a planar surface. This generated a smooth sidescan texture with moderate to low backscatter (Fig. 10iii). Dredges over similar smooth, low backscatter areas also recovered sediment (Fig. 6xi).

As high relief pillow flows grow older and move off axis their hummocky nature becomes less pronounced (Fig. 6xii). This is a result partly of tectonic dismemberment and partly of sediment cover. As the hummocks are progressively covered by sediment, the steep, rounded edges of the flows remain brightly backscattering, but the sizes and number of the bright patches are reduced as the poorly backscattering interstitial sediment ponds increase in thickness and lateral extent. The overall effect is one of both smoothing and reduction of average backscatter, although the volcanic hummocky terrain still remains clearly identifiable.

Sediment covered flat-topped seamounts are identified by the strongly backscattering arcuate forms of their steep flanks, which stand out against the darker circle of the sedimented summit.

(ii) Talus and erosional gullies. Talus forms uniform, smooth, highly backscattering areas at the bases of fault scarps or other steeply dipping surfaces. Sometimes the talus forms an obvious fan which has prograded out into a convex or scalloped front (Fig. 7iii). Streaks of high backscatter are often seen trending downslope on the otherwise smooth talus. They occur on the walls of the inside corner high where Karson & Dick (1983) have observed rock debris slides forming curvilinear trails of loose fragments hundreds of metres long. On the southern transform wall similar streaks contain some barely resolvable specks. The ANGUS camera traverse across this area showed reactivated debris slides which contain blocks of consolidated talus from older deposits (Karson & Dick 1983). The sonar specks may result from such individual larger blocks, which are most likely to occur when older consolidated deposits are broken up or if deeper crustal levels are exposed. In contrast, talus composed solely of recently broken pillow lavas will tend to have a more homogeneous texture. When active deformation on a fault ceases, the talus becomes 'inactive' and increasingly poorly backscattering as it is covered by sediment. A progression of

fault scarp evolution from immature to draped stage is described in Allerton *et al.* (1995a).

The material which forms the talus ramps and debris slides has mostly come from mass wasting of the exposed upper areas of the scarp. The erosional gullies that develop as a result produce shadowed depressions between the brighter streaks and knobbles of rock butresses. This gives the top of the fault scarp a crenulated appearance with the gullies occurring approximately every 100–200 m along strike. The talus fans often emerge from the base of the gullies suggesting that the degraded material has been channelled down them. McAllister *et al.* (1993) have studied the progressive change in fault morphology off axis and concluded that the dominant degrading mechanism is tectonic activity on either the parent fault or adjacent ones.

(iii) Mass flows: landslides and sediment slumping. Sediment slumping and landslips are evident in the transform and on the scarp faces of rift mountains. On the transform median ridge, and occasionally along the median valley wall, we see a terrain characterized by irregular, diffuse, moderately backscattering, low amplitude domes, which downlaps with an abrupt lobate contact onto the terrain below (Fig. 9vii). Cruise CD65 produced TOBI images of nine areas along the MAR between the Kane and Atlantis transforms with similar acoustic characteristics and also interpreted as landslides (Cann *et al.* 1992); the largest of these was the major landslide described by Tucholke (1992). Sampling from these landslide sites has recovered material ranging from basalt blocks and mud to metabasalt, gabbro and serpentinite (Cann *et al.* 1992). A sediment slump which was identified in our TOBI 7.5 kHz profiler record corresponds to an area of valley wall with very low intensity backscatter on the sidescan records (Fig. 9viii) and a convex bulge in the contours of the valley wall. The amplitude of the backscatter increases towards the steeper front of the slump which has a disrupted diffuse texture compared to the sharply defined fault scarps to the south (Fig. 9ix).

Along-axis variations in median valley floor geology

The following section will focus on the changing character of the volcanic and tectonic terrains seen along the inner valley. This includes variations in the volcanic morphologies, the faulting that cuts the NVZ, and the boundary between the neovolcanic and tectonic fissured terrains. The varying character of the wall faulting is discussed in Allerton *et al.* (this

volume) and will only be mentioned in passing. For the purposes of this description, the MAR-NOK area has been divided into four regions: Segment 2 (the northern half of which is shown in Fig. 7); the third-order non transform offset (NTO) and the area near the junction of the two segments (Fig. 9); the north half of Segment 1 (Fig. 10); and the southern half of Segment 1 which includes the RTI (Fig. 11). The account will start in the north and continue southwards towards the RTI.

Segment 2

The robust AVR of Segment 2 is composed almost entirely of hummocky terrain in a broad composite pile. There is a strong degree of linearity near the crest while the flanks have a more isotropic texture. This suggests that larger structures (probably faults) dominate at the crest, while there is less evidence of prominent fault (or fissure) control on the flanks. The eastern flank is more extensively faulted than the western flank and the southern end of the AVR is more tectonized and less voluminous than the northern end. The faults cutting the hummocky terrain range in length from a couple of hundred metres to 5 km. In some places they show a remarkable congruence with the valley wall faults while in others there seems to be very little relationship between the wall faulting and the tectonism on axis.

The northern half of the AVR is dominated by a large, poorly backscattering, flat-topped seamount (Fig. 7iv). This sits astride a partly faulted and partly constructional ridge that extends for over 15 km along most of the northern half of Segment 2 (Fig. 7v–vii). The seamount is the largest in the study area, being 2.5 km in diameter and at least 200 m high. The low backscatter amplitudes of the summit plateau and the smooth degraded texture of the western flank imply that it is relatively old. The highest point of the AVR is a 3 km-long, strongly backscattering, broad, hummocky area of recent volcanism (Fig. 7viii). It is located immediately to the northeast of the older seamount and directly underlies another partly faulted, partly constructional ridge (Figure 7ix–x). These observations confirm that there is a close relationship between faulting and volcanic construction. The AVR seems to be built up by eruptions associated with closely spaced, subparallel structures (faults and also probably fissures) which almost certainly act as conduits for magma. The marginal basin between the west side of the AVR and the valley wall is very poorly backscattering, indicating that sediment cover is more than 1 m thick (Fig. 7xi), while the

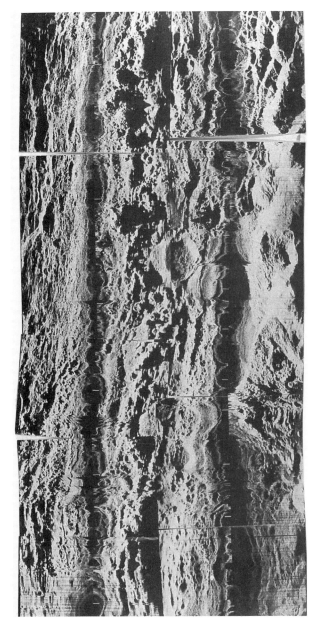

Fig. 11. (a) Sidescan mosaic of the southern half of Segment 1. Features i–xiii are identified in (b). (b) Interpretation of (a) to illustrate various terrains observed on the sidescan sonar. Positions of dredge sites 3D, 4D, 5D, 6D, 7D, 8D, 9D, 10D, 11D, 12D, 13D, 14D & 27D in the southern half of Segment 1 are shown. Features i–xiii are indicated on (a) and discussed in the text: (i) median valley widens as western valley-wall faults decrease in throw and deviate to the west; (ii) expanse of fissured terrain on western side of median valley increases as valley widens; (iii) hummocky mound; (iv) Thatcher's Nose; (v) high backscattering smooth-textured pillow field or sheet flows; (vi) flat-topped seamount; (vii) tectonised smooth- textured seamount on walls of nodal basin; (viii) oblique hummocky ridge; (ix) neovolcanic zone curves out to the west and then strikes N–S towards the transform, following the edge of the nodal basin; (x) sharply defined hummocky ridge trends into the transform valley; (xi) fine blibble ridges onlap fissured terrain as neovolcanic zone widens; (xii) eastern wall abruptly deviates to strike of 120°; (xiii) northern wall of nodal basin trends at 100°; (xiv) terrace of older crust uplifted into valley wall between fault splays.

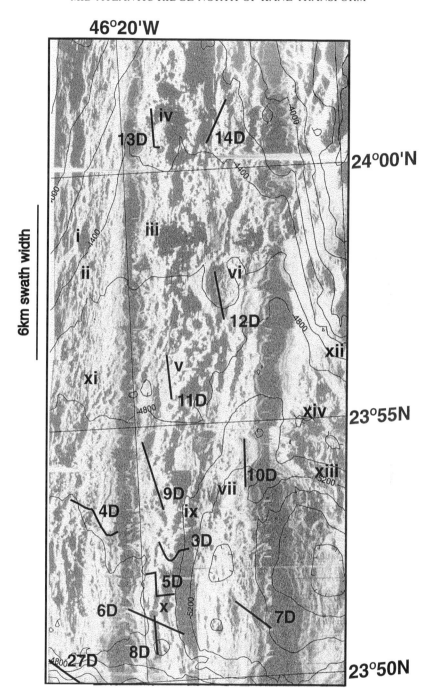

eastern marginal basin contains much younger, highly backscattering terrain which reflects the oblique trend of the ridge at the northern end of the segment. To the south of the area shown in Fig. 7 both eastern and western marginal basins comprise poorly backscattering, diffusely hummocky terrain. However, the marginal basin is wider on the eastern side and contains more debris from sediment slumps and landslides reflecting the greater maturity of the larger eastern wall faults.

The contact of the AVR with the older crust in

the marginal basins seems relatively sharp, but for most of Segment 2 the actual contact is not observed as it falls within the nadir of the sidescan. The presence of large syn-magmatic faults which dominate the ridge crest may imply that the construction of the robust AVR is controlled at least partly by these structures. These large faults also point to a relatively thick lithosphere across the median valley of this segment. The well defined age contrast between the AVR and most of the marginal basin suggests that the AVR was built after a relatively long period of magmatic quiescence.

The junction of segments 1 and 2

The most striking feature of the area around the 24°22′N NTO is the large number of distinctive seamounts (Fig. 9i–iii, vi, x). Most of these seamounts are flat-tops with a wide range of apparent ages. A cratered cone (Fig. 9vi) and flat-topped seamount immediately to the north (Fig. 9x) are both situated on the W flank of the AVR of segment 2. However, many other seamounts are situated at the edges of the median valley adjacent to the wall faults, so it seems likely that some of these segment-end seamounts were erupted at the valley edges next to the median valley wall.

The NVZs of the two segments do not overlap, but the hummocky terrain is continuous across the area between the tips of the AVRs, and is cut by a series of small-throw, ridge-parallel, en-echelon faults (Fig. 9xi). There is also on the median valley floor a set of minor structures, parallel to the oblique trend of the eastern wall fault, that are probably transfer faults (Fig. 9iv, v). The eastern valley-wall fault increases in throw towards the NTO while on the western side of the ridge there is a series of small throw fault terraces. The eastern fault has thick sediment cover and displays features resulting from mass wasting and slope failure, indicative of a highly mature fault scarp. Landslides have occurred on both the east and west walls of the NTO. Such features may preferentially occur at the NTO because of the increased fault throw and scarp dissection across the median valley walls there, in part perhaps resulting from the intersection of faults with different orientations.

Northern end of Segment 1

The median valley of the northern half of Segment 1 is notable for its uniformity. The NVZ comprises a very uniform, low relief, hummocky AVR. Elongated hummocky ridges dominate and hummocky mounds are rare,

while other distinctive seamount morphologies are entirely absent from the sidescan images. However, the bathymetry data indicate that there are two seamounts adjacent to the western wall at 24°18′N and 24°11′N that have fallen in the nadir of the sidescan swaths. The one at 24°18′N (relatively close to the NTO) appears to be a flat-top and that at 24°11′N to be a hummocky mound.

In contrast to the AVR of Segment 2, the hummocky ridges here are much shorter, and at no point along the crest does a single one dominate. There is a well-defined sharp contact between the hummocky neovolcanic terrain and older fissured and faulted terrain imaged to the west of the AVR (Fig. 10ii). The trace of this contact shows that the NVZ is actually offset slightly to the east of the centre of the valley floor. The sharpness of this contact implies that volcanism is highly focused and/or has a relatively long-period episodicity.

The width of the AVR narrows gradually towards the bathymetric saddle (Fig. 10iii). Thus, in contrast to orthodox views of segmentation, the bathymetric centre of the segment is actually where the NVZ is least extensive. The dredge stations in this area (D15, D16 & D17) recovered the freshest basalts in the sampling program (pristine glass crusts up to 3 cm thick were recovered in D16) showing that although the NVZ is relatively narrow, it has been recently active. The extent to which the hummocky terrain is tectonized is relatively constant along axis and is less than that seen at the NTO and over parts of the AVR of Segment 2. Those faults that do cut the hummocky AVR are ridge-parallel, as are the valley wall faults.

Southern end of Segment 1: the RTI

The morphological effect of the transform extends as far as 24°04′N, 26 km from the Principal Transform Displacement Zone. This is similar to the width of other 'transform domains' (Fox & Gallo 1986) and to the estimated length scale of the TFE. The clearest manifestation of the change in the stress regime is seen in the median valley walls: the eastern, linear wall breaks down into a complex structure (Fig. 10iv) and the western wall starts to deviate to the west (Fig. 11i), causing the valley to widen. At the same time, the width of the fissured terrain increases on the western side of the NVZ (Fig. 11ii), although the width of the hummocky terrain remains relatively constant.

The hummocky terrain also widens out westwards as fine hummocky ridges onlap onto the fissured terrain (Fig. 11xi). Adjacent to the transform, this terrain is 10 km wide and comprises two ridges, each about 5 km wide. The easternmost, adjacent to the nodal basin, is the present NVZ.; the western one (in the western basin) may be a relict NVZ.

The morphological diversity of the NVZ dramatically increases in the transform domain. In addition to the hummocky ridges there are hummocky mounds (Fig. 11iii), chains of hummocky seamounts with summit cones (e.g. Thatcher's Nose – Fig. 11iv), and smooth-textured areas of sheet flows or low-relief pillow fields (Fig. 6v). There are also at least two flat-topped seamounts imaged within the NVZ (e.g. Fig. 11vi).

At 23°57′N there is a prominent, oblique-trending hummocky ridge (Fig. 11viii), whose trend parallels both the edge of the nodal basin and the oblique faults in the valley walls. At the same latitude the orientation and width of the NVZ change: the NVZ curves first to the west, and then back to follow an approximately N–S strike towards the transform. It thus parallels the edge of the nodal basin (Fig. 11ix). At the southern tip of the segment there are two narrow, sharply crested hummocky ridges (Fig. 11x), one of which extends right into the transform valley where it terminates against the talus ramp of its southern wall.

The boundary between the hummocky terrain and the fissured terrain is gradational along most of the southern half of Segment 1, although in places a clear onlapping relationship is seen. This suggests that volcanism and tectonism were more-or-less simultaneous, or at least succeeded one another on a short time scale. The heterogeneous character of the NVZ implies that volcanism is taking place in small discrete episodes, probably from individual magma bodies separated in space and time. The westward widening of the valley near the transform and the presence of the relict NVZ in the western basin are consistent with strongly asymmetric valley wall faulting across the RTI and a reduction in the half spreading rate to the east as deduced from our near-bottom magnetic anomalies.

Geological map and summary

A geological map of the MARNOK area (Fig. 12) provides a summary of the sidescan interpretations. Note the clusters of seamounts as opposed to more uniform hummocky terrain at the discontinuity between segments 1 and 2 and in the RTI area. This increase in volcanic complexity toward segment ends is generally accompanied by an increase in tectonic complexity in terms of the faulting patterns observed in the valley walls. The NVZs of the two segments are quite different in plan view. That of Segment 2 tapers monotonically towards its ends, and its constituent hummocky ridges are relatively laterally extensive and have strongly pronounced crests. The NVZ of Segment 1 also tapers into the discontinuity, but in contrast it narrows at the bathymetric saddle of the segment centre. It then widens as the valley itself widens towards the transform. Adjacent to the nodal basin there are two subparallel volcanic ridges which curve slightly, mirroring the edge of the nodal basin.

Photoanalysis

We ran two traverses of the WASP deep tow camera system across the AVR of Segment 2 and the bathymetric saddle of Segment 1 (stations 26W and 28W, respectively). Not only did these prove useful in groundtruthing TOBI's sidescan sonar, but they also provide a comparison of the tectonic and volcanic style and age variations across the centres of the two segments. The photographs were described systematically in terms of lava morphology, tectonic features and sediment cover (Figs 4 and 5; Lawson 1996).

Relative and absolute age estimates

The ages of the pillow flows were estimated by using the extent of degradation of the surface and the amount of pelagic sediment cover. Frames with the complete spectrum of sediment cover were arranged in order of increasing sedimentation and assigned values on a relative age scale to correlate as closely as possible with the sketches of Ballard et al. (1981). Figure 13 shows a sample from this relative age scale. During systematic logging of the photos, each frame was assessed by visually matching against this scale.

The variations in microtopography between different flow surfaces can make relative age assessment difficult, particularly when trying to estimate the relative ages of sheet flows from a scale which is based primarily on the cover of pillow surfaces. Often, although various sheet flow surfaces seem, on morphological evidence, to have formed contemporaneously, they can give the appearance of slightly different ages because of systematic differences in their apparent sediment cover. Figures 4 and 5 show profiles of the relative age estimates for both traverses. Repeat analysis of several sections of film gave a maximum variation of 0.4 on the relative age scale, which can be taken as an approximate error for the method.

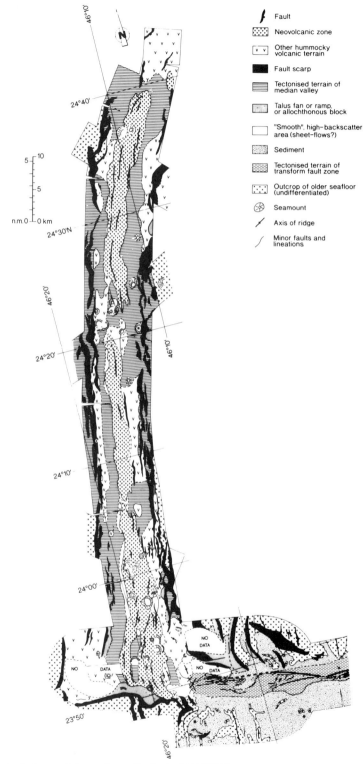

Fault

Neovolcanic zone

Other hummocky volcanic terrain

Fault scarp

Tectonised terrain of median valley

Talus fan or ramp, or allochthonous block

"Smooth", high-backscatter area (sheet-flows?)

Sediment

Tectonised terrain of transform fault zone

Outcrop of older seafloor (undifferentiated)

Seamount

Axis of ridge

Minor faults and lineations

Fig. 12. Geological map of the median valley in the MARNOK area.

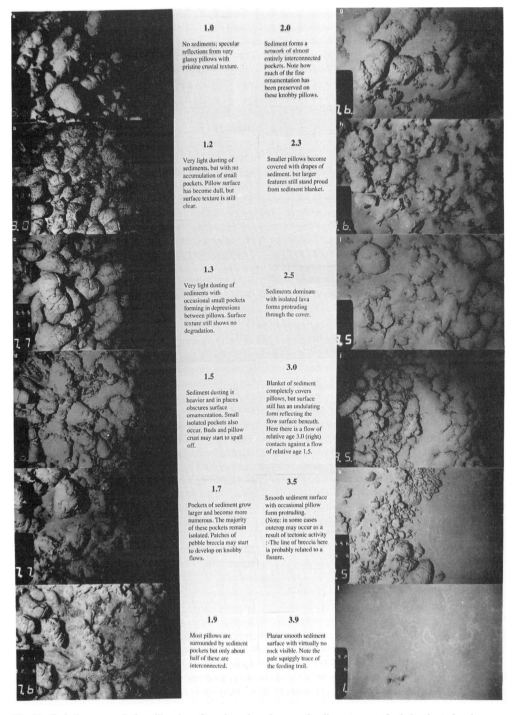

1.0

No sediments; specular reflections from very glassy pillows with pristine crustal texture.

2.0

Sediment forms a network of almost entirely interconnected pockets. Note how much of the fine ornamentation has been preserved on these knobby pillows.

1.2

Very light dusting of sediments, but with no accumulation of small pockets. Pillow surface has become dull, but surface texture is still clear.

2.3

Smaller pillows become covered with drapes of sediment, but larger features still stand proud from sediment blanket.

1.3

Very light dusting of sediments with occasional small pockets forming in depressions between pillows. Surface texture still shows no degradation.

2.5

Sediments dominate with isolated lava forms protruding through the cover.

1.5

Sediment dusting is heavier and in places obscures surface ornamentation. Small isolated pockets also occur. Buds and pillow crust may start to spall off.

3.0

Blanket of sediment completely covers pillows, but surface still has an undulating form reflecting the flow surface beneath. Here there is a flow of relative age 3.0 (right) contacts against a flow of relative age 1.5.

1.7

Pockets of sediment grow larger and become more numerous. The majority of these pockets remain isolated. Patches of pebble breccia may start to develop on knobby flows.

3.5

Smooth sediment surface with occasional pillow form protruding. (Note: in some cases outcrop may occur as a result of tectonic activity :-The line of breccia here is probably related to a fissure.

1.9

Most pillows are surrounded by sediment pockets but only about half of these are interconnected.

3.9

Planar smooth sediment surface with virtually no rock visible. Note the pale squiggly trace of the feeding trail.

Fig. 13. Relative age scale for pillow lava flows based on degree of sediment cover. Scale is adapted and extended from that of Ballard *et al.* (1981).

There are large errors involved in relating the areal sediment cover to the absolute age of pillows. These include uncertainties about the sedimentation rate, the degree of sediment redistribution, and variations in the mode of sediment cover, with draping dominant in some areas and ponding in others. The age variations included in Figs 4 and 5 were estimated using a sedimentation rate of 1–2 cm per 1000 years for the centre of the North Atlantic (Marks 1981), and for sediment cover >50% are based on the exponential curve of Lichtman *et al.* (1984) who estimate that a terrain of pillows averaging 1 m in diameter will be completely obscured in 40 000 years. For smaller degrees of sediment cover, this curve is deemed to be inaccurate, as it assumes complete sediment ponding into the inter-pillow pockets whereas observations show that a sprinkling of sediment over the entire flow surface comprises a relatively large proportion of the total sediment cover on young pillows. We estimate that flows with relative ages of 1.2, 1.5 and 1.7 have absolute ages of about 100, several hundred and 1000 years, respectively.

The phototraverse across the AVR of Segment 2

Figure 7 shows the location of station 26W which ran for about 5 km across the AVR of Segment 2, from 24°35.1′N, 46°11.1′W to 24°34.6′N, 46°08.4′W. The AVR at this locality consists of two peaks, of which the western is slightly higher (Fig. 4). 550 frames were exposed, showing the AVR to be composed entirely of pillowed flows of varying ages. The flows commonly exhibit frontal breccias. Even the flatter area of the western marginal basin is floored by recent pillow flows (several 100 years old), which overlie older areas of crust with an age of several thousands of years. The sedimented and de-graded nature of the eastern flank of the AVR contrasts with the pristine pillows on the main crest and the western flank, which probably all erupted within the last several 100 years. The oldest flows exposed on the eastern flank are at least 20 000 years old and volcanism ceased there about 10 000 years ago as the focus of volcanism moved progressively to the west.

Fissures are seen throughout the traverse, even on relatively young crust. The western crest of the AVR is coincident with a fault on which there has been syn-magmatic tectonism. Three faults on the eastern flank all have displacements down to the west, with a cumulative throw of about 60 m. This faulting is probably related to volcano collapse, as documented in the FAMOUS area by Ballard & Van Andel (1977), where volcanoes were thought to subside by as much as 100 m shortly after formation by faulting around their bases. Before collapse, the pillow volcano which forms the eastern summit of the AVR would have been the same height as the younger crest to the west (Fig. 4).

The phototraverse across the narrowgate of Segment 1

Station 28W traverses the inner valley about 1.5 km north of the bathymetric saddle of the narrowgate of Segment 1 between 24°07.3′N, 46°18.6′W and 24°06.3′N, 46°6.4′W (Fig. 10). In this 5 km long deployment, 800 frames were exposed though there are occasional data gaps owing to the difficulty of maintaining a suitable altitude over the sharply dipping scarps.

The photos revealed a highly variable terrain of sheet and pillow flows with a wide range of ages and in varying states of tectonic dismem-berment. The NVZ lies in the deepest part of the valley's cross-section and is poorly developed in terms of axial topography, achieving a relief of only 70 m compared with the 500 m of Segment 2's AVR. The NVZ is only 2 km wide and contains a number of sheet flows confined to flat-lying depressions between adjacent pillow mounds. Some of the pillow flows have large, extensive tubes that were not observed on the northern traverse. The presence of tubes and especially sheet flows indicates higher effusion rates (Griffiths & Fink 1992*a*, *b*) than for Segment 2. Such lavas are thought to occur during the early phase of a volcanic cycle when channelling is relatively undeveloped (Ballard *et al.* 1989). None of the narrowgate pillow flows have frontal breccias. Some may be buried by surrounding sheet flows, but probably the pillow fronts in this lower relief terrain did not become steep enough for extensive slumping and brec-ciation to take place.

The majority of NVZ flows seem to be little more than a few hundred years old, but faulting and fissuring, which is confined to the pillow mounds, has accelerated the brecciation of the surface. The tectonic disruption of younger pillow flows within the narrowgate traverse is more extensive than that within the AVR of Segment 2, suggesting that the Segment 1 NVZ is being dismembered relatively rapidly. If this faulting is related to the subsidence of the pillow volcanoes then it seems that the lithosphere at the narrowgate can support only a small con-structional edifice compared with that of Seg-ment 2.

The sequence of extrusion is quite complex, with pillow flows downlapping onto older sheet flows, but also younger sheet flows onlapping older pillow fronts. The youngest flows appear to be at the centre of the NVZ, but there are also inlicrs of older sheet and pillow flows in this central intrusion zone. There is no evidence for systematic movement of the focus of volcanism across the NVZ, but in 80% of cases the flow immediately to the east is younger, suggesting a tendency of local volcanic fronts to build eastwards. Nearly all of the lavas in the NVZ have westerly flow directions, implying that a westerly dipping slope has been maintained. This is easily explained by the presence of easterly dipping faults which truncate the NVZ on both east and west sides, producing a half graben tilted down to the base of the western fault.

There is a well defined transition from the NVZ to the tectonized zone, marked by a sudden increase in the extent of tectonism from about 3 to 8 faults per kilometre across strike, in a way that was not observed in the traverse of Segment 2. The age contrasts between adjacent areas are also more dramatic on the narrowgate traverse. At the western end of the profile a number of horizontal terraces have been uplifted by inward (east)-dipping normal faults. Immediately to the west of the NVZ the tectonized crust has variable sediment cover no more than a few centimetres deep, indicating a maximum age of several thousand years, but the adjacent terrace at the western end of the profile is heavily sedimented and at least 40 000 years old.

The outward (also east)-dipping normal fault which truncates the NVZ at its eastern end forms the western wall of the first of two small fault-controlled grabens. The talus fan from the eastern bounding fault of this graben downlaps onto a recent pillow mound or 'haystack' which is at least several thousand years younger than the surrounding lavas. Similar syntectonic haystacks were observed in the FAMOUS area where they are thought to represent the final remnant of a magma body which has been squeezed up the fault plane (Stakes *et al.* 1984). The second graben is bounded to the east by a composite fault which has a cumulative throw of nearly 300 m and an overall dip of 35°, although the actual fault planes are much steeper. The relative maturity of this composite fault is reflected by the welded breccia, fault gouge, and ponds of dark, hydrothermal sediment which have collected towards its base.

Petrography and geochemistry of dredged samples

Rocks from the 28 dredges were described petrographically, minerals and glass inclusions analysed by microprobe, and bulk rocks/glasses analysed for major and trace elements and Sr, Nd and Pb isotope ratios. The full data set and its detailed interpretation will be published elsewhere. Here we focus on just the key observations required to understand the volcanic geology.

Petrography

The main phenocryst phases are olivine and plagioclase, which occur in both quench and intratelluric forms. Clinopyroxene is rare and occurs only in the most fractionated samples. Radiating clusters of plagioclase laths are sometimes intergrown with olivine, while clinopyroxene, when present, commonly encloses the radiating plagioclase microlites with an optically continuous microphenocryst. These textural relationships suggest that, within the quench environment, olivine and plagioclase nucleated simultaneously while clinopyroxene was the last major phase to form.

Common features of the rocks are plagioclase megacrysts similar to those previously described in Atlantic MORB (Aumento 1968; Melson & Thompson 1971; Donaldson & Brown 1977; Dungan & Rhodes 1978; Kuo & Kirkpatrick 1982). On rare occasions olivine is found partially included in the plagioclase megacrysts, suggesting that it formed at the same time, and was possibly first in the crystallisation order. The plagioclase megacrysts are up to 15 mm in diameter and, with aspect ratios sometimes less than 1:2 and usually no greater then 1:3, they are more stubby than the typical lath-form of plagioclase. They are sub-equant with rounded margins which indicate a previous phase of resorption, are commonly zoned (usually in the normal sense), twinned (usually lamellar but sometimes sector-twinned) and are commonly rimmed by a quench overgrowth.

Melt inclusions in megacrysts can be uniformly distributed or concentrated towards the centre or, rarely, occur in concentric zones near the margins. The inclusions take many forms including a sieve-like texture of round dots, larger amorphous blebs, and sometimes elongated forms with a consistent preferred elongation that is usually parallel to cleavage. In some of the more fractionated rocks, the edges of the megacrysts appear to be preferentially resorbed along the cleavage planes to a zone boundary allowing the surrounding glass to be injected into the megacryst. The similarity of these resorbed margins to the melt inclusions seen in other phenocrysts suggests that some of the inclusions may have formed as a result of resorption to form a crystal with a sieve-like texture followed by subsequent growth which entrapped the melt.

A sample from dredge 21D at the northern tip of Segment 1 (21D-2-1b) contains a xenolithic crystal clot comprising an agglomeration of sub-equant plagioclase phenocrysts. The clot was clearly in active disequilibrium with the host glass as the rims of the megacrysts have reacted to produce a serrated, almost vermicular texture. Other large single plagioclase phenocrysts in the same sample are also variably zoned. These are distinct from the quench growth of the groundmass, suggesting that there are at least three generations of plagioclase within this sample.

The olivine microphenocrysts exhibit a continuous spread of compositions (Fo$_{88-81}$). Plagioclase compositions are, by contrast, distinctly bimodal, with phenocrysts and microlites of An$_{87-68}$ and refractory calcic megacrysts of An$_{93-84}$. Although plagioclase megacrysts occur throughout the whole area, there is a tendency for greater concentrations in samples from the RTI area. Textures of intratelluric single crystals, clusters and clots showing variable degrees of disequilibrium, and a wide range in the degree and style of zoning highlight the complex history of fractionation, magma-mixing, and resorption within the intratelluric environment of the Mid-Atlantic Ridge.

Glass compositions

Over 500 electron microprobe analyses were obtained from glass separates and the rims of whole rock specimens. Although the degree of devitrification varies, it was possible by hand picking to select sufficient unaltered material to obtain an analysis from most of the dredge localities.

By MORB standards the major element chemistry of the glasses is unremarkable: for example, MgO varies from 8.6 to 6.2 wt%, TiO$_2$ from 1.3 to 2.2 wt%, and Na$_2$O from 2.7 to 3.4 wt%. Five Fourier Transform InfraRed spectroscopy (FTIR) determinations on samples bracketing the full range in MgO indicate that H$_2$O content varies from 0.17 to 0.38 wt%.

Least-squares modelling indicates that the range in glass chemistry for the more compatible elements can be adequately explained by about 25% crystallization of olivine and plagioclase in the proportion of *c.* 1 : 2, with no need to invoke clinopyroxene. For incompatible elements such as TiO$_2$ and H$_2$O, higher estimates of 50% crystallization for the equivalent samples may

underline the rôle of open-system fractionation. The experimental work of Tormey *et al.* (1987), on basalts from north and south of the Kane transform, indicates that olivine and plagioclase dominate low-pressure crystallization, but that olivine, plagioclase and clinopyroxene crystallization may take place at elevated pressure. Hence we may be seeing only the effects of very shallow processes in our dredged samples. The explanation of Dungan & Rhodes (1978), that combined magma chamber replenishment and crystallization drive the composition from clinopyroxene saturation onto the olivine-plagioclase cotectic, may equally apply.

Geochemistry of melt inclusions

The melt inclusions in the calcic plagioclase megacrysts overlap (in terms of MgO) the compositional range of the separated glasses, but also extend to compositions approaching 18 wt% MgO. However, the melt inclusions have experienced the post-entrapment loss of a plagioclase component since the kinetics of the nucleation process favour precipitation onto the surrounding megacrysts. It is therefore necessary to adjust the melt inclusion compositions for this process, in order to attempt to determine the initial parental composition. This is illustrated by a plot of TiO$_2$ against Al$_2$O$_3$ (Fig. 14). The glass inclusion data plot directly away from plagioclase, as plagioclase precipitation from the inclusions has depleted them in plagioclase components such as Al$_2$O$_3$ and enriched them in other components such as TiO$_2$. When the compositions are extrapolated along plagioclase control lines to intersect the glass trend, they give a composition more primitive than any of the erupted glasses. The MgO content of the glass inclusions prior to entrapment is about 11.5 wt%, which is close to many estimates of primary MORB magmas. We can thus infer that the plagioclase megacrysts formed early in the fractionation history of magmas having this level of MgO.

Correlation of glass chemistry with ridge segmentation

One of the principal results of this study is the striking relationship between the MgO content of the dredged glasses and their location within the ridge segment (Fig. 15). Over the 40 km length of Segment 1, the MgO content of the basalt glasses attains a maximum of about 8.5 wt% just to the south of the narrowgate saddle point, and declines to values of around 6 wt% at the transform intersection and 6.5 wt%

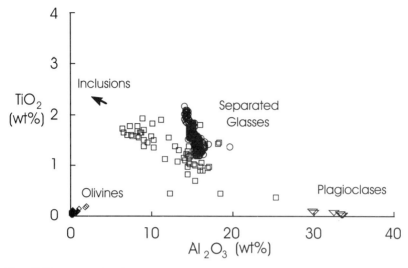

Fig. 14. Plot of TiO$_2$ versus Al$_2$O$_3$ from electron microprobe analyses, showing how the trend of glass separates (circles) can be explained by fractionation of a mixture of olivine (diamonds) and plagioclase (triangles). The data for fluid inclusions squares plot directly away from plagioclase.

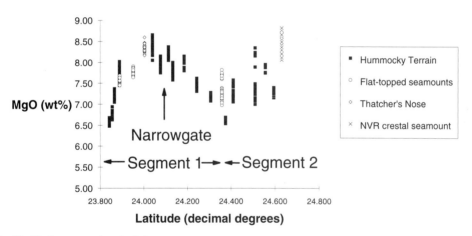

Fig. 15. MgO content of probed glasses plotted against latitude showing that the least fractionated glasses occur in the segment centres.

at the boundary with Segment 2. Complete sample coverage of the second segment was not achieved, but a trend of increasing MgO towards the segment centre is still apparent. The inferred along-axis variation in extent of fractionation is similar to that seen at intermediate spreading rates, where highly fractionated lavas are sampled at the tips of propagating rifts (Christie & Sinton 1981).

Fractionation-corrected Na$_2$O ('Na$_{8.0}$') which

is commonly used as a measure of bulk mantle melting (Klein & Langmuir 1987; Niu & Batiza 1991; Plank & Langmuir 1992), has a relatively uniform value of about 2.8 across the entire MARNOK area (Fig. 16). This corresponds to about 13% batch melting (Plank & Langmuir 1992) and suggests that there is no variation in the degree of melting along either segment. This observation favours focused melt migration models over focused mantle flow models for

MARNOK Glasses

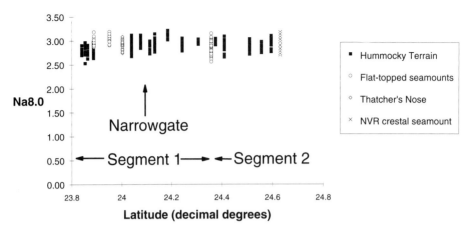

Fig. 16. $Na_{8.0}$ of MARNOK glasses (calculated using the equation of Plank & Langmuir 1992), shows little variation when plotted against latitude. This implies that the bulk degree of melting along the length of the segments is relatively uniform.

Linear volcanic ridge at the RTI

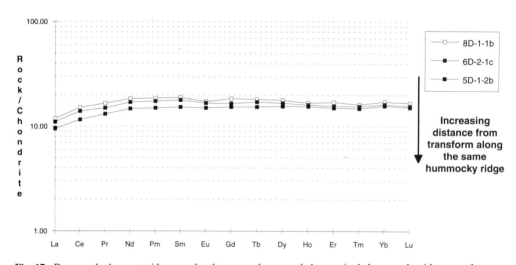

Fig. 17. Rare earth element spidergram for three samples spaced along a single hummocky ridge near the ridge-transform intersection. The patterns are parallel implying a common parental magma, and there is a trend to more fractionated compositions towards the transform.

creating the crustal thickness variations along axis.

Trace element geochemistry

Whole-rock trace element data confirm that many samples can be simply related by fractional crystallization, but also allow basalts with different parental magmas to be discriminated by subtle variations in the highly incompatible trace elements. This helps to define the scale of petrological segmentation along the ridge crest. Figure 17 shows REE patterns for dredges 5D, 6D and 8D all of which sampled the same long hummocky ridge near the RTI. The parallel REE patterns show that these basalts are all related to the same parental magma but vary in the extent of fractionation, with more evolved

Hummocky ridges across narrowgate and northern half of segment 1

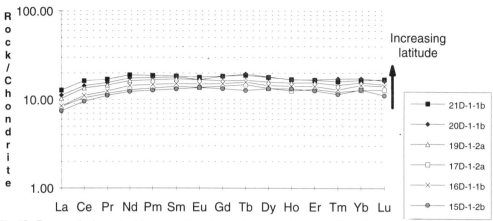

Fig. 18. Rare earth element spidergram of basalt samples across the very uniform hummocky axial volcanic ridge of the narrowgate and the northern half of Segment 1. The patterns are parallel implying a common parental magma, and there is a clear trend towards more fractionated compositions at the non-transform offset.

MORB normalised extended element plot

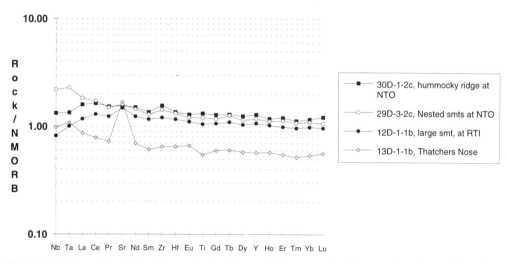

Fig. 19. Multielement variation diagrams for basalts sampled from near the ridge-transform intersection and the non-transform offset. The crossing patterns imply that distinctive features on the sidescan sonar images have differing parental magmas.

lavas occurring at the tip nearest the transform. Systematic variations in the degree of fractionation have been reported in very long dykes (Upton & Thomas 1980), and Rickwood (1990) suggests that such variation can be used as a weak correlator of distance from the source (less evolved compositions occurring closer to the source). It is therefore possible that the variations along this hummocky ridge reflect lateral dyke injection with movement of magma to-

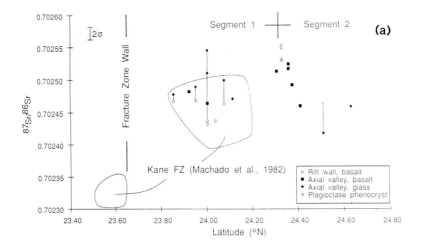

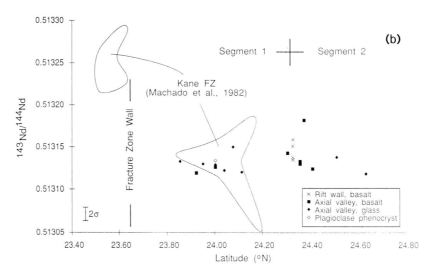

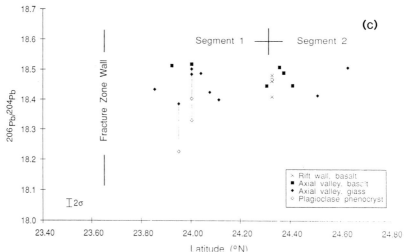

wards the transform. Such dyke injection towards a transform is predicted from seismic studies of crustal thickness variations, particularly in layer 3 (Cormier *et al.* 1984; White *et al.* 1984).

The basalts sampled across the narrowgate and along the northern half of Segment 1 tell a similar story (Fig. 18). Here the AVR, which has a very uniform hummocky morphology, comprises lavas which have parallel patterns on normalised REE plots and show consistently more fractionated compositions towards the end of the segment.

The principal exception to this pattern is found at the boundary of segments 1 and 2 (Fig. 19). Here, the nested seamounts (Fig. 9i) have distinctly flatter REE patterns than most samples from the two segments, including hummocky terranes from similar latitudes. This results in patterns that cross within the LREE. Whereas the other, subparallel, patterns in Figure 19 can be explained by fractional crystallization, this pattern requires a different explanation.

The multi-element pattern in figure 19 shows examples of basalts that are not related by common parents. This plot compares some moderately to highly incompatible element values for the nested seamounts (29D-3-2c) with those for a nearby basalt of similar MgO content from the tip of the hummocky AVR of Segment 1 (30D-1-2c). The principal feature is the crossing pattern and divergence towards the most highly incompatible elements such as Nb and Ta.

Two prominent features from the RTI region are also compared in Fig. 19: Thatcher's Nose (13D-1-1b, Figs 8 and 11iv) and the large flat-topped seamount (12D-1-1b, Figs 6iv and 11vi). The pattern for Thatcher's Nose has particularly low concentrations of incompatible elements and a prominent peak at Sr, which is related to the high concentration of plagioclase megacrysts in this sample. The pattern shows a slight enrichment trend towards the highly incompatible elements (La, Ta and Nb) compared with the nearby flat-topped seamount, which shows a downward trend towards the left-hand side of the plot.

This type of differential enrichment cannot be explained by differences in fractional crystallization, nor by differences in bulk degree of melting if all melt fractions are pooled. It may,

however, be explained by derivation from mantle sources with varying degrees of enrichment or by selective tapping of small melt fractions from a melting column, invoking dynamic melting of an adiabatically upwelling source (Devey *et al.* 1994). In the dynamic melting process, the final erupted magma is a mix of melts from the entire melting column and this mix is very sensitive to the physical processes that allow deep melts to separate and migrate. Small variations in the proportions of melts from deeper sources can cause an increase in the concentrations of the highly incompatible elements without affecting the bulk melting signature.

The petrological segmentation, as defined by the sampling of discrete parental magmas, is greater in the southern half of Segment 1 than in the northern half. This difference is also reflected in the morphological diversity of the two regions, the RTI area having many discrete edifices on the sidescan images, while the AVR north of the narrowgate is very uniform. The sampling density is not great enough to gain an accurate idea of the scale of petrological segments in Segment 2, but from the data available it seems to be longer than those of the RTI area, but with a less uniform petrological trend than for the northern half of Segment 1.

Isotope geochemistry

A representative suite of whole rock basalts, glasses and separated plagioclase phenocrysts were analysed for Sr, Nd and Pb isotope ratios. Although appearing fresh, all samples were leached in 6M HCl for 30 minutes to minimise the effects of seawater alteration; some unleached basalts and leachates were also analysed. Although leaching has not affected Nd isotope ratios, Sr isotope ratios are clearly reduced, leached residues having lower values than both unleached whole rocks and leachates. This is particularly true for the older samples dredged from the median valley wall. These samples have the highest $^{87}Sr/^{86}Sr$ ratios of the MARNOK leached samples, suggesting that they had a slightly more radiogenic source or that, even when leached, they retained a small signature of interaction with seawater not seen in the other samples analysed. Since the Nd and Pb isotopic compositions of both leached and

Fig. 20. Isotope variations against latitude for (**a**) $^{87}Sr/^{86}Sr$, (**b**) $^{143}Nd/^{144}Nd$, (**c**) $^{206}Pb/^{204}Pb$. The isotopic variation within the MARNOK area is not much greater than the analytical error. There is no systematic variation within or between segments, but the data do confirm the isotopic anomaly across the fracture zone, as reported by Machado *et al.* (1982).

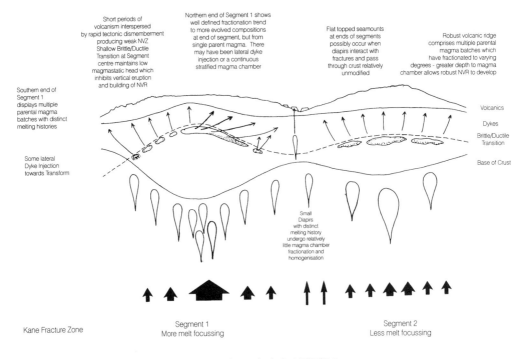

Model of crustal accretion in the MARNOK Area

Fig. 21. Cartoon showing petrogenetic model for crustal formation in the MARNOK area.

unleached samples are similar to other basalts from the area, we prefer the latter interpretation.

Sr, Nd and Pb isotope plots for leached residues are shown as a function of latitude in figure 20. MARNOK basalts show a relatively restricted range in isotopic values (^{143}Nd/^{144}Nd = 0.513118–0.513182, ^{87}Sr/^{86}Sr = 0.70242–0.70255, ^{206}Pb/^{204}Pb = 18.227–18.518) and lie within the field of other North Atlantic MORB. Compared with data published by Machado *et al.* (1982), our new data fall in the range of his Sr and Nd data from the MARNOK area, but are distinct from those dredged from the eastern end of the transform, which have higher Nd and lower Sr isotope ratios (Fig. 20). Although the total range of values is not large, there is a small isotopic variation outside the range of reproducibility for Sr, Pb and possibly Nd. The range of ratios is no greater between segments than within a segment, suggesting that the magmas from the two segments are derived from mantle of comparable isotopic range, but that the source area is slightly heterogeneous on the scale of tens of kilometres. The isotopic variations of the leached basalts and glasses do not seem to be correlated with any other aspect of the geochemical data. Thus the long-term mantle source for the MARNOK area is assumed to be relatively homogeneous. We observe isotopic disequilibrium between glass and associated plagioclase phenocrysts in several samples, but particularly in the plagioclase-rich material recovered from Thatcher's nose (CD57-13D-1-1). In all but one case (CD57-31D-1-1), plagioclase has lower ^{87}Sr/^{86}Sr and ^{206}Pb/^{204}Pb ratios than the associated glass. Possible reasons for this are discussed in the next section.

Discussion: a magmatic model of spreading segments

We begin by discussing the unusual plagioclase megacrysts discovered here, and go on to develop a model for the magmatic aspects of mid-ocean ridge segmentation. In our model, crustal thinning toward segment ends reflects a combination of focused melt delivery to the crust and redistribution within it by lateral dyke injection. Increased fractionation towards segment ends, and distinct geochemistries of segment-end seamounts, are explained by the balance between crustal residence time and magma replenishment (Fig. 21). Ultimately, these may be largely controlled by variations in the depth to the brittle-ductile transition, which also explains the observed variations in volcanism and tectonism on the median valley floor.

Significance of the plagioclase megacrysts and their melt inclusions

The cores of the plagioclase megacrysts from MARNOK are out of equilibrium with their basalt hosts, both isotopically (lower Sr and Pb isotope ratios) and in terms of their anorthite content (An_{92}). They also trap melt inclusions with original MgO = 11.5 wt%. Similar megacrysts from other Atlantic MORB (e.g. Donaldson & Brown 1977; Dungan & Rhodes 1978; Kuo & Kirkpatrick 1982) have raised the questions of why the megacryst cores are so calcic, and why the trapped melt is so magnesian yet trapped by plagioclase rather than olivine. Our study raises the additional question of why the megacrysts and their host glass are not in isotopic equilibrium.

The calcic nature of the plagioclase is significant because this composition is too refractory to have crystallized from the magma parental to the MARNOK basalts. In order to achieve compositions of An_{92}, the plagioclase must have been in equilibrium with magmas either with much higher Ca/Na ratios than MORB, or with H_2O contents in excess of 5 wt%, for there to have been sufficient depression and steepening of the liquidus. The former explanation requires crystallization from magma derived from depleted (high Ca/Na ratio) mantle; the latter requires addition of water well beyond that normally found in MORB mantle or magma. The low Pb and Sr isotopic ratios in the megacrysts do not indicate any seawater addition, and water contents of MARNOK glasses are normal for MORB. We therefore favour crystallization of melts derived from depleted mantle as the cause of the calcic nature of the megacrysts. This conclusion corresponds to that reached by Donaldson & Brown (1977), Dungan & Rhodes (1978) and others, where it is further supported by analysis of inclusions in olivine. It provides further evidence for the presence of refractory melts at mid-ocean ridges and hence for dynamic, or multistage, melting models (e.g. Duncan & Green 1980; Elliot et al. 1991).

Given this interpretation, the low Pb and Sr isotope ratios in the megacrysts still require explanation. There are two principal possibilities (1) the melted mantle may have had Sr and Pb isotope ratios that were less radiogenic than those of the glasses, as well as high Ca/Na ratios; or (2) the higher Sr and Pb isotope ratios in the

glasses may reflect assimilation of hydrothermally altered rocks from the walls of the magma chamber. In the latter case the megacrysts would have grown in deeper chambers unaffected by such assimilation. Future stable-isotope analyses may resolve this issue.

Although there is little dispute that the megacrysts and their inclusions provide evidence for magma mixing, the region where the megacryst cores grow and incorporate melt inclusions has already prompted debate (e.g. Stakes et al. 1984). We support models in which such melts (with high Ca/Na ratios) separated at high levels in the melting column and crystallised calcic plagioclase (together with olivine and spinel) in magma reservoirs just beneath or at the base of the main crustal axial magma chamber. After a period of growth sufficient to allow formation of the megacryst cores, the mush (melt plus crystals) eventually mixed with the main magma reservoir. During this mixing process, the megacrysts trapped samples with the composition of the ambient liquid available in the chamber. The megacrysts may then have risen to the top of the chamber suspended in plumes of magma, to be erupted as the crystal-rich pillow lavas that characterise the prominent seamounts of Thatcher's Nose.

Comparison with Icelandic volcanism and implications for crustal magma transport

Studies of crustal accretion during the Krafla rifting episode in Iceland showed that magma was injected along the length of a fissure swarm from a central crustal magma chamber, while simultaneously a small shield erupted a chemically distinct, primitive lava at the swarm tip (Björnsson 1985). However, there is a discrepancy between the volume of intruded magma (expelled from the central magma chamber) and the volume increase deduced from widening of the fissures. Björnsson proposed that dykes that are formed close to a magma chamber are fed by horizontal magma movement along a propagating fissure but that, farther from the chamber, magma ascends vertically. He suggested that fissures that formed in the uppermost few kilometres propagate down to the base of the crust where they meet a partially molten layer, facilitating withdrawal of magma from this layer and producing the chemically distinct lavas at the tip of the fissure swarm.

A similar process may be operating in the north of MARNOK Segment 1. Here, the uniform hummocky AVR comprises lavas that have similar parents and show consistently more fractionated compositions towards the end of the segment. The varying degree of fractionation along the ridge may be related to lateral dyke intrusion from a density-stratified magma chamber similar to the Krafla model. The nested seamounts, which are located beyond the northern tip of the hummocky ridge, are relatively primitive and come from a distinct, unrelated parental magma. It is conceivable that their formation may be similar to that of the shield at the tip of the Krafla fissure swarm.

Interpretation of rift valley evolution

The gross morphology of a ridge segment and the chemical evolution of the magma during the crustal accretion process will depend on the complex interplay between (a) magmatic processes (magma supply, eruption episodicity, melt migration, mixing and fractionation within and below the crust), (b) tectonic processes (strain partitioning, periodicity and symmetry of faulting, and (c) hydrothermal circulation (and particularly the part it plays in increasing the brittle-ductile transition under the ridge crest).

Episodicity and focusing of volcanism

The relative age contrasts and the sharpness of the contact between the NVZ and the older tectonized crust of the valley floor can be used as a measure of the relative episodicity and degree of focusing of the volcanism within a segment.

The AVR of Segment 2, especially on the western side, has a strong age contrast with the marginal basins, and the boundary between the two terrains is relatively sharp. Over the northern half of Segment 1 the contact is sharper still, but the degree of sedimentation and thus age of the marginal basin seem relatively small compared with those observed over Segment 2. However, stronger age contrasts are seen towards the bathymetric saddle of Segment 1. In the transform domain the contact is neither sharp nor suggestive of a dramatic age contrast, though there are some places where individual hummocky ridges can be seen onlapping the highly fissured terrain.

These observations, together with the varying heterogeneity of the NVZ, imply that volcanism in the north has occurred, at least recently, in long, widely separated episodes, whereas in the south it has been either more continuous or episodic on a much shorter time scale. Such short, discrete episodes in the transform domain probably derive from individual magma bodies separated in space and time. Compared to Segment 2, the northern half of Segment 1 probably has a shorter periodicity of AVR construction and the lifespan of the wall faults seems shorter.

Fractionation and crustal residence time

A principal result of this study is our discovery of the increasingly basic composition of the rocks toward the centres of the segments. Building on ideas proposed for propagating ridges (Christie & Sinton 1981; Sinton *et al.* 1983), we propose that the along-axis fractionation systematics ultimately reflect the residence time of magmas within the crust, buffered by the supply rate of primitive magma from the mantle. The correlation of basalt composition with the crustal structure inferred from bathymetry suggests that a balance between magma supply, crystallization rate and eruption rate is operating on the scale of a single ridge segment. In the centre of the segment, where the crust is thicker, magma supply dominates the open system and a relatively primitive steady-state composition is erupted. The higher supply rate will tend to maintain the magmastatic pressure within a chamber and promote regular eruptions that prevent the composition in the chamber from becoming too evolved. Conversely, crystallization dominates in the thinner and cooler crust at the ends of the segments, leading to the eruption of more fractionated lavas. Between segments, where the supply is low, not all the rising melt may feed a central crustal chamber, but may instead form distinctive eruptive edifices, such as the nested seamounts, which display their own individual melting histories.

The TOBI and WASP observations have demonstrated how the episodicity of NVZ construction may vary across the MARNOK area. They have emphasised the close relationship between volcanism and tectonism that occurs at slow spreading ridges, even showing some volcanic episodes that have apparently utilized pre-existing faults in the NVZ as conduits. Crustal fractures that are not directly controlled by magma overpressure may also have an effect on the crustal residence time. At the centre of the narrowgate segment the continuous dismemberment of the NVZ may allow magma batches to be tapped immediately on arrival in the crust. At the ends of the segment the brittle-ductile transition is likely to be deeper and strain may be more partitioned into the valley wall faulting, so magma batches may not be able to erupt so frequently. Segment 2 has had a more prolonged and robust volcanic episode which has produced a 'noisier' fractionation profile. We suggest that the construction of the thick volcanic pile has inhibited the immediate eruption of magma at the centre of the segment, allowing a broader range in the MgO contents than in Segment 1.

The magma chamber model of Sinton & Detrick (1992) suggests that, at slow spreading ridges, fractionation takes place almost entirely during the migration of melts through a crystal mush zone, with little or no discrete melt lenses accumulating in the crust. In their model, eruptions are closely coupled, in time, with episodes of injection of new magma from the mantle. The multiple parental magma batches with distinct melting histories observed in our study seem to preclude extensive mixing and along-axis transport within the crust, except in the northern half of Segment 1, in agreement with the model of Sinton and Detrick. However, this model suggests that the mush zone is thickest in the centre of the segment. If fractionation occurs by vertical percolation through the mush, then the magmas at the centre of the segment might be expected to be more fractionated than those at the ends. Thus if this model is to stand, a mechanism must be found that will increase the interaction of magmas erupting at the ends of the segment compared with those in the centre. The additional buoyancy caused by the greater melt fraction at the segment centre may produce an increase in the percolation rate and hence less fractionation. Alternatively, or additionally, lateral, along-strike transport of melt through the mush zone would cause those magmas erupting at the ends of the segment to have been exposed to the mush for longer than those erupting at the centre.

Brittle-ductile transition related to magmastatic head and faulting

To some extent the differences in morphology of the rift valley may be considered a consequence of the varying width of the inner valley floor: segment 1 has a narrow valley so the volcanism can fill the entire floor; conversely, Segment 2 has a wide floor so the constructional activity builds a ridge which stands apart from the valley walls. However, this does not account for all the differences between the two segments: segment 2 has an AVR with a higher summit than the narrowgate, has older, more mature walls, and has greater asymmetry at the segment centre. The narrowness of the NVZ at the centre of the narrowgate segment has to be explained.

The small height and width of the NVZ, the relatively brief volcanic episodes, and the rapid tectonic dismemberment at the centre of the narrowgate segment imply either that there is only a narrow and short lived magma chamber at the centre of this segment, or that there is some process operating which is inhibiting the surficial expression of volcanism there. Possible processes are low magmastatic head and/or low relative buoyancy forces, which may inhibit eruption at the surface and/or promote lateral dyke injection. The fast effusion rates indicated by the presence of sheet flows may also be a consequence of shorter (and/or wider) conduits, and consequent prolongation of the early, effusive, phase in the volcano building cycle. These processes are all promoted by a thin brittle lid. We suggested above that the period between major magmatic pulses may be greater for Segment 2 than Segment 1. In Segment 1, the time between episodes of magmatic activity (which elevate the isotherms) and episodes of rapid tectonism (which cause crustal thinning) may be small compared to the time required for hydrothermal circulation to penetrate to deep levels, inhibiting formation of a thick brittle lid. The tectonic dismemberment of the NVZ occurs quickly, so a thick volcanic pile is unable to develop. This prevents the formation of extensive pillow mounds which are associated with the higher degree of channelisation and slower flow rates of the later stages of a volcano-building episode.

There are several observations that are consistent with the hypothesis of a thinner brittle lid under the centre of Segment 1 than Segment 2. First, the faults along the centre of the robust AVR of Segment 2 are larger and more continuous than those which cut the NVZ of Segment 1. Secondly, a robust AVR must itself be supported by thick lithosphere. On the other hand, the rapid dismemberment of the NVZ of Segment 1 implies that the lithosphere is very thin and weak so that it can only support a small constructional edifice. The asymmetric wall faulting in Segment 2, identified by Allerton *et al.* (1995), is also indicative of a thick brittle lid according to the hypothesis of Harper (1985).

These differences between the segments can be related to the observation of Sempéré *et al.* (1993) that the group 1, narrowgate-type segments show strongly focused mantle Bouguer anomalies, while the group 2 segments have no strong, well-defined central anomaly. This implies that the narrowgate segments have greater along-axis crustal thickness variations and hence are likely to have experienced more focused melt delivery. It seems probable that a highly focused sub-crustal melt flow will tend to produce frequent, relatively regular injection events in the centre of the segment, thus maintaining a shallow brittle lid.

A model for crustal accretion in the MAR-NOK area, combining the features discussed above, is shown in Fig. 21. This model tends to favour a magmatic control of second-order

segmentation, with melt migration as the preferred cause of crustal thickness variations along
axis. The model is consistent with our observation that the petrological effects of the
second-order segmentation outweigh those of
the transform segmentation. However, Allerton
et al. (1995) propose that it is the episodicity of
the valley-wall faulting that drives the tectonic
thinning of the lithosphere and in turn produces
the pulses of upwelling asthenosphere that drive
local adiabatic melting and melt focusing. Their
tectonically controlled model is also consistent
with the observations, though further modelling
is necessary to deduce whether such shallow
processes could produce the variations in melting required.

Conclusions

The two second-order segments differ on the
basis of their morphology. Segment 1, adjacent
to the transform, is a narrowgate-type segment
with a V-shaped inner valley at the segment
centre. The floor deepens away from the
bathymetric saddle point as the valley becomes
increasing asymmetric towards the ends of the
segment. The low relief AVR is narrowest at the
saddle point in the segment centre, whose recent
history is characterised by short periods of
volcanism interspersed by extensive tectonism,
implying a low flux of melt to the surface. The
present NVZ represents a focused, but meagre
volcanic episode lasting not much longer then
1000 years. Quiescent periods of at least several
thousand years seem to separate volcanic episodes allowing the NVZ to become rapidly
dismembered. The northern half of Segment 1
has a very uniform, hummocky AVR, on which
there are no single dominant structures, and the
lavas all seem to be related by a common parent.
The contact between the AVR and older
fissured crust is sharp and the valley wall faults
are relatively immature. There is a striking
variation in the MgO content of glasses along
Segment 1, with increasing fractionation towards the segment ends.

The transform domain extends up to 26 km
along most of the south of Segment 1. This is
reflected by widening of the untectonized zone
and median valley towards the transform,
development of duplex-style wall faulting towards the inside corner, and development of
considerable morphological diversity correlat-
ing to a relatively wide range of parental basalt
compositions.

The inner valley of Segment 2 is on average
wider than that of Segment 1, and the NVZ takes
the form of a robust, 400–600 m high AVR. This
seems to have been built relatively steadily in
several prolonged phases, with only short
periods of quiesence which caused little tectonic
dismemberment of the AVR as a whole. The
crest of the AVR is dominated by long syn-
magmatic faults and a crestal flat-topped seamount. There is a large age contrast between the
AVR and the marginal basins implying that the
major 'ridge building' volcanic episodes are
infrequent. The along-segment trend of increasing MgO towards the bathymetric crest of the
AVR is noisier than, but similar to, that of the
narrowgate segment, being less well constrained
because of low sampling density. Near the NTO
there is an abundance of flat-topped seamounts,
many of which are situated at the edges of the
median valley adjacent to the wall faults. They
show a wide range of apparent ages and different
parental compositions to the adjacent hummocky terrain. There is no overlap in the
neovolcanism of the two segments, but the
hummocky terrain is cut by a series of small-
throw ridge parallel en-echelon faults.

One of the main objectives of this research
was to test the different mechanisms proposed
for decreasing crustal thickness towards the
segment boundaries. The major element geochemistry of basalts sampled along the strike of
the ridge crest suggests that there is little
variation in the degree of bulk melting along the
segments. Hence melt migration is favoured
over focused mantle upwelling as the main cause
of the crustal thickness variations that define
second-order segmentation. The extrusion of
more primitive compositions at the segment
centres indicates that more magma feeds into the
centres of segments. Similarly, at the segment
ends, the basalts are more fractionated indicating a high ratio of crystallization to replenishment; there is also more variability in the highly
incompatible element compositions, suggesting
that there is less homogenization of melt
packages from the mantle. Radiogenic isotopes
show only very small variations, with no systematic pattern emerging within or between the
ridge segments. The isotopic irregularities are
not related to the variations in incompatible
elements, implying that the latter result from the
action of dynamic melting processes rather than
from long-lived source heterogeneity. The
highly anorthitic plagioclase megacrysts and
their primitive melt inclusions provide further
evidence for the presence of refractory melts at
mid-ocean ridges and hence for multistage (or
dynamic) melting models. The petrological
segmentation of the NVZ as defined by variations in parental compositions of basalts correlates well with the identification of discrete

volcanic edifices on the sonar images. The petrological effect of the transform is minor compared to the manifestation of the second-order segmentation.

This work was funded by the Natural Environment Research Council, who also supported K.L. through a PhD studentship. We gratefully acknowledge the assistance of the officers, crew and shipboard scientific party of RRS *Charles Darwin* cruise 57. S. Hussenoeder, M. Tivey and H. Schonten interpreted the TOBI magnetic data. This work greatly benefited from discussions with S. Allerton and C. Mével. We thank R. Batiza and C. MacLeod for helpful reviews of the manuscript.

References

ABRAMS, L. J., DETRICK, R. S. & FOX, P. J. 1988. Morphology and crustal structure of the Kane fracture zone transverse ridge. *Journal of Geophysical Research*, **93**, 3195–3210.

ALLERTON, S., MURTON, B. J., SEARLE, R. C. & JONES, M. 1995. Extensional faulting and segmentation of the Mid-Atlantic Ridge north of the Kane fracture zone (24°00'N to 24°40'N). *Marine Geophysical Researches*, **17**, 37–61.

——, SEARLE, R. C. & MURTON, B. J. 1996. Bathymetric segmentation and faulting on the Mid-Atlantic Ridge, 24°00'N to 24°40'N. *This volume*.

AUMENTO, F. 1968. The Mid-Atlantic Ridge near 45°N. II. Basalts from the area of Confederation Peak. *Canadian Journal of Earth Science*, **5**, 1–21.

BALLARD, R. D. & VAN ANDEL, T. H. 1977. Morphology and tectonics of the inner rift valley at lat. 36°50'N on the Mid-Atlantic Ridge. *Geological Society of America Bulletin*, **88**, 507–530.

——, FRANCHETEAU, J., JUTEAU, T., RANGAN, C. & NORMARK, W. 1981. East Pacific Rise at 21°N: The volcanic, tectonic and hydrothermal processes of the central axis. *Earth and Planetary Science Letters*, **55**, 1–10.

——, HOLCOMB, R. T. & VAN ANDEL, T. H. 1989. The Galapagos Rift at 86°W: Sheet Flows, Collapse Pits and Lava Lakes of the Rift Valley. *Journal of Geophysical Research*, **84**, 5407–5422.

BATIZA, R. & VANKO, D. 1983. Volcanic development of small conic central volcanoes on the flanks of the East Pacific Rise: inferred from narrow beam echo sounder surveys. *Marine Geology*, **54**, 53–90.

——, FOX, P. J., VOGT, P. R., CANDE, S. C., GRINDLAY, N. R., MELSON, W. G. & O'HEARN, T. 1989. Morphology, abundance, and chemistry of near-ridge seamounts in the vicinity of the Mid-Atlantic Ridge c. 26°S. *Journal of Geology*, **97**, 209–220.

BENDER, J. F., LANGMUIR, C. H. & HANSON, G. N. 1984. Petrogenesis of basalt glasses from the Tamayo Region, East Pacific Rise. *Journal of Petrology*, **25**, 213–254.

BJÖRNSSON, A. 1985. Dynamics of crustal rifting in NE Iceland. *Journal of Geophysical Research*, **90**, 10151–10162.

——, JOHNSEN, S., SIGURDSSON & THORBERGSSON, G. 1979. Rifting of the Plate Boundary in North Iceland 1975–1978. *Journal of Geophysical Research*, **84**, 3029–3038.

CANN, J. R., SMITH, D. K., BROOKS, B., DOUGHERTY, M. E., GARLAND, S., KEETON, J., LIN, J., MCALLISTER, E., MACLEOD, C., PASCOE, R. & SPENCER, S. 1992. *Building the crust at the Mid Atlantic Ridge*. RRS Charles Darwin Cruise 65. University of Leeds, UK.

CARBOTTE, S., WELCH, S. & MACDONALD, K. C. 1990. Spreading rates and offset histories of the southern MAR 31°–34°30'S, 25°–27°30'S. *Marine Geophysical Researches*, **13**, 51–80.

CHRISTIE, D. & SINTON, J. M. 1981. Evolution of abyssal lavas along propagating segments of the Galapagos spreading center. *Earth and Planetary Science Letters*, **56**, 321–335.

CORMIER, M. H., DETRICK, R. S. & PURDY, G. M. 1984. Anomalously thin crust in oceanic fracture zones: New seismic constraints from the Kane fracture zone. *Journal of Geophysical Research*, **89**, 10 249–10 266.

CRANE, K. 1985. The spacing of rift axis highs: dependence upon diapiric processes in the underlying asthenosphere? *Earth and Planetary Science Letters*, **72**, 405–414.

DELANY, P. & POLLARD, D. D. 1982. Solidification of basaltic magma during flow in a dike. *American Journal of Science*, **282**, 856–885.

DEMETS, C., GORDON, R. G., ARGUS, D. F. & STEIN, S. 1990. Current Plate Motions. *Geophysical Journal International*, **101**, 425–478.

DEPLUS, C., MAIA, M., ASLANIAN, D. & GENTE, P. 1992. Segmentation of the Mid-Atlantic Ridge south of the Kane Fracture Zone revealed by gravity anomalies. *EOS, Transactions American Geophysical Union*, **73**, 568.

DETRICK, R. S., MUTTER, J. C., BUHL, P. & KIM, I. I. 1990. No evidence from multichannel reflection data for a crustal magma chamber in the MARK area on the Mid-Atlantic Ridge. *Nature*, **347**, 61–64.

——, NEEDHAM, H. D. & RENARD, V. 1995. Gravity anomalies and crustal thickness variations along the Mid-Atlantic Ridge between 33°N and 40°N. *Journal of Geophysical Research*, **100**, 3767–3787.

DEVEY, C. W., GARBE-SCHONBERG, C. D., STOFFERS, P., CHAUVEL, C. & MERTZ, D. F. 1994. Geochemical effects of dynamic melting beneath ridges: Reconciling major and trace element variations in Kolbeinsey (and global) mid-ocean ridge basalt. *Journal of Geophysical Research*, **99**, 9077–9095.

DONALDSON, C. H. & BROWN, R. W. 1977. Refractory megacrysts and magnesium rich melt inclusions within spinel in oceanic tholeiites: Indicators of magma mixing and parental magma composition. *Contributions to Mineralogy and Petrology*, **37**, 81–89.

DRAGONI, M. & TALLARICO, A. 1994. The effects of crystallisation on the rheology and dynamics of lava flows. *Journal of Volcanology and Geothermal Research*, **59**, 241–252.

DUNCAN, R. A. & GREEN, D. H. 1980. Role of

multistage melting in the formation of oceanic crust. *Geology*, **8**, 22–26.

DUNGAN, M. A. & RHODES, J. M. 1978. Residual glasses and melt inclusions in basalts from DSDP legs 45 and 46: evidence for magma mixing. *Contributions to Mineralogy and Petrology*, **67**, 417–431.

ELLIOT, T. R., HAWKESWORTH, C. J. & GRONVOLD, K. 1991. Dynamic melting of the Iceland Plume. *Nature*, **351**, 201–206.

FORNARI, D. J., BATIZA, R. & LUCKMAN, A. M. 1987. Seamount abundances and distribution near the East Pacific Rise 0–24°N based on Seabeam data. *In*: KEATING, B. H. *et al.* (eds) *Seamounts, islands and atolls.* American Geophysical Union, Geophysical Monographs, **43**, 13–21.

——, RYAN, W. B. F. & FOX, P. J. 1984. The evolution of craters and calderas on young seamounts: Insights from Sea MARC I and SeaBeam sonar surveys of a small group near the axis of the East Pacific Rise at c. 10°N. *Journal of Geophysical Research*, **89**, 11 069–11 083.

FORSYTH, D. W. 1992. Geophysical constraints on mantle flow and melt generation beneath mid-ocean ridges. *In*: PHIPPS MORGAN, J., BLACKMAN, D. K. & SINTON, J. M. (eds) *Mantle flow and melt generation beneath mid-ocean ridges.* American Geophysical Union, Geophysical Monographs, **31**, 1–65.

FOX, P. J., DETRICK, R. S. & PURDY, G. M. 1980. Evidence for crustal thinning near fracture zones: Implications for ophiolites. PANAYIOTOU, A. (ed.) *Ophiolites. Proceedings of the International Ophiolite Symposium*, 1979, Cyprus Geological Survey Dept, 161–168.

—— & GALLO, D. G. 1984. A tectonic model for ridge-transform-ridge plate boundaries: implications for the structure of oceanic lithosphere. *Tectonophysics*, **104**, 205–242.

—— & —— 1986. The geology of North Atlantic transform plate boundaries and their aseismic extensions. *In*: VOGT, P. R. & TUCHOLKE, B. E. (eds) *The Western North Atlantic Region.* The Geology of North America, **M**, Geological Society of America, 157–172.

FUJIMOTO, H., BRYAN, B., KOBAYASHI, K., KINOSHITA, H., TIVEY, M., KELEMEN, P., TAKEUCHI, A., MATSUMOTO, T., ISHIZUKA, H., FURUTA, T., FUJIWARA, T., LIN, J., HOTTA, H. & PURDY, M. 1994. Diving and surface surveys of the western part of the Kane Transform Fault. *InterRidge News*, **3**, 20.

GENTE, P., POCKALNY, R. A., DURAND, C., DEPLUS, C., MAIA, M., CEULENEER, G., MAVEL, C., CANNAT, M. & LAVERNE, C. 1995. Characteristics and evolution of the segmentation of the Mid-Atlantic Ridge between 20°N and 24°N during the last 10 million years. *Earth and Planetary Science Letters*, **129**, 55–71.

GRIFFITHS, R. W. & FINK, J. H. 1992a. The morphology of lava flows in planetary environments: predictions from analogue experiments. *Journal of Geophysical Research*, **97**, 19 739–19 748.

—— & —— 1992b. Solidification and morphology of submarine lavas: A dependence on extrusion rate. *Journal of Geophysical Research*, **97**, 19 729–19 737.

GRINDLAY, N. R., FOX, P. J. & MACDONALD, K. C. 1991. Second-order ridge axis discontinuities in the South Atlantic: Morphology, structure, and evolution. *Marine Geophysical Researches*, **13**, 21–49.

——, —— & VOGT, P. R. 1992. Morphology and tectonics of the Mid-Atlantic Ridge (25°–27°30'S) from Sea Beam and magnetic data. *Journal of Geophysical Research*, **97**, 6983–7010.

HARPER, G. D. 1985. Tectonics of slow spreading mid-ocean ridges and consequences of a variable depth to the brittle/ductile transition. *Tectonics*, **4**, 395–409.

HEKINIAN, R. & THOMPSON, G. 1976. Comparative geochemistry of volcanics from rift valleys, transform faults and aseismic ridges. *Contributions to Mineralogy and Petrology*, **57**, 145–162.

HUGGETT, Q. J. & MILLARD, N. W. 1992. Towed Ocean Bottom Instrument TOBI: A new deep-towed platform for side-scan sonar and other geophysical surveys. *24th Annual Ocean Technology Conference*, Houston, 349–354.

JOHNSON, G. L. & VOGT, P. R. 1973. Mid-Atlantic Ridge from 47° to 51°N. *Geological Society of America Bulletin*, **84**, 3443–3462.

KARSON, J. A. & DICK, H. J. B. 1983. Tectonics of ridge-transform intersections at the Kane Fracture Zone. *Marine Geophysical Researches*, **6**, 51–98.

KLEIN, E. M. & LANGMUIR, C. H. 1987. Lobal correlations of ocean ridge basalt chemistry with axial depth & crustal thickness. *Journal of Geophysical Research*, **92**, 8089–8115.

KONG, L. S., DETRICK, R. S., FOX, P. J., MAYER, L. A. & RYAN, W. B. F. 1988/89. The morphology and tectonics of the MARK area from Sea Beam and Sea MARC I observations (Mid-Atlantic Ridge 23°N). *Marine Geophysical Researches*, **10**, 59–90.

KUO, B. Y. & FORSYTH, D. W. 1988. Gravity anomalies of ridge-transform system in South Atlantic. *Marine Geophysical Researches*, **10**, 205–232.

KUO, L.-C. & KIRKPATRICK, R. J. 1982. Pre-eruption history of phyric basalts from DSDP legs 45 and 46: Evidence from morphology and zoning patterns in plagioclase. *Contributions to Mineralogy and Petrology*, **79**, 13–27.

LANGMUIR, C. H. & BENDER, J. F. 1984. The geochemistry of ocean basalts in the vicinity of transform faults: observations and implications. *Earth and Planetary Science Letters*, **69**, 107–127.

——, —— & BATIZA, R. 1986. Petrological and tectonic segmentation of the East Pacific Rise, 5°30'–14°30'N. *Nature*, **322**, 422–429.

LAUGHTON, A. S. & SEARLE, R. C. 1979. Tectonic processes on slow spreading ridges. *In*: TALWANI, M., HARRISON, C. G. & HAYES, D. E. (eds) *Deep drilling results in the Atlantic Ocean: Ocean crust.* Maurice Ewing Series, American Geophysical Union, **2**, 15–32.

LAWSON, N. K. 1996. *Crustal accretion near ridge–transform intersections: Kane Fracture Zone,*

Mid-Atlantic Ridge. PhD thesis, University of Durham.

LE ROEX, A. P., DICK, H. J. B. & FISHER, R. L. 1989. Petrology and geochemistry of MORB from 25°E to 46°E along the Southwest Indian Ridge: Evidence for contrasting styles of mantle enrichment. *Journal of Petrology*, **30**, 947–986.

LICHTMAN, G. S., NORMARK, W. R. & SPIESS, F. N. 1984. Photogeologic study of a segment of the East Pacific Rise axis near 21°N latitude. *Geological Society of America Bulletin*, **95**, 743–752.

LIN, J. & PHIPPS MORGAN, J. 1990. A mechanical investigation of axial rifting and discrete faulting processes at slow-spreading mid-ocean ridges. *EOS, Transactions of the American Geophysical Union*, **71**, 627.

—— & —— 1992. The spreading rate dependence of three-dimensional mid-ocean ridge gravity structure. *Geophysical Research Letters*, **19**, 13–16.

LOUDEN, K. E., WHITE, R. S., POTTS, C. G. & FORSYTH, D. W. 1986. Structure and seismotectonics of the Vema Fracture Zone, Atlantic Ocean. *Journal of the Geological Society, London*, **143**, 795–805.

MACDONALD, G. A. & ABBOTT, A. T. 1970. *Volcanoes In The Sea.* University of Hawaii Press, Honolulu.

MACDONALD, K. C. & LUYENDYK, B. P. 1977. Deep-tow studies of the structure of the Mid-Atlantic ridge crest near 37°N (FAMOUS). *Geological Society of America Bulletin*, **88**, 621–636.

MACHADO, N., LUDDEN, J. N. & BROOKS, C. 1982. Fine scale isotopic heterogeneity in the sub-Atlantic mantle. *Nature*, **295**, 226–228.

MAGDE, L. S. & SMITH, D. K. 1995. Seamount volcanism at the Reykjanes Ridge: Relationship to the Iceland hot spot. *Journal of Geophysical Research*, **100**, 8449–8468.

MARKS, S. N. 1981. Sedimentation on new ocean crust: The Mid-Atlantic Ridge at 37°N. *Marine Geology*, **43**, 65–82.

MARSH, B. D. 1989. Magma chambers. *Annual Review of Earth and Planetary Sciences*, **17**, 439–474.

MCALLISTER, E. & CANN, J. R. 1996. Initiation and evolution of boundary-wall faults along the Mid-Atlantic Ridge, 25–29°N. *This volume.*

——, SPENCER, S. & DOUGHERTY, M. 1993. Observations of mass-wasting in the crestal mountains of the MAR: implications for the seismic life of a crestal mountain fault. *BRIDGE Newsletter*, 7–8.

MCBIRNEY, A. R. & MURASE, T. 1984. Rheological properties of magmas. *Annual Review of Earth and Planetary Sciences*, **12**, 337–357.

MELSON, W. G. & THOMPSON, G. 1971. Petrology of a transform fault zone and adjacent ridge segments. *Philosophical Transactions of the Royal Society, London, A*, **268**, 423–441.

MITCHELL, N. C. 1993. A model for attenuation of backscatter due to sediment accumulations and its application to determine sediment thickness with GLORIA sidescan sonar. *Journal of Geophysical Research*, **98**, 22 477–22 493.

MURTON, B. J. & PARSON, L. M. 1993. Segmentation, volcanism and deformation of oblique spreading centres: a quatitative study of the Reykjanes Ridge. *Tectonophysics*, **222**, 237–257.

MUTTER, J. C. & KARSON, J. A. 1992. Structural

processes at slow-spreading ridges. *Science*, **257**, 627–634.

NIU, Y. & BATIZA, R. 1991. An empirical method for calculating melt compositions produced beneath mid-ocean ridges: Application for axis and off-axis (seamounts) melting. *Journal of Geophysical Research*, **96**, 21 753–21 777.

PARMENTIER, E. M. & PHIPPS MORGAN, J. 1990. Spreading rate dependence of three-dimensional structure in oceanic spreading centres. *Nature*, **348**, 325–328.

PARSON, L. M., MURTON, B. J., SEARLE, R. C., BOOTH, D., EVANS, J., FIELD, P., KEETON, J., LAUGHTON, A., MCALLISTER, E., MILLARD, N., REDBOURNE, L., ROUSE, I., SHOR, A., SMITH, D., SPENCER, S., SUMMERHAYES, C. & WALKER, C. 1993. En echelon volcanic ridges at the Reykjanes Ridge: a life cycle of volcanism and tectonics. *Earth and Planetary Science Letters*, **117**, 73–87.

PERFIT, M. R., FORNARI, D. J., MALAHOFF, A. & EMBLEY, R. W. 1983. Geochemical studies of abyssal lavas recovered by DSRV Alvin from Eastern Galapagos Rift, Inca Tranform, and Ecuador Rift 3. Trace element abundances and petrogenesis. *Journal of Geophysical Research*, **88**, 10 551–10 572.

PHILIPS, J. D. & FLEMING, H. S. 1978. *Multi-beam sonar study of the Mid-Atlantic rift valley 36°–37°N FAMOUS.* Geological Society of America, MC-19.

PHIPPS MORGAN, J., PARMENTIER, E. M. & LIN, J. 1987. Mechanisms for the origin of Mid-Ocean Ridge axial topography: implications for the thermal and mechanical structure of accreting plate boundaries. *Journal of Geophysical Research*, **92**, 12 823–12 836.

PLANK, T. & LANGMUIR, C. H. 1992. Effects of the melting regime on the composition of the oceanic crust. *Journal of Geophysical Research*, **97**, 19 770–19 794.

POCKALNY, R. A., DETRICK, R. S. & FOX, P. J. 1988. The morphology and tectonics of the Kane Transform from Sea Beam bathymetry data. *Journal of Geophysical Research*, **93**, 3179–3193.

PURDY, G. M., SEMPÉRÉ, J.-C., SCHOUTEN, H., DUBOIS, D. L. & GOLDSMITH, R. 1990. Bathymetry of the Mid-Atlantic Ridge, 24°–31°N: A map series. *Marine Geophysical Researches*, **12**, 247–252.

RABINOWICZ, M., ROUZO, S., SEMPÉRÉ, J.-C. & ROSEMBERG, C. 1993. Three-dimensional mantle flow beneath mid-ocean ridges. *Journal of Geophysical Research*, **98**, 7851–7869.

RICHTER, D. H., EATON, J. P., MURATA, K. J., AULT, W. A. & KRIVOY, H. L. 1970. *Chronological narrative of the 1955–60 eruption of the Kilauea volcano, Hawaii.* United States Geological Survey Professional Paper, **537-E**.

RICKWOOD, P. C. 1990. The anatomy of a dyke and the determination of proportion and magma flow directions. *In:* PARKER, A. J., RICKWOOD, P. C. & TUCKER, D. H. (eds) *Mafic dykes and emplacement mechanisms.* Balkema, Rotterdam, 81–100.

ROEST, W. R., SEARLE, R. C. & COLLETTE, B. J. 1984.

Fanning of fracture zones and a three-dimensional model of the Mid-Atlantic Ridge. *Nature*, **308**, 527–531.

Rouzo, S., Rabinowicz, M. & Briais, A. 1995. Segmentation of mid-ocean ridges with an axial valley induced by small-scale mantle convection. *Nature*, **374**, 795–798.

Schouten, H., Klitgord, K. D. & Whitehead, J. A. 1985. Segmentation of mid-ocean ridges. *Nature*, **317**, 225–229.

Searle, R. C. 1983. Submarine central volcanoes on the Nazca Plate – high-resolution sonar observations. *Marine Geology*, **53**, 77–102.

——, Field, P. R. & Owens, R. B. 1994. Segmentation and a nontransform ridge offset on the Reykyanes Ridge near 58°N. *Journal of Geophysical Research*, **99**, 24 159–24 172.

Sempéré, J.-C., Lin, J., Brown, H. S., Schouten, H. & Purdy, G. M. 1993. Segmentation and morphotectonic variations along a slow spreading center: The Mid-Atlantic Ridge (24°00′N–30°40′N). *Marine Geophysical Researches*, **15**, 153–200.

——, Purdy, G. M. & Schouten, H. 1990. Segmentation of the Mid-Atlantic Ridge between 24°N and 30°40′N. *Nature*, **344**, 427–431.

Shen, Y. & Forsyth, D. W. 1992. The effects of temperature- and pressure-dependant viscosity on three-dimensional passive flow of the mantle beneath a ridge-transform system. *Journal of Geophysical Research*, **97**, 19 717–19 728.

Sinton, J. M. & Detrick, R. S. 1992. Mid-ocean ridge magma chambers. *Journal of Geophysical Research*, **97**, 197–216.

——, Wilson, D. S., Christie, D. M., Hey, R. N. & Delaney, J. R. 1983. Petrologic consequences of rift propagation on oceanic spreading ridges. *Earth and Planetary Science Letters*, **62**, 193–207.

Smith, D. K. & Cann, J. R. 1990. Hundreds of small volcanoes on the median valley floor of the Mid-Atlantic Ridge at 24–30°N. *Nature*, **348**, 152–155.

—— & —— 1992. The role of seamount volcanism in crustal construction at the Mid-Atlantic Ridge (24°–30°N). *Journal of Geophysical Research*, **97**, 1645–1658.

—— & —— 1993. Building the crust at the Mid-Atlantic Ridge. *Nature*, **365**, 707–715.

——, ——, Dougherty, M. E., Lin, J., Spencer, S., MacLeod, C., Keeton, J., McAllister, E., Borroks, B., Pascoe, R. & Robertson, W. 1995a. Mid-Atlantic Ridge volcanism from deep-towed side-scan sonar images, 25°–29°N. *Journal of Volcanology and Geothermal Research*, **67**, 233–262.

——, Humphris, S. E. & Bryan, W. B. 1995b. A comparison of volcanic edifices at the Reykjanes Ridge and the Mid-Atlantic Ridge at 24°–30°N. *Journal of Geophysical Research*, **100**, 22 485–22 498.

Sparks, D. W. & Parmentier, E. M. 1991. Melt extraction from the mantle beneath spreading centers. *Earth and Planetary Science Letters*, **105**, 368–377.

Stakes, D. S., Shervais, J. W. & Hopson, C. A. 1984. The volcano-tectonic cycle of the FAMOUS and AMAR valleys, Mid-Atlantic Ridge (36°47′N): Evidence from basalt glass and phenocryst compositional variations for a steady state magma chamber beneath the valley midsections, AMAR 3. *Journal of Geophysical Research*, **89**, 6995–7028.

Thorarinsson, S. 1969. The Lakagigar eruption of 1783. *Bulletin of Volcanology*, **33**, 910–929.

——, Steinthorsson, S., Einarsson, T. H., Kristmannsdottir, H. & Oskarsson, N. 1973. The eruption on Heimay, Iceland. *Nature*, **241**, 372–75.

Tolstoy, M., Harding, A. J. & Orcutt, J. A. 1993. Crustal thickness on the Mid-Atlantic Ridge – Bull's-eye gravity-anomalies and focused accretion. *Science*, **262**, 726–729.

Tormey, D. R., Grove, T. L. & Bryan, W. B. 1987. Experimental petrology of normal MORB near the Kane Fracture Zone: 22°–25°N Mid-Atlantic Ridge. *Contributions to Mineralogy and Petrology*, **96**, 121–139.

Tucholke, B. E. 1992. Massive submarine rockslide in the rift-valley wall of the mid-Atlantic Ridge. *Geology*, **20**, 129–132.

Upton, B. G. J. & Thomas, J. E. 1980. The Tugtutoq Younger Giant Dyke Complex, South Greenland: fractional crystallization of transitional olivine basalt magma. *Journal of Petrology*, **21**, 167–198.

White, R. S., Detrick, R. S., Sinha, M. C. & Cormier, M. H. 1984. Anomalous seismic crustal structure of oceanic fracture zones. *Geophysical Journal of the Royal Astronomical Society*, **79**, 779–798.

Whitehead, J. A., Dick, H. J. B. & Schouten, H. 1984. A mechanism for magmatic accretion under spreading centres. *Nature*, **312**, 146–148.

Wilson, J. T. 1965. A new class of faults and their bearing on continental drift. *Nature*, **207**, 343–347.

Zonenshain, L. P., Kuzmin, M. I., Lisitsin, A. P., Bogdanov, Y. A. & Baranov, B. V. 1989. Tectonics of the Mid-Atlantic rift valley between the TAG and MARK areas (26–24°N): evidence for vertical tectonism. *Tectonophysics*, **159**, 1–23.

Magmatic segmentation of mid-ocean ridges: a review

RODEY BATIZA

Department of Geology and Geophysics and Hawaii Center for Volcanology, University of Hawaii, 2525 Correa Road, Honolulu, HI, 96822, USA

Abstract: Tectonic and magmatic segmentation of ridges occurs at several scales that overlap or form a continuum. A good working hypothesis is that segmentation is hierarchical and that magmatic and tectonic segmentation are linked and related by mantle flow and upwelling patterns. The largest scale of magmatic segmentation is isotopic, reflecting differences in mantle history and composition spanning a range of scales as large as individual ocean basins or larger. Superimposed, is the most conspicuous scale of segmentation, defined by large transforms and/or non-transform offsets. Magmatically, this scale of segmentation seems to reflect differences in the melting process together with the effects of smaller scale mantle heterogeneity. There appear to be differences between slow- and fast-spreading magmatic segments, although other factors such as magma supply, plume influence and ridge obliquity may also have important effects. Recurring patterns at individual segments, at least partly independent of spreading rate, include: constancy v. variability in inferred melting parameters, regular v. irregular spatial distribution of enriched basalts, MgO values, and other parameters, and correlation vs. no correlation of chemistry with axial depth.

The smallest scales of magmatic segmentation require the most closely spaced and detailed sampling. Since there have been relatively few studies of this type, there are many questions remaining. It is likely that this scale of segmentation is controlled by deep crustal and shallow upper mantle processes. However, the superimposed effects of mantle heterogeneity and time-dependent processes such as volcanic-magmatic cycles complicate the picture.

The global mid-ocean ridge system is about 50 000 km long, segmented at various length scales by triple junctions, transform faults, non-transform offsets and other small features along its length (e.g. Macdonald 1986; Macdonald *et al*. 1988; Lonsdale 1994). It is widely believed that this tectonic segmentation is related to magmatic segmentation of the ridge because, commonly, the boundaries of petrological and geochemical provinces correspond with tectonic elements of segmentation. More generally, the undulating along-axis depth profile of the ridge, with topographic humps separated by deeps at ridge-offsets, seems to define spreading 'cells' or units (LeDouaran & Francheteau 1981; Francheteau and Ballard 1983) that correlate, at least roughly, with petrological or geochemical characteristics of basalts erupted at the axis. Though by no means perfect, this correlation between large-scale tectonic and magmatic segmentation has led to the widespread belief that both may be the result of mantle thermochemical characteristics and upwelling, possibly via feedback mechanisms.

However segmentation occurs on several length scales and available evidence suggests that different segmentation scales may have different fundamental causes (Forsyth 1992). Current models favor a hierarchy of tectonic segmentation scales (e.g. Macdonald *et al*. 1988; Fig. 1) linked with a corresponding depth hierarchy of magmatic processes such as mantle upwelling, melting, and melt segregation. At the same time, there is evidence that there may be significant differences in segmentation style due to differences in spreading rate, magma supply, tectonic history, obliquity and perhaps other factors, though these issues are controversial.

This review first briefly discusses tectonic segmentation, and ideas on its causes as a function of length scale, including possible differences with spreading rate and other variables. After this, and in more detail, the available evidence for magmatic segmentation at several length scales and the ideas that have been developed to explain these observations is discussed. Along-axis transport of magma in the mantle and crust has been suggested as playing an important role at both fast and slow-spreading centres, and this is also briefly reviewed. Finally, some of the outstanding questions regarding magmatic segmentation and its links to tectonic segmentation processes is discussed.

From: MacLeod, C. J., Tyler, P. A. & Walker, C. L. (eds) 1996, *Tectonic, Magmatic, Hydrothermal and Biological Segmentation of Mid-Ocean Ridges*, Geological Society Special Publication No. 118, pp. 103–130.

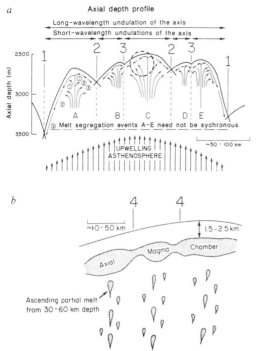

Fig. 1. The hierarchical model of magma supply and ridge segmentation of Macdonald *et al.* (1988). The numbers apply to first order (1°) to fourth order (4°) segments with the lower panel being an enlargement of the circled area of the top panel. Models such as this suggest a fundamental link between tectonic and magmatic segmentation of ridges.

Tectonic segmentation

Tectonic segmentation of the mid-ocean ridge system and recent ideas about its causes have been reviewed recently by Forsyth (1992). One difficulty is that since the mid-ocean ridge system is not completely mapped at the scale needed to fully define its segmentation, we are working with a highly incomplete data set. Nevertheless, portions of the system, particularly portions of the East Pacific Rise (EPR) and Mid-Atlantic Ridge (MAR) are well-mapped. While large portions of the Pacific–Antarctic (Lonsdale 1994; Castillo *et al.* 1995) and Indian Ocean ridges are currently being studied, most ideas about tectonic segmentation are based principally on data from the EPR and MAR.

The largest possible length scale of tectonic segmentation is the ocean basin scale, though this seems to be less significant than the smaller scale defined by the plate boundaries of the major lithospheric plates (DeMets *et al.* 1990). LeDouaran & Francheteau (1981) documented a large-scale ($c.$ 10^3 km) segmentation in axial

depth and gravity along the northern MAR that seems to correspond well with the distribution of mantle plumes or hotspots along the ridge (e.g. Schilling 1986, 1991). This they attributed to regional mantle flow structure (upwellings and downwellings) and/or differences in mantle temperature, composition, or mineralogy. The EPR has a much smoother along-axis depth profile and this scale of segmentation is less obvious, presumably because there are fewer hotspots near the ridge, although Bonatti *et al.* (1993) have proposed an equatorial along-axis deep for the EPR, indicating downwelling. Regional downwelling or cold mantle temperature may also be present at the very deep Australian–Antarctic Discordance (AAD) (e.g. Klein *et al.* 1991; Pyle *et al.* 1992; West *et al.* 1994).

Kane & Hayes (1992, 1994) and Hayes & Kane (1994) called attention to a large ($c.$ 10^3–10^2 km) scale of tectonic segmentation that seems to define long-lived tectonic corridors (Fig. 2). The length scale and longevity of these corridors, identified in the southern MAR, Pacific–Antarctic Ridge (PAR) and parts of the Southeast Indian Ridge (SEIR), are analogous to those defined by the large north and equatorial Pacific fracture zones (e.g. Atwater 1989). Though Kane and Hayes find no single model that adequately explains all their observations, they suggest a combination of regional and local mantle temperature variations as a possible cause.

The next smaller scale of segmentation ($c.$ 10^2 km) is that defined by transform faults and large non-transform offsets at both fast and slow ridges. This is the scale that Macdonald *et al.* (1988) called first order (1°) and is the most obvious scale of segmentation in both map view and along-axis profile for most of the ridge system (Fig. 1). The first-order segmentation also has the greatest temporal stability, with transforms lasting stably for $c.$ 10^7 years or more. Superimposed on the first-order are smaller scales of segmentation (2° to 4°) dividing the ridge into progressively smaller and shorter-lived segments (Macdonald *et al.* 1988). Macdonald *et al.* (1988) argued that the first-order segmentation reflects aesthenospheric upwelling cells, with the 2° and 3° segmentation due to melt segregation events and the 4° segmentation (Langmuir *et al.* 1986) caused by the pinching and swelling of sub-axial crustal magma chambers or melt lenses (Sinton *et al.* 1991; Sinton & Detrick 1992).

Possible causes of segmentation

The parallelism of the MAR to the rifted coastlines of Africa, South America, Europe and North America and its location in the middle

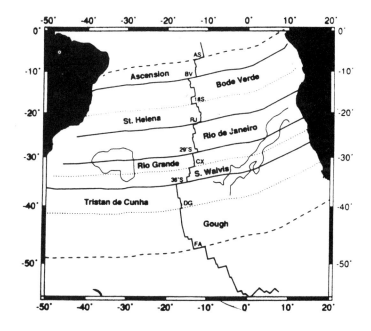

SOUTHEAST INDIAN

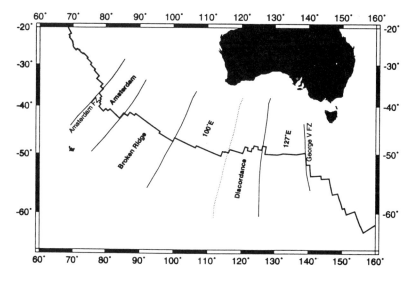

Fig. 2. Examples of long-lived spreading corridors of the (top) south Atlantic (Kane & Hayes 1992) and the Pacific Antarctic Ridge (Hayes & Kane 1994). This is the largest scale of tectonic segmentation observed at ridges.

of the Atlantic basin suggests that the largest scale segmentation of the MAR is controlled by initial continental rifting (Forsyth 1992). This is supported by the fact that many of the present offsets of the MAR can be traced back to the initial rifting episode (e.g. Klitgord & Schouten 1986). In contrast, the Pacific (Atwater 1989) and the Indian Ocean (e.g. Royer et al. 1992) have more complicated histories of opening, with several major reorganizations of spreading

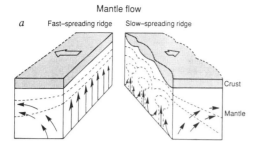

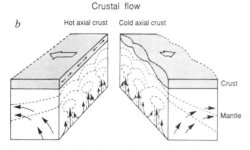

Fig. 3. (a) Model of spreading rate dependence of mantle flow geometry of Lin & Phipps Morgan (1992). (b) Alternative suggestion of Bell & Buck (1992) invoking rapid ductile flow in the lower crust at fast spreading rate. The figure is from Bell & Buck (1992).

centers. In the Indian Ocean basin it is possible to reconstruct the Gondwana continents, thereby closing the basin, although the early rifting is not well constrained (Royer *et al.* 1992); similar closing of the Pacific basin has not been possible so far. Thus, while continental rifting may be the cause of initial segmentation in some cases, it is clearly possible for such initial patterns to be later modified and even completely replaced by new spreading center systems with new segmentation patterns.

The relative lack of temporal stability of segmentation patterns in the Pacific (e.g. Hey 1977; Menard 1978; Atwater 1989; Lonsdale 1991), contrasts with the stable segmentation of the Atlantic. Though transforms in the Pacific have greater temporal stability than non-transform offsets, they are apparently shorter-lived on average than Atlantic counterparts. Possibly this is due to the faster spreading of the Pacific with, in general, smaller age offsets across the transforms (Forsyth 1992).

The temporal stability of Atlantic transforms (Schouten & White 1980; Schouten & Klitgord 1982) lead Schouten *et al.* (1985) and Crane (1985) to suggest an asthenospheric cause of regular and stable segmentation. They suggested Rayleigh–Taylor instabilities in a partly molten sub-axial asthenospheric layer and showed that the spacing of instabilities should increase with spreading rate, consistent with data available at that time. New multibeam data generally support the notion that fast ridges have, on average, longer 1° segment lengths; however, because there are several scales or orders of segmentation, measurements of segment length are not necessarily simple (Forsyth 1992). Further, while the mean segment lengths may be different, there is a large variability of lengths along many ridges (e.g. Lonsdale 1994). Despite this uncertainty about the spreading rate dependence of the mean lengths of tectonic segments, Sinton *et al.* (1991) showed that the lengths of magmatically defined segments also seems to increase with spreading rate, strengthening the association between tectonic and magmatic segmentation.

It is worth noting that despite the apparent stability of 1° segmentation and spreading corridors in the Atlantic, detailed studies in both the north Atlantic (Cannat *et al.* 1995; Tucholke & Lin 1994) and the south Atlantic (Grindley *et al.* 1992; Michael *et al.* 1994) indicate that first-order segmentation patterns in the Atlantic can change with time as they do in the Pacific and elsewhere. This means that if segmentation is controlled by mantle instabilities, their number and spacing changes over relatively short (10^5–10^6 years) time periods.

A series of studies (Phipps Morgan & Forsyth 1988; Lin *et al.* 1990; Parmentier & Phipps Morgan 1990; Lin & Phipps Morgan 1992; Sparks & Parmentier 1993; Sparks *et al.* 1993; Jha *et al.* 1994) exploring buoyant mantle flow beneath ridges indicates that segmented (along-axis) flow, termed three dimensional or 3D is favoured by a number of factors including slow spreading rate. In contrast, more 2D flow is favoured by faster spreading and lower viscosity. While buoyancy-driven flow (Sotin & Parmentier 1989; Su & Buck 1993) seems to be favoured at slow-spreading ridges where there is strong geophysical evidence for along axis geophysical changes linked with segmentation, the case for buoyant versus passive upwelling (Phipps Morgan 1987) at fast-spreading centres is less clear and remains controversial (e.g. Batiza *et al.* 1989; Wilson 1992; Wang & Cochran 1993). This issue will be discussed later. Overall, theoretical studies of mantle flow indicate that segmentation effects can be enhanced by buoyant mantle flow. Jha *et al.* (1994) propose that the scale length of 3D flow controlled by the initial depth of mantle melting might explain the large-scale segmentation noted by Kane &

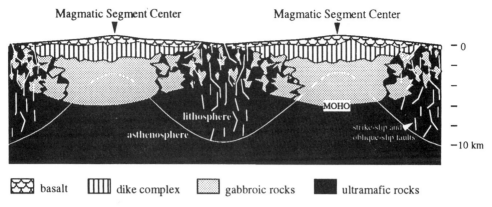

Fig. 4. Crustal and lithosphere structure inferred for slow-spreading ridges by Cannat *et al.* (1995).

Hayes (1994) and that the smaller wavelength of 1° segmentation might be due to changes in asthenosphere thickness and viscosity distribution.

Available evidence indicates that spreading rate exerts an important control on segmentation patterns, although this is probably not the only significant factor. Bell & Buck (1992) (Fig. 3), using data from the Reykjanes ridge, propose that along-axis segmentation in crustal thickness is controlled not by spreading rate-dependent mantle flow processes, but rather by differences in the thermal state of the crust controlled by magma supply. They show that the Reykjanes ridge, despite slow spreading, has a small mantle Bouguer anomaly. However, the Reykjanes ridge differs from other portions of the MAR in other characteristics, some of which may also be significant. For example, it lacks conventional plan-view first-order segmentation (Appelgate & Shor 1994). Furthermore, it is spreading obliquely, a factor that may affect the style of segmentation via the structural balance of forces at the ridge (Taylor *et al.* 1995). Another possible complication along the Reykjanes ridge is the possibility of lateral aesthenospheric flow from the Iceland hotspot (Vogt 1971). Such flow might interfere with the vertical 3D mantle flow pattern that might otherwise be present (Lin & Phipps Morgan 1992). The enhanced melt supply and lateral mantle flow may also explain why, in contrast with most of the North Atlantic, the tectonic history of the Reykjanes ridge has been marked by large changes in segmentation patterns with time (Klitgord & Schouten 1986), more characteristic of fast spreading.

The extent to which the 1° tectonic segmentation of ridges is a result of spreading rate-dependent mantle flow patterns, lithosphere characteristics, magma supply, asthenospheric layering, crustal processes, and/or plate driving forces is currently unresolved. A widely accepted idea (Fig. 3) is that first order segmentation is, in fact, a result of mantle flow processes, with more 2D-like upwelling below fast ridges and more 3D flow beneath slow ones (Lin & Phipps Morgan 1992). Figure 4 shows the crust and lithosphere structure inferred for slow ridges with gravity 'bulls-eye' patterns resulting from along-axis gradients in crustal thickness (e.g. Kuo & Forsyth 1988; Tolstoy *et al.* 1993). Figure 5 shows the relationships among the mantle Bouguer anomaly (MBA) gradient (along-axis), spreading rate and ridge morphology (Wang & Cochran 1995). Wang & Cochran argue that the consistent low value of the MBA gradient for ridges with axial highs suggests that axial morphology reflects the efficiency of along-axis melt transport. This suggestion, consistent with that of Bell & Buck (1992), implies that both fast and slow spreading ridges may have localized 3D upwelling (Wang & Cochran 1993) but that ridges with axial highs (fast-spreading ridges and the Reykjanes ridge) have efficient along-axis melt redistribution to build crust of relatively constant thickness along axis. Thus, while the 2D-fast and 3D-slow dichotomy seems to explain many of the first-order morphological and geophysical differences between fast and slow ridges, it is probably too simple to explain all the observations. One possibility is that mantle flow beneath all ridges is somewhat localized or segmented along axis (3D), with fast ridges having a larger component of passive flow and less efficient localization and slow ridges having more efficient localization and a larger component of buoyancy-driven flow. It is to be hoped that future studies will allow testing this idea as well as the others discussed above.

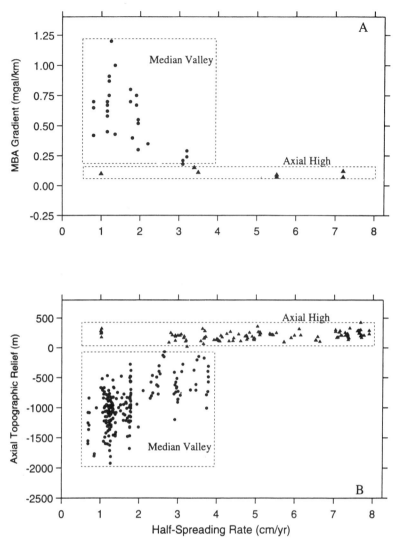

Fig. 5. Plots of mantle Bouguer anomaly gradient and topographic relief vs. spreading rate, from Wang & Cochran (1995). These data are interpreted to indicate efficient along-axis melt transport below fast-spreading ridges and the Reykjanes Ridge.

Magmatic segmentation

Having briefly discussed some of the issues of tectonic segmentation at various scales, and having touched on the connection between tectonic and magmatic segmentation, magmatic segmentation is now discussed. As with tectonic segmentation, it occurs at several length scales, discussed in order below from largest to smallest.

A working hypothesis that associates the various length scales of magmatic segmentation with diverse magmatic processes in a hierarchi-

cal fashion is given by Sinton *et al.* (1991) (Fig. 6). In this model, differences in mantle history and composition are responsible for segmentation at the largest geographic length scale. Superimposed on this is smaller-scale magmatic segmentation due to melting processes. At the smallest scale, Sinton *et al.* (1991) find boundaries between MgO-value domains that they· interpret as the result of crustal magma chamber processes at shallow level. This working hypothesis, adopted here, does not require a one-to-one correspondence between tectonic offsets and magmatic boundaries, as mantle com-

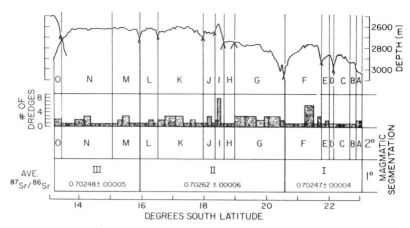

Fig. 6. A model for magmatic segmentation along the southern East Pacific Rise, with mantle isotopic domains (roman numerals) comprising the largest scale of segmentation. Superimposed on this is magmatic segmentation (lettered segments) due to differences in melting, defined by 1° and 2° tectonic elements (from Sinton *et al.* 1991).

positional provinces and regions of similar melting conditions may in some cases be larger than tectonically-defined segments. The Sinton *et al.* (1991) model is roughly analogous to the model of Macdonald *et al.* (1988) in its hierarchical scheme.

Large-scale segmentation

The largest scale of magmatic 'segmentation' or domain size observable along active ridges and mid-plate eruptions, comparable to the 1° segmentation of Sinton *et al.*, (1991) or larger, is isotopically defined and is thought to reflect significant differences in the history and composition of mantle sources feeding the ridge. Examples include large (10^4–10^3 km) geographic provinces of distinctive isotopic composition, such as the DUPAL and SOPITA anomalies (Hart 1984; Smith *et al.* 1989), including basinwide isotopic provinces (Zindler & Hart 1986; White *et al.* 1987; Pyle *et al.* 1992; Mahoney *et al.* 1992). In some cases, these isotopic provinces or portions thereof can be attributed more or less directly to known mantle plumes (e.g. Weis *et al.* 1992; Schilling *et al.* 1994) especially when the plume is directly below or near a ridge (Schilling 1991). As an example, Fig. 7 shows the mantle source provinces of the north Atlantic (Schilling 1986). Presumed mantle source effects also may appear as steady gradients in trace element concentration decoupled from isotopic ratios (Sinton *et al.* 1991; Mahoney *et al.* 1994). Figure 8 shows such a gradient in K/Ti found along the southern EPR (Sinton *et al.* 1991).

Additional evidence for large scale geo-

graphic differences in chemistry were shown by Klein & Langmuir (1987) (Fig. 9) who showed that different basins plot in different areas of a Na_8 v. depth plot. Recently, Batiza *et al.* (1995) showed that the Pacific Ocean does not apparently obey the global chemistry–depth systematics of Klein & Langmuir (1987) and attribute this not to mantle composition directly, but rather to a lack of dynamic equilibrium between mantle upwelling, melting, and seafloor depth. Castillo *et al.* (1995) find a similar lack of chemistry–depth systematics along the newly sampled Pacific Antarctic Ridge and ascribe this to the effects of mantle heterogeneity on melting (e.g. Natland 1989). The Klein & Langmuir systematics were based on regionally averaged depth and chemistry, whereas the studies of Batiza *et al.* (1995) and Castillo *et al.* (1995) use raw, unaveraged data. More data, allowing regional averages, will be needed to test fully the global systematics in the future.

First-order magmatic segmentation

As the availability of multibeam bathymetric mapping limits our understanding of the tectonic segmentation of the ridge system, so limited sampling at the axis constrains our knowledge and understanding of magmatic segmentation. Much of the EPR has been sampled at a spacing of *c.* 10–15 km (e.g. Langmuir *et al.* 1986, 1992; Sinton *et al.* 1991; Batiza 1989, 1991 and references therein) and much of the MAR has been sampled at *c.* 50 km spacing (e.g. Schilling 1986; Hanan *et al.* 1986). Along both these ridges, some study areas have been sampled

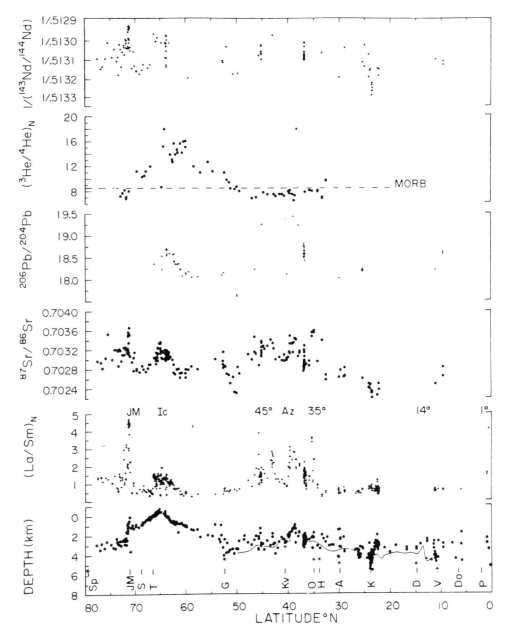

Fig. 7. Variations of trace element and isotope ratios and depth in the north Atlantic from Schilling (1986). Note the general correspondence between depth and geochemical anomalies.

more intensively and in some cases, much more intensively. Available global sampling is partly summarized by Klein & Langmuir (1987), Langmuir *et al.* (1992), Niu & Batiza (1993), and additional references given in this section.

The scale of sample spacing is an important consideration for magmatic segmentation be-

cause adequate sample spacing is needed to detect existing chemical variations at each spatial length scale. Equally important, at all length scales, is good sample location. The advent of routine GPS positioning of ships, as well as pinpoint sampling techniques (e.g. Reynolds *et al.* 1992) have helped increase the

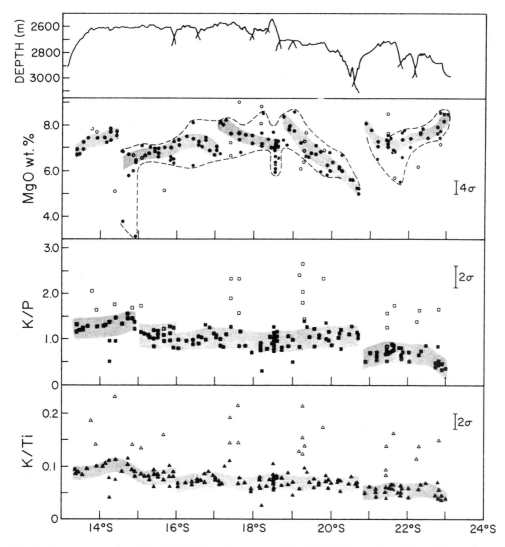

Fig. 8. Long wavelength gradient in K/Ti along the southern East Pacific Rise and examples of along-axis discontinuities in MgO value (at 17°05′ and 14°30′S) from Sinton *et al*. (1991). Dots are NMORB and circles are TMORB samples. The shading indicates 2σ and 4σ values to show general trends in the data.

accuracy and precision of sampling. Precise sampling and good geochronologic age control (e.g. Goldstein *et al*. 1994; Rubin *et al*. 1994) are important at all scales for sorting out spatial and temporal variability.

The regional averages of Klein & Langmuir (1987) and the groupings of Niu & Batiza (1993) involve ridge segments of length roughly comparable (in some cases longer due to sample limitations) to the 1° tectonic segmentation of the ridge. The results of these studies indicate that indeed, magmatic variations at this scale are due principally to differences in the inferred

melting conditions among first-order segments, possibly due to upwelling mantle domains with distinctive temperatures (Klein & Langmuir 1987; McKenzie & Bickle 1988). Figures 10 and 11, for portions of the northern and southern EPR show that transforms and large nontransform offsets, 1° or 2° tectonic elements, commonly serve as boundaries separating domains of more-or-less homogeneous values of various inferred melting parameters (Langmuir *et al*. 1986; Niu & Batiza 1991; Sinton *et al*. 1991 Bach *et al*. 1994). However, in some cases, melting parameters vary gradually or are

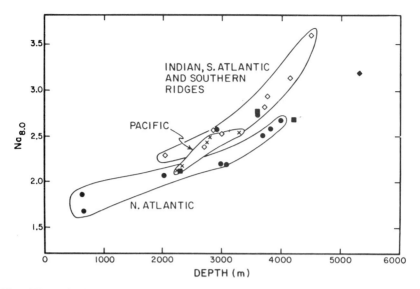

Fig. 9. Plot of Na$_8$ v. depth showing distinctions among different ocean basins, from Klein and Langmuir (1987). Dots are Atlantic averages; crosses are Pacific; solid squares are back arc basins and hollow diamonds are Indian and southern Ridges.

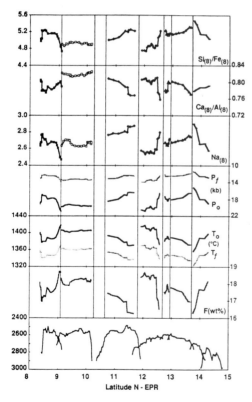

Fig. 10. Magmatic segmentation along the northern East Pacific Rise defined by discontinuities in inferred melting parameters at 1° and 2° tectonic features, from Niu & Batiza (1991). 'F' is extent of melting and 'T' and 'P' are temperatures and pressures of melting, respectively.

scattered significantly along a segment. Further, not all 1° and 2° tectonic elements are boundaries, for example lavas north and south of the Clipperton transform have the same inferred parental melt composition and thus apparently formed by melting under similar conditions (Langmuir et al. 1986; Barth et al. 1994). Also, there does not seem to be a correlation between the length of offset at transform and non-transform offsets and the magnitude of the difference in melting parameters across the offset. Nevertheless, significant offsets in melting parameters, where they do occur, are located at transforms and non-transform offsets (Bender et al. 1994).

For medium and slow-spreading ridges, 1° to 2° magmatic segmentation involving melting effects is apparently less simple, in general, than for the EPR. There are at least two important reasons for this. First, there seems to be more significant *intrasegment* variation in inferred melting conditions at slow and medium spreading segments than at fast ones (e.g. Niu & Batiza 1993). This was noted by Christie et al. (1988) who pointed out the difference between 'fractionation-dominated' (one or a few parent melt types) and 'melt-dominated' (many parent melt types) ridge segments. This difference has not yet been fully quantified. However, many segments of the EPR show only very small differences in Na$_8$, Fe$_8$, and other fractionation-corrected oxide values and ratios of values, whereas most medium and slow segments with comparable sampling usually show much larger differences. Secondly, many of the medium and

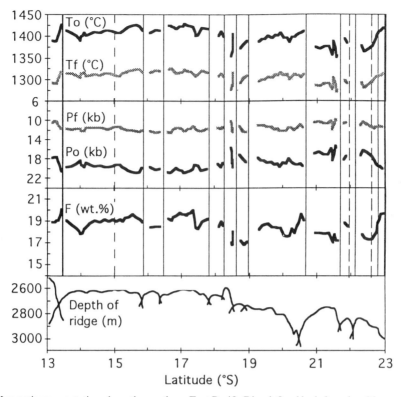

Fig. 11. Magmatic segmentation along the southern East Pacific Rise defined by inferred melting parameters. Solid lines are the magmatic segment boundaries of Sinton *et al.* (1991) (from Bach *et al.* 1994).

slow segments analysed so far contain fairly common enriched basalts whose melting parameters are more difficult to calculate with existing models (e.g. Niu & Batiza 1991; Kinzler & Grove 1993; Shen & Forsyth 1995; but see also Langmuir *et al.* 1992), though admittedly, many of these are affected by hot spots. Enriched basalts are also present at the EPR (e.g. Langmuir *et al.* 1986, 1992; Sinton *et al.* 1991) but in general appear to be more dispersed (Langmuir *et al.* 1992), at least along the southern EPR (Sinton *et al.* 1991; Mahoney *et al.* 1994).

Segments of the northern Juan de Fuca Ridge (e.g. Karsten *et al.* 1990; Michael *et al.* 1989) and the Chile ridge (Klein & Karsten 1995) are good examples of medium-spreading segments containing abundant enriched (EMORB and TMORB) basalts (Figs 12 and 13). In contrast, the southern Juan de Fuca segments (e.g. Smith *et al.* 1994) seem to have mostly normal depleted mid-ocean ridge basalt (NMORB). Examples of segments with significant intrasegment variation in melting parameters from the northern and southern MAR (Langmuir *et al.* 1992; Niu & Batiza 1994; Michael *et al.* 1994) are shown in Figs 14 and 15. Niu & Batiza (1993) list many others.

Magmatic segmentation at the scale of the first- or second-order tectonic segmentation can be quite complex, and several distinct types of behaviour are observed.

Constancy of melting parameters. This was discussed above. Along the southern and northern EPR, many 1° segments show relatively homogeneous inferred melting parameters. This is another way of saying that the lavas within a segment are commonly derived from very similar parental melt compositions. This is not always the case, as some segments show significant gradients in melting parameters (Figs 10 and 11). Nevertheless, constancy seems to be more common than at slow and intermediate ridges. Along segments with constant melting parameters, it is perhaps reasonable to infer a relatively 2D mantle upwelling pattern, although another possibility is efficient along-axis distribution of melt.

Eruption of enriched basalts in the center of a segment. This is a fairly common pattern observed along many 1° to 2° segments, though it may be more common for medium and slow spreading segments. The enrichment can either

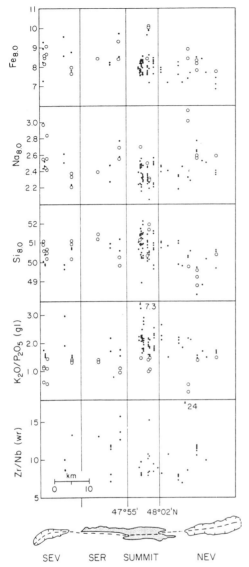

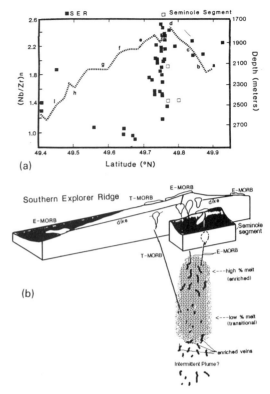

Fig. 13. (a) Nb/Zr v. latitude and (b) petrogenetic cartoon for the southern Explorer Ridge from Michael *et al.* (1989) showing enrichment in the centre of the segment and at its southern end. Note the interpretation of along-axis dyke transport of EMORB melt from the alon-axis high to the end of the segment.

Fig. 12. Magmatic segmentation of the Endeavour segment of the Juan de Fuca ridge, from Karsten *et al.* (1990). Circles are samples with $K_2O/P_2O_5 < 1.5$; SEV is Southern Endeavour Valley, SER is Southern Endeavour Ridge and NEV is Northern Endeavour Valley.

be in incompatible trace elements such as found in many types of TMORB and EMORB (Hekinian *et al.* 1989), and/or radiogenic isotope ratios. In some cases, the trace elements and isotope ratios correlate among themselves but in others not. An examples of the former sort of behavior is the EPR between 11°30′N and the Orozco transform (Castillo *et al.* 1988). This ridge segment has a fairly simple humped

topographic profile (see Fig. 10) and isotopic values tend to follow the topography, with high values for Sr and Pb isotopes near the broad top of the dome. In this area and in much of the northern EPR, there is a reasonable correlation between trace element and isotope ratios (e.g. Hekinian *et al.* 1989; Prinzhofer *et al.* 1989), although it is not clear how commonly these correlate with axial depth changes. The dispersed distribution of T and EMORB along much of the northern EPR (Langmuir *et al.* 1986) may indicate that the depth correlation is relatively weak, although the EPR segment immediately north of the Orozco transform, the shallowest segment of the northern EPR, appears to erupt only enriched basalts, possibly due to a hotspot-like influence in the region (Langmuir *et al.* 1990).

The southern EPR has isotopic differences at a comparable geographic scale, but there is no correlation with axial depth, and trace element abundances and ratios are decoupled from isotopic values (Mahoney *et al.* 1994), as also

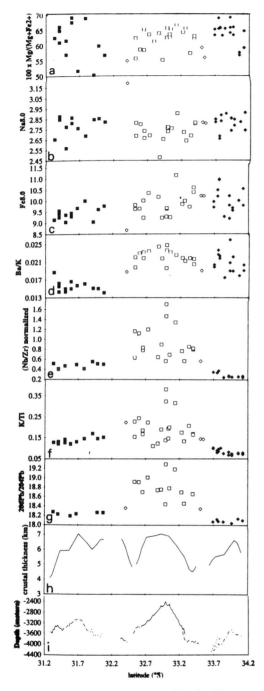

found along the Kolbeinsey ridge (Devey *et al.* 1994), in the FAMOUS and AMAR areas of the northern MAR (Frey *et al.* 1993), and other places.

At a smaller geographic length scale, the Endeavour Ridge (Karsten *et al.* 1990) shows a general enrichment in trace element ratios near the center of the segment (Fig. 12), but this is not correlated with isotopic abundances (J. Karsten, pers. comm.). The Explorer Ridge (Michael *et al.* 1989) shows a similar pattern, but enriched basalts are also found at the southern extremity of the segment (Fig. 13), interpreted to indicate along-axis dyke injection. The 33°S segment of the MAR (Michael *et al.* 1994) shows enrichment of both trace elements and isotopes in the center of the segment (Fig. 14), while the 26°S segment of the MAR shows a central enrichment of isotopes alone (P. Castillo, pers. comm.). The pattern at 26°S is actually more complicated, with Sr and Pb isotopes showing a 'W' shaped pattern of enrichment at the centre of the segment but also at the segment ends similar to the pattern in Mg# (Fig. 15). Nd isotope values show an inverted 'W' ('M') pattern. Along the recently-sampled MAR between 37° and 41°N (C. Langmuir & B. Bourdon, pers. comm.), one segment (at *c.* 39°N) shows a central enrichment in Ba and Ba/Ti.

While superficially similar, these enrichments in trace elements, isotopes or both are removed from known hotspots (except for the 39°N MAR segment) and are probably distinct from the better-understood along-axis enrichments due to the migration of plume material toward active ridges (Schilling 1991). They may be due to melting of mantle heterogeneities as suggested by several workers (e.g. Michael *et al.* 1989, 1994; Karsten *et al.* 1990) and merely reflect differences in mantle history and composition. But if so, it is difficult to understand why the heterogeneities should so commonly be located in the central, shallowest portion of the segment. One possible explanation, at least for medium to fast-spreading reidges, may be provided by the model of Bideau & Hekinian (1995), invoking a multistage melting and melt extraction cycle. If the T and EMORB phase of melt extrusion is restricted mostly to the central part of a segment, the observed patterns could be produced. Possible support for the idea that the central parts of segments behave differently than the ends comes from recent studies of fissuring by Wright *et al.* (1995), indicating a non-uniform distribution of fissuring within segments of the 9°–10°N region of the East Pacific Rise. More closely-spaced sampling studies are needed to determine how widespread these patterns of central enrichment are along the ridge. What-

Fig. 14. Magmatic segmentation and depth of the southern Mid-Atlantic ridge from Michael *et al.* (1994). The different symbols are for samples from different segments defined tectonically and bathymetrically.

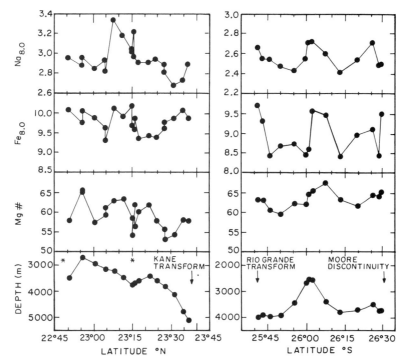

Fig. 15. Chemistry and depth along two segments of the Mid-Atlantic Ridge: the area south of the Kane transform and the 26°S segment (Niu & Batiza 1994), from Langmuir *et al.* (1992). Note the crude 'W' pattern of Mg# for the 26°S segment.

ever their cause, these sorts of patterns do seem to preclude very efficient along-axis melt redistribution from a central feeder, as this process would be expected to reduce or eliminate along-axis heterogeneity and regular along-axis depth–chemistry correlations.

Regular patterns or gradients in MgO. There are some cases of apparent regular variation in MgO, apart from clear propagating rifts and thermal edge effects, encompassing 1° or 2° segments. Even so, regular variation in MgO values, a proxy for eruption temperature, seem to be associated mainly with the smallest (3° and 4°) scales of ridge segmentation. One example is the central portion of the 26°S segment of the MAR (Niu & Batiza 1994) (Fig. 15), with MgO (for the segment as a whole) showing a 'W' pattern along axis. Fairly smooth gradients in MgO are found along several segments of the southern EPR (Fig. 8; Sinton *et al.* 1991) and along segments of the Chile ridge (S. Sherman & J. Karsten, pers. comm.). It has been speculated that such patterns may be caused by along-axis dyke injection, however, as discussed below, Icelandic and Hawaiian rifts, where along-axis dyke injection is a well-documented process,

show no such regular geographic patterns of MgO variation.

Complex temporal and spatial patterns of variation. Some well-studied segments exhibit great chemical variability, but in complex geographic patterns inferred to reflect the interplay of complex petrogenetic processes and/or mixed temporal and spatial variation. Segments showing relatively large but spatially complex chemical variations include the FAMOUS and AMAR segments (Stakes *et al.* 1984), the Kolbeinsey ridge (Devey *et al.* 1994), and the MAR segment south of the Kane transform (Langmuir *et al.* 1992) (Fig. 15). Similar complexity is found along the southern and northern EPR in the form of sporadic occurrences of EMORB and TMORB (e.g. Langmuir *et al.* 1986, 1992; Sinton *et al.* 1991; Batiza & Niu 1992; Hekinian *et al.* 1989; Castillo *et al.* 1988). At the Endeavour segment (Karsten *et al.* 1990) there is evidence that the enriched basalts in the centre of the segment may be relatively recent. Other possibilities, including temporal variation, mantle heterogeneity on a variety of scales, possible mixing in the mantle and crust, and along-axis melt transport all add complexity to

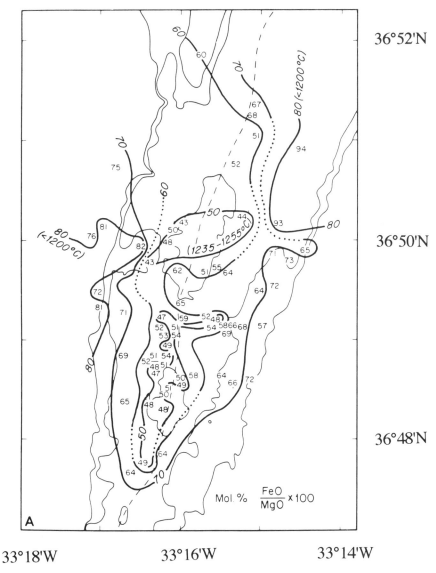

Fig. 16. FeO/MgO* 100 contours in the FAMOUS area from Hekinian (1982). This zoned pattern may be due to temporal variation in the chemistry of erupted products, or possibly, to eruption from a zoned magma chamber.

the interpretation of irregular along-axis chemical variation patterns. Irregular patterns of chemical variation probably indicate the geological complexity that is possible from place to place at various times in the history of a segment and pose a challenge to testing globally applicable and simple conceptual models of mantle upwelling and melt supply.

Simple patterns of temporal variability. In contrast with the complex patterns of temporal and spatial variability above, some segments exhibit relatively simple patterns of off-axis variation. An example from the FAMOUS area is given by Bryan & Moore (1977) and Hekinian (1982) (Fig. 16) and from the EPR at 21°N by Hekinian & Walker (1987). These patterns are thought to be due to real temporal variation in the composition of erupted basalts, though an alternative in the FAMOUS area may be eruption from a zoned magma chamber (Bryan & Moore 1977). Reynolds *et al.* (1992) show

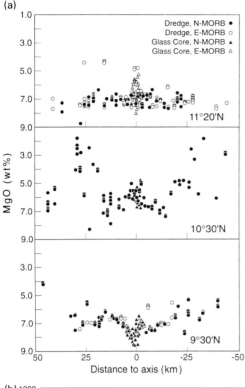

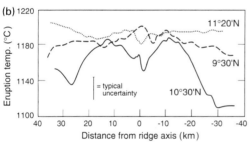

Fig. 17. MgO v. distance from the EPR axis for off-axis samples at three latitudes. Also shown is average smoothed eruption temperature, from Batiza *et al.* (1996).

tation patterns. For example, it has been argued that the migration of many 1° to 2° offsets is controlled in many cases by purely tectonic forces (see Lonsdale 1994 and references therein). If so what are the implications for models of asthenosphere-controlled segmentation and indeed, the widely accepted fundamental correlation between tectonic and magmatic segmentation?

Edge effects. Geochemical and petrological edge effects at transforms (e.g. Bender *et al.* 1984), non-transform offsets (e.g. Langmuir *et al.* 1986) and propagating rifts (Sinton *et al.* 1983) have been well known for many years. It is not clear how such edge effects are related to magmatic segmentation, partly because edge effects are not always present when expected on the basis of simple models of mantle flow (e.g. Phipps Morgan & Forsyth 1988). The chemical effects of ridge propagation may represent time-dependent v. steady state behaviour, as suggested by Sinton *et al.* (1983). A different sort of effect, occurring over *c*. 500 km of the MAR axis approaching the St Paul transform from south to north (Fig. 18), is a chemical gradient interpreted to represent a gradient in the extent of melting and crustal thickness that progresses along a half dozen ridge segments despite the presence of numerous long-offset transforms (Schilling *et al.* 1995). The gradient comes to an abrupt end at the St Paul fracture zone, known also to be a boundary of interaction between the MAR and Sierra Leone plume (Schilling *et al.* 1994). The authors suggest that the gradient may be due to the presence of cold mantle either due to local factors linked to the opening of the equatorial Atlantic or larger scale convective motions. An interesting feature of this phenomenon is that, while it occurs on a large geographic length scale, its properties, especially its monotonic nature and abrupt end at a fracture zone, resemble edge effects that usually occur on much smaller scales.

Magmatic segmentation and spreading rate. Langmuir *et al.* (1992) discuss the effects of spreading rate on MORB chemistry and find that in general, slower spreading ridges have more chemical variability than faster ridges. They also discuss the contrast between the distribution of enriched basalts between slow ridges like the MAR, with plumes located at or near the ridge and fast ridges like the EPR with a more dispersed distribution of enriched basalts. Another interesting contrast between slow and fast ridges was discussed by Niu & Batiza (1993), who showed that fast and slow ridge segments exhibit very

temporal variation of basalt chemistry at the EPR near 12°N and Perfit *et al.* (1994) from the EPR at *c*. 9°30′N. Batiza *et al.* (1996) (Fig. 17) show temporal variation suggesting that some magma chambers at the EPR are steady state whereas others, associated with relatively deep portions of the EPR, show large variations in average temperature with time. Data from off-axis studies such as these will be needed to address the question of the temporal stability of magmatic segmentation for comparison with tectonic studies of temporal changes in segmen-

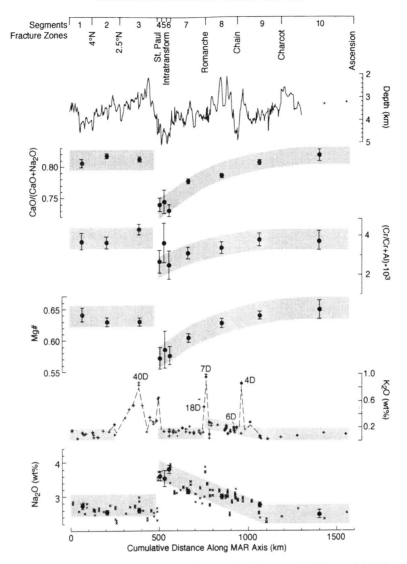

Fig. 18. Chemistry and depth along the equatorial Mid-Atlantic Ridge from Schilling *et al*. (1995). The bottom two panels show raw data and the middle three show only segment averages. Note the apparent edge effect associated with the St Paul transform and the long distance of its effect on the chemistry of axial eruptives.

different chemical systematics (Fig. 19). Interpretation of this observation relates to the origin of the chemical patterns originally called by Klein & Langmuir (1989) the 'local trend'. Niu & Batiza (1993, 1994) interpret the trend to be due to a melting reaction favored by partial reaction between melt and solid during ascent and melting of mantle. This reaction, dissolving pyroxene into the melt and precipitating olivine as a solid (Keleman *et al*. 1990) may be favoured by diapiric 3D upwelling, with slower separation of liquid from solid (Niu & Batiza 1993) and is consistent

with some of the geophysical evidence for 3D upwelling at slow ridges discussed earlier. This interpretation is controversial and alternative explanantions are available. For example, Langmuir *et al*. (1992) propose an origin related to volatiles in the mantle, while Kinzler & Grove (1992*a*, *b*, 1993) suggest the local trend is due to polybaric fractionation.

One difference between the petrological interpretations of Niu & Batiza (1993, 1994) and the geophysical models is that Niu & Batiza need more than one diapir to account for the distinct

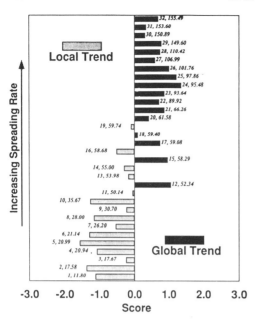

Fig. 19. Chemical systematics for slow (local trend) and fast (global trend) ridges from Niu & Batiza (1993). The numbers for each bar show the segment ID number (from Niu & Batiza, 1993), followed by the full spreading rate in mm/a^{-1}.

geochemical differences of samples along the axis (Fig. 20). Along-axis differences in melting parameters (Fig. 15) and isotopes require differences in the mantle source and melting processes, difficult to accomplish with a single, homogeneous, centrally-located, buoyant upwelling.

Lower order (3° and 4°) segmentation

The final and smallest scale of segmentation considered is the segmentation defined by small 3° and 4° tectonic offsets (Macdonald et al. 1988). This scale of segmentation is on the order of c. 10 km, one order of magnitude or so smaller than the first-order (1° to 2°) segmentation. This scale of segmentation is considered by Sinton et al. (1991) to reflect boundaries in the along-axis magma lens, as some low-order segment boundaries in the southern EPR separate provinces of differing MgO content, related to differences in eruption temperature rather than differences in melting (Fig. 7). Along the northern EPR, very tiny discontinuities, such as at 12°09' and 12°15'N, may apparently serve as boundaries between segments characterized by very different trace element ratios and inferred melting history (Langmuir et al. 1986). On the other

hand, the EPR between 12°00'N and 12°30'N is not continuously underlain by a magma lens (Detrick et al. 1987) and this supposed segmentation could actually be due to the sporadic occurrence of enriched basalts (Reynolds et al. 1992). Enriched basalts do sometimes occur preferentially at small-order discontinuities (e.g. Langmuir et al. 1986; Batiza & Niu 1992). Additional work in this area of the EPR (Reynolds et al. 1992), including much off-axis sampling, shows fairly subtle chemical changes in NMORB along the axis, together with apparent large changes in magmatic segmentation at the 10–30 km scale over the last 10^5 years.

The 3° to 4° tectonic and magmatic segmentation is not yet well understood. Partly this is because very detailed sampling is needed for good spatial resolution (e.g. Perfit et al. 1994) and there are very few areas that have been adequately sampled. In addition, good geochronological data is needed in order to sort out purely spatial variation from temporal variation in chemistry. For example, Goldstein et al. (1994) showed that in the 9°–10°N area of the EPR, off-axis eruptions may occur up to several kilometers away from the narrow axial summit depression, leading to a petrologically complex region in the vicinity of the axis (Fig. 21). It is thus clear that additional detailed sampling and dating studies will be needed to fully clarify the nature and causes of 3° to 4° magmatic segmentation. As shown in Fig. 21, it is likely that the dynamics of the magma lens and mush zone of the axial magma chamber (Sinton & Detrick 1992) play an active role in this scale of magmatic segmentation; at present, however, this role is not well understood.

The axial magma lens is thought to play an important role, not only petrologically, but also in terms of crustal structure and axial morphology (e.g. Phipps Morgan & Chen 1993). Along the EPR, where a majority of the axis is underlain by a melt lens (Detrick et al. 1987, 1993), it is reasonable to expect small-order tectonic and magmatic segmentation to be linked with processes in the lens. For example, steady state magma chambers (Fig. 17) might be expected to lead to a different sort of magmatic segmentation than non-steady state chambers. Another possible consequence of steady or near-steady state magma chambers might be the observed rough correlation along the EPR between axial depth and MgO (Scheirer & Macdonald 1993) (Fig. 22). This correlation is observed for both the northern (Langmuir et al. 1986) and southern EPR and is also observed at a small scale as in the 9°–10°N region (Batiza &

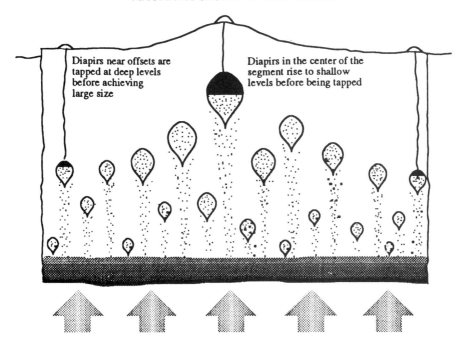

Fig. 20. Petrogenetic cartoon model for the slow spreading segment at 26°S MAR from Niu & Batiza (1994). Note that many small blobs are needed to explain the major element and isotopic differences observed along-strike.

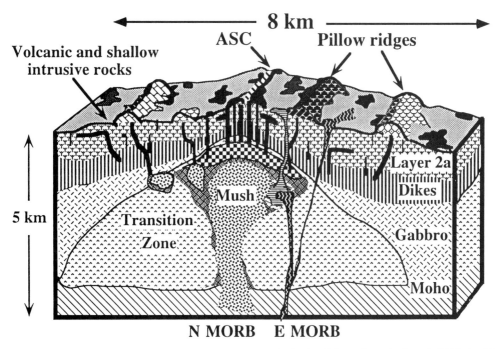

Fig. 21. Petrogenetic cartoon model for the East Pacific Rise in the 9–10°N area from Perfit *et al*. (1994). In this model, NMORB and some EMORB may have separate conduits to the surface.

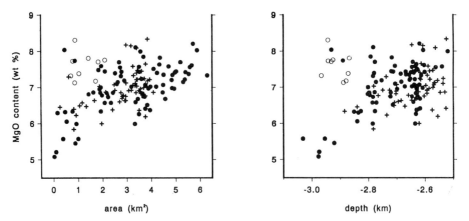

Fig. 22. Plot of MgO v. cross-sectional area and axial depth for the northern EPR (crosses) and the southern EPR (dots). The circles are for the Easter microplate area, from Scheirer & Macdonald (1993).

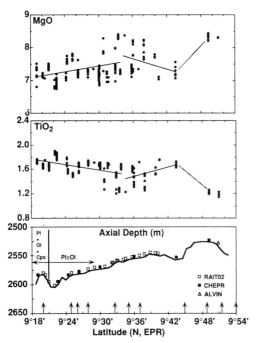

Fig. 23. MgO, TiO_2 and axial depth in the 9–10° area of the EPR, from Batiza & Niu (1992). Note the crude correlation between axial depth and chemistry.

to complications such as off-axis eruptions, lateral along-axis transport, and mantle heterogeneity and mixing (or lack thereof) in the mantle prior to delivery of melt to the magma chamber (Holness & Richter 1989). Existing data for the Atlantic, especially the greater chemical diversity in small regions of the axis (e.g. Langmuir *et al.* 1992; Stakes *et al.* 1984; Frey *et al.* 1993), suggests that a similar relationship between MgO values and axial depth is probably absent, although the high MgO lavas at Mt Pluto and Mt Venus (Fig. 16) may suggest otherwise. Since magma chambers at slow ridges are thought to be more ephemeral and smaller that those below the EPR (e.g. Sinton & Detrick 1992), the absence of an MgO-axial depth correlation would perhaps not be surprising.

Lateral magma transport

Careful monitoring of seismicity, ground deformation and eruption on Iceland (see Sigurdsson 1987) (Fig. 25) and Hawaiian rift zones (e.g.Ryan 1988; Rubin & Pollard 1987) shows that magma transport by lateral dyke injection is a common process in these volcanoes. It has been widely assumed that such a process also occurs along the ridge system (e.g. Macdonald *et al.* 1988), partly because of the strong evidence for lateral dyking in ophiolites such as Oman (MacLeod & Rothery 1992) and Troodos (Staudigel *et al.* 1992) In addition to lateral dyke injection, other forms of deeper along-axis melt distribution have also been invoked at ridges. In this section the existing evidence for these processes is reviewed briefly.

Definitive evidence for along-axis dyke in-

Niu 1992) (Fig. 23). This correlation may be due to the pinching and swelling of the magma lens (Fig. 24), with the hottest, highest MgO values found over the thickest, newest (?) and topographically highest parts of the lens (Sinton & Detrick 1992; Scheirer & Macdonald 1993). If so, then the scatter in the correlation may be due

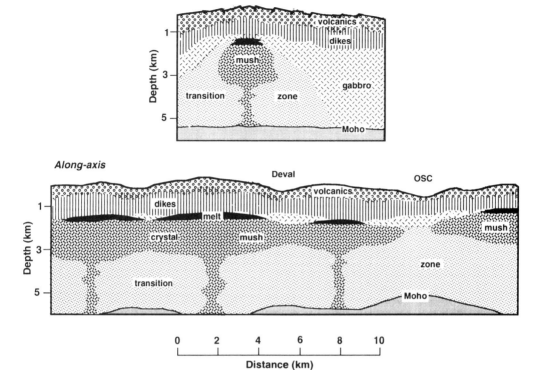

Fig. 24. Model of the 4° segmentation at fast ridges from Sinton & Detrick (1992). Note that the smallest segment boundaries correspond to gaps in the melt lens and that segmentation is defined by the spacing of vertical conduits.

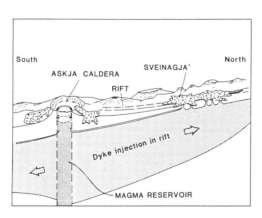

Fig. 25. Conceptual diagram of lateral dyke injection in Icelandic rifts, from Sigurdsson (1987).

jection at ridges, as in Iceland and Hawaii, will probably require monitoring of seismicity, ground deformation and eruption to track progressive movement of the magma down-rift and its appearance at the surface. Since this has

not yet been done, the best available evidence for lateral dyke injection comes from more fragmentary but still rather convincing evidence provided by the study of well-documented recent eruptions such as those at the EPR at 9° to 10°N (Haymon *et al.* 1993), the Cleft segment of the Juan de Fuca ridge (Embley & Chadwick 1994; Smith *et al.* 1994) and the progressive seismic swarm and eruption at the Co-Axial segment of the Juan de Fuca (Fox *et al.* 1995; Dziak *et al.* 1995; Embley *et al.* 1995; Chadwick *et al.* 1995). At the Cleft segment (Fig. 26), evidence from geological and structural mapping and the distribution of hydrothermal activity suggests that recent eruptions at two separate sites along the segment were supplied by dykes migrating along-axis. The petrology and geochemistry of lavas from these sites (the Young Sheetflow Site and the Young Pillow Mounds site) are consistent with this process. Interestingly, the young sheet flow samples differ chemically from the pillow mounds, seeming to require two dyking events. The young sheet flow samples are very homo-

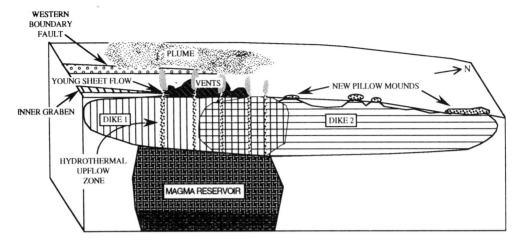

Fig. 26. Cartoon of lateral dike injection along the Cleft segment of the Juan de Fuca ridge from Embley & Chadwick (1994). Note that two dykes are required to explain the chemical differences between young lavas at the Young Sheet Flow site versus the multiple New Pillow Mounds sites.

geneous, whereas the pillow mound samples show some chemical diversity (see also Rubin & Macdougall 1990), interpreted by Smith *et al.* (1994) to perhaps indicate mixing of newly supplied primitive magma with older magma as occurs along Hawaiian rift zones (e.g. Wright & Fiske 1971; Garcia *et al.* 1992).

As noted by Dziak *et al.* (1995), in Hawaiian and Icelandic rift zones, magma injected down rift initially rises below caldera complexes. Although the topographic gradients at many mid-ocean ridges are similar to those of Hawaii and Iceland, most ridges lack circular caldera systems at their along-axis summit. A significant exception is Axial seamount on the Juan de Fuca ridge. Although it is still not certain where the Co-Axial dyke and flow originated, it is possible that the dyke originated near the summit of Axial seamount (Dziak *et al.* 1995). For most other ridges lacking circular caldera complexes, it is possible that the locus of initial dyke injection is not long-lived and shifts up and down the axis with time. The gravity study of Wang & Cochran (1993) suggests the presence of a locus of central upwelling at a fast segment of the southern EPR; additional studies will be needed to determine whether such loci are temporally stable or not. By analogy with subaerial volcanoes (e.g. Ryan 1987) these vertical conduits should, with time, develop into circular collapse structures. One possible explanation for the the fact that these are not observed at most ridge segments is that the loci of vertical supply are temporally unstable.

It is worth noting that the petrological

consequences of lateral dyke injection for patterns of magmatic segmentation are uncertain. With a central conduit providing magma for down-rift injection, one would perhaps expect a simple pattern of decreasing eruption temperature at progressively greater distances from the conduit. Interestingly no such patterns are observed in Icelandic and Hawaiian rift zones. Instead, Hawaiian rift zones generally show great petrologic diversity along their lengths and they lack simple spatial–chemical patterns (D. Clague & M. Garcia pers. comm.). This may be partly because along-axis dykes can stop at variable distances from the caldera and isolated magma batches may then cool for variable times prior to eventual eruption or mixing with fresh magma batches in subsequently-injected dykes (e.g. Wright & Fiske 1971). Another possibility is that complex temporal variations may obscure simple short-term patterns (M. Garcia & D. Clague pers. comm.). It is thus interesting that the EPR exhibits reasonably good correlation between MgO content and axial depth. Possibly this could be due to lateral dyking if the vertical injection sites are the along-axis highs. On the other hand, it may be expected that temporal shifts in the locus of vertical supply would quickly overprint and obliterate simple along-axis chemical patterns. If such shifts do not occur, then why are central calderas not present? Another possibility is that the chemistry–depth relations are not due to lateral dyking but are simply controlled by the chemistry and eruptive patterns of the most recent flows. For example, the hottest most primitive magmas may be supplied most voluminously and with greatest magma pressure, thus building the highest volcanic constructs along the ridge.

Available evidence suggests that along-axis dyke injection is an important process at ridges, although the details are still not well known. If so, this process probably operates at depths within the crust. However, this may not be the only process of lateral melt migration that occurs at mid-ocean ridges. Wang & Cochran (1993, 1995) argue in favour of along-axis melt transport and their gravity data from the southern EPR suggest that such transport may occur in the mantle. It is unclear what, if any, petrological consequences may result from such a process although it is obvious that if such a process is important, it invalidates inferences about mantle composition and melting based on the assumption of vertical transport alone. Many ridges show isotopic heterogeneity along-axis (see previous sections) and it is widely assumed that such differences reflect the composition of pooled melts derived from mantle directly underlying each portion of the axis. With along-axis melt transport in the mantle this is not necessarily true; however such a process, if important, must be such as to prevent extensive mixing, otherwise the observed along-axis chemical and isotopic differences would not be seen. As lateral mass transport in the mantle apparently can occur in situations of plume–ridge interaction (Schilling 1991) and perhaps along the southern EPR (Mahoney et al. 1994), it should not be ruled out as a potentially important process along most ridges. Nevertheless, there is presently little direct evidence that it occurs.

Some outstanding questions

The most fundamental question regarding magmatic segmentation at ridges is whether it is causally linked with the tectonic/structural segmentation and, if so, how? Prevailing opinion favours the idea that segmentation is controlled by the physical and chemical properties of the mantle, especially as they affect the patterns of mantle upwelling and melting. On the other hand, the evidence that tectonic forces alone may be responsible for the creation and migration of some offsets (e.g. Lonsdale 1994; B. Taylor pers. comm.), may indicate otherwise. It is likely that the answer to this question is not simple, and may involve feedback not only between lithosphere and asthenosphere, but also between crust and mantle.

There is strong qualitative evidence for the

models of Langmuir et al. (1986), Macdonald et al. (1988) and Sinton et al. (1991) of a hierarchical system linking magmatic and tectonic segmentation (see also Hart 1993). On the other hand, there is insufficient rock sampling at present to test definitively whether the scales of magmatic and tectonic segmentation truly correspond in detail on a global basis. These models seem to work best for fast spreading ridges, raising the question of whether processes are very different at slow ones, as implied by available petrological data and models of mantle flow (e.g. Lin & Phipps Morgan 1992)?

Finally there is the question of the role of horizontal versus vertical mantle flow and melt delivery. It seems very likely that along-axis magma transport occurs both in the crust, by lateral dyke injection, and deeper in the mantle. How do processes of lateral mantle flow interact with inferred passive and buoyant upwelling? Are the processes of lateral transport apparently needed to explain some of the gravity and seismic data incompatible with vertical transport schemes that seem to explain a lot of the available geophysical and geochemical data?

It is to be hoped that new mapping and sampling efforts along the mid-ocean ridge system, together with new insights gained from laboratory and theoretical studies will begin to provide answers to some of these questions.

I thank the following individuals for discussion and/or providing data or preprints: B. Taylor, J. Sinton, J. Mahoney, P. Johnson, M. Perfit, J. Karsten, M. Garcia, D. Clague, B. Bourdon, C. Langmuir, and P. Castillo. I thank K. Rubin, E. Klein, D. Christie, P. Browning and C. MacLeod for very helpful reviews. Supported by the NSF and ONR. SOEST Contribution No. 3930.

References

APPELGATE, B. & SHOR, A. N. 1994. The northern Mid-Atlantic and Reykjanes Ridges: spreading center morphology between 55°50′N and 63°00′N. Journal of Geophysical Research, **99**, 17935–17956.

ATWATER, T. 1989. Plate tectonic history of the northeast Pacific and western North America. In: WINTERER, E. L., HUSSONG, D. M. & DECKER, R. W. (eds.) The Eastern Pacific Ocean and Hawaii. The Geology of North America, **N**, Geological Society of America, 21–72.

BACH, W., HEGNER, E., ERZINGER, J. & SATIR, M. 1994. Chemical and isotopic variations along the superfast spreading East Pacific Rise from 6° to 30°S. Contributions to Mineralogy and Petrology, **116**, 365–380.

BARTH, G. A., KASTENS, K. A. & KLEIN, E. M. 1994. The origin of bathymetric highs at ridge-transform intersections: a multi-disciplinary case study at the Clipperton fracture zone. *Marine Geophysical Researches,* **16**, 1–50.

BATIZA, R. 1989. Petrology and geochemistry of eastern Pacific spreading centers. *In:* WINTERER, E. L., HUSSONG, D. M. & DECKER, R. W. (eds) *The Eastern Pacific Ocean and Hawaii*: The Geology of North America, **N**. Geological Society of America, Boulder, Colorado, 145–160.

—— 1991. Pacific Ocean Crust. *In:* FLOYD, P. (ed.) *Oceanic Basalts.* Blackie and Son, Glasgow, 264–288.

—— & NIU, Y. 1992. Petrology and magma chamber processes at the East Pacific Rise ~9°30′N. *Journal of Geophysical Research,* **97**, 6779–6797.

——, SMITH, T. & NIU, Y. 1989. Geologic and petrologic evolution of seamounts near the EPR based on submersible and camera study. *Marine Geophysical Researches,* **11**, 169–236.

——, HEKINIAN, R., BIDEAU, D. & FRANCHETEAU, J. 1995. Chemistry of deep (3500–5600m) Pacific zero-age MORB- why is the Pacific anomalous?, *Geophysical Research Letters,* **22**, 3067–3070.

——, NIU, Y., KARSTEN, J. L., BOGER, W., POTTS, E., NORBY, L. & BUTLER, R. 1996. Steady and non-steady state magma chambers below the East Pacific Rise. *Geophysical Research Letters,* **23**, 221–224.

BELL, R. E. & BUCK, R. 1992. Crustal control of ridge segmentation inferred from observations of the Reykjanes Ridge. *Nature,* **357**, 583–586.

BENDER, J. F., LANGMUIR, C. H. & HANSON, G. N. 1984. Petrogenesis of basalt glass from the Tamayo region, East Pacific Rise. *Journal of Petrology,* **25**, 213–254.

BIDEAU, D. & HEKINIAN, R. 1995. A dynamic model for generating small-scale heterogenieities in ocean foor basalts. *Journal of Geophysical Research,* **100**, 10141–10162.

BONATTI, E., SEYLER, M. & SUSHEVSKAYA, N. 1993. A cold suboceanic mantle belt at the Earth's equator. *Science,* **261**, 315–320.

BRYAN, W. B. & MOORE, J. G. 1977. Compositional variations of young basalts in the Mid-Atlantic Ridge rift valley near lat. 36°49′N. *Geological Society of America Bulletin,* **88**, 556–570.

CANNAT, M., MEVEL, C., MAIA, M., DEPLUS, C., DURANT, C., GENTE, P., AGRINIER, P., BELAROUCHI, A., DUBUISSON, G., HUMLER, E. & REYNOLDS, J. 1995. Thin crust, ultramafic exposures, and rugged faulting patterns at the Mid-Atlantic Ridge (22°–24°N). *Geology,* **23**, 49–52.

CASTILLO, P. R., KLEIN, E., SHIREY, S. B., BENDER, J. F., BATIZA, R. & LANGMUIR, C. H. 1988. Relationship of Sr-Nd-Pb isotopes of EPR and MAR basalt glasses to bathymetry and ridge segmentation. *EOS, Transactions of the American Geophysical Union,* **69**, 1477.

——, NATLAND, J. H. & NIU, Y. 1995. Petrology and Sr, Nd, and Pb isotope geochemistry of MORB from the fast-spreading ridges in the south Pacific. *EOS, Transactions of the American Geophysical Union,* **76**, F530.

CHADWICK, W. W. JR., EMBLEY, R. W. & FOX, C. G.

1995. SeaBeam depth changes associated with recent lava flows, CoAxial segment, Juan de Fuca ridge: evidence for multiple eruptions between 1981–1993. *Geophysical Research Letters,* **22**, 167–170.

CHRISTIE, D. M., PYLE, D. G., SEMPERE, J.-C., PHIPPS MORGAN, J. & SHOR, A. 1988. Petrologic and tectonic observations in and adjacent to the Australia–Antarctic Discordance. *EOS, Transactions of the American Geophysical Union,* **69**, 1426.

CRANE, K. 1985. The spacing of rift-axis highs: dependence upon diapiric processes in the underlying asthenosphere? *Earth and Planetary Science Letters,* **72**, 405–414.

DETRICK, R. S., BUHL, P., VERA, E., MUTTER, J., ORCUTT, J., MADSEN, J. & BROCKER, T. 1987. Multichannel seismic imaging of a crustal magma chamber along the East Pacific Rise. *Nature,* **236**, 35–41.

——, HARDING, A. J., KENT, G. M., ORCUTT, J. A., MUTTER, J. C. & BUHL, P. 1993. Seismic structure of the Southern East Pacific Rise. *Science,* **259**, 499–503.

DEMETS, C., GORDON, R. G., ARGUS, D. F. & STEIN, S. 1990. Current plate motions. *Geophysical Journal,* **101**, 425–478.

DEVEY, C. W., GARB-SCHONBERG, C.-D., STOFFERS, P., CHAUVEL, C. & MERTZ, D. F. 1994. Geochemical effects of dynamic melting beneath ridges: Reconciling major and trace element variations in Kolbeinsey (and global) mid-ocean ridge basalt. *Journal of Geophysical Research,* **99**, 9077–9095.

DZIAK, R. P., FOX, C. G. & SCHREINER, A. E. 1995. The June-July 1993 seismo-acoustic event at CoAxial segment, Juan de Fuca ridge: evidence for a lateral dike injection. *Geophysical Research Letters,* **22**, 135–38.

EMBLEY, R. W. & CHADWICK, W. W. JR. 1994. Volcanic and hydrothermal processes associated with a recent phase of seafloor spreading at the northern Cleft segment: Juan de Fuca ridge. *Journal of Geophysical Research,* **99**, 4741–4760.

——, CHADWICK, W. W. JR., JONASSON, I. R., BUTTERFIELD, D. E. & BAKER, E. T. 1995. Initial results of the rapid response to the 1993 CoAxial event: relationships between hydrothermal and volcanic processes. *Geophysical Research Letters,* **22**, 143–146.

FORSYTH, D. W. 1992. Geophysical constraints on mantle flow and melt generation beneath mid-ocean ridges. *In:* PHIPPS MORGAN, J., BLACKMAN, D. K. & SINTON, J. M. (eds) *Mantle flow and melt generation at Mid-Ocean Ridges.* AGU Geophysical Monographs, **71**, 1–65.

FOX, C. G., RADFORD, W. E., DZIAK, R. P., LAU, T.-K., MATSUMOTO, H. & SCHREINER, A. E. 1995. Acoustic detection of a seafloor spreading episode on the Juan de Fuca ridge using military hydrophone arrays. *Geophysical Research Letters,* **22**, 131–134.

FRANCHETEAU, J. & BALLARD, R. D. 1983. The East Pacific Rise near 21°N, 13°N and 20°S: inferences for along-strike variability of axial processes of the Mid-Ocean Ridge. *Earth and Planetary Science Letters,* **64**, 93–116.

FREY, F. A., WALKER, N., STAKES, D., HART, S. R. & NIELSEN, R. 1993. Geochemical characteristics of basaltic glasses from the AMAR and FAMOUS axial valleys, Mid-Atlantic Ridge (36°–37°N): petrogenetic implications. *Earth and Planetary Science Letters,* **115,** 117–136.

GARCIA, M. O., RHODES, J. M., WOLFE, E. W., ULRICH, G. E. & HO, R. A. 1992. Petrology of lavas from episodes 2–47 of the Pu'u O'o eruption of Kilauea volcano, Hawaii: evaluation of magmatic processes. *Bulletin of Volcanology,* **55,** 1–16.

GOLDSTEIN, S. J., PERFIT, M. R., BATIZA, R., FORNARI, D. J. & MURRELL, M. T. 1994. Off-axis volcanism at the East Pacific Rise based on U-series dating of basalts. *Nature,* **367,** 157–159.

GRINDLAY, N. R., FOX, P. J. & VOGT, P. R. 1992. Morphology and tectonics of the Mid-Atlantic Ridge (25°–27°30'S) from Sea Beam and magnetic data. *Journal of Geophysical Research,* **97,** 6983–7010.

HANAN, B. B., KINGSLEY, R. H. & SCHILLING, J.-G. 1986. Pb isotope evidence in the South Atlantic for migrating ridge-hotspot interactions. *Nature,* **322,** 137–144.

HART, S. R. 1984. A large-scale isotope anomaly in the southern hemisphere mantle. *Nature,* **309,** 753–757.

—— 1993. Equilibration during mantle melting: a fractal tree model. *Procedings of the National Academy of Science USA,* **90,** 11914–11918.

HAYES, D. E. & KANE, K. A. 1994. Long-lived mid-ocean ridge segmentation of the Pacific-Antarctic ridge and the Southeast Indian Ridge. *Journal of Geophysical Research,* **99,** 19679–19692.

HAYMON, R. M., FORNARI, D. J., VON DAMM, K. L., LILLEY, M. D., PERFIT, M. R., EDMOND, J. M., SHANKS, W. C. III, LUTZ, R. A., GREBMEIER, J. M., CARBOTTE, S., WRIGHT, D., MCLAUGHLIN, E., SMITH, M., BEEDLE, N. & OLSON, E. 1993. Volcanic eruption of the mid-ocean ridge along the East Pacific Rise crest at 9°45'–52'N: Direct submersible observations of seafloor phenomena associated with an eruption event in April 1991. *Earth and Planetary Science Letters,* **119,** 85–101.

HEKINIAN, R. 1982. *Petrology of the Ocean Floor.* Elsevier, Amsterdam.

——, THOMPSON, G. & BIDEAU, D. 1989. Axial and off-axial heterogeneity of basaltic rocks from the East Pacific Rise at 12°35'N–12°51'N and 11°26'N–11°30'N. *Journal of Geophysical Research,* **94,** 17 437–17 464.

—— & WALKER, D. 1987. Diversity and spatial zonation of volcanic rocks from the East Pacific Rise near 21°N. *Contributions to Mineralogy and Petrology,* **96,** 265–280.

HEY, R. N. 1977. Tectonic evolution of the Cocos-Nazca spreading center. *Geological Society of America Bulletin,* **88,** 1404–1420.

HOLNESS, M. B. & RICHTER, F. M. 1989. Possible effects of spreading rate on MORB isotopic and rare earth composition arising from melting of a heterogeneous source. *Journal of Geology,* **97,** 247–260.

JHA, K., PARMENTIER, E. M. & PHIPPS MORGAN, J. 1994. The role of mantle-depletion and melt-retention bouyancy in spreading center segmentation. *Earth and Planetary Science Letters,* **125,** 221–234.

KANE, K. A. & HAYES, D. E. 1992. Tectonic corridors in the South Atlantic: evidence for long-lived Mid-Ocean Ridge segmentation. *Journal of Geophysical Research.* **97,** 17317–17330.

—— & —— 1994. Long-lived mid-ocean ridge segmentation: constraints on models. *Journal of Geophysical Research,* **99,** 19693–19706.

KARSTEN, J. L., DELANEY, J. R., RHODES, J. M. & LIIAS, R. A. 1990. Spatial and temporal evolution of magmatic systems beneath the Endeavour segment, Juan de Fuca Ridge: Tectonic and petrologic constraints. *Journal of Geophysical Research.* **95,** 19 235–19 256.

KELEMEN, P. B., JOYCE, D. M., WEBSTER, J. D. & HOLLOWAY, J. R. 1990. Reaction between ultramafic rock and fractionating basalt magma, II, experimental investigations of reactions between olivine tholeiite and harzburgite at 1150–1050°C and 5 kb. *Journal of Petrology,* **31,** 99–134.

KINZLER, R. J. & GROVE, T. L. 1992a. Primary magmas of mid-ocean ridge basalts, 1. experiments and results. *Journal of Geophysical Research,* **97,** 6885–6906.

—— & —— 1992b. Primary magmas of mid-ocean ridge basalts, 2, Applications. *Journal of Geophysical Research,* **97,** 6907–6926.

—— & —— 1993. Corrections and further discussion of the primary magmas of mid-ocean ridge basalts 1 and 2. *Journal of Geophysical Research,* **98,** 22 339–22 348.

KLEIN, E. M. & LANGMUIR, C. H. 1987. Global correlations of ocean ridge basalt chemistry with axial depth and crustal thickness. *Journal of Geophysical Research.* **92,** 8089–8115.

—— & —— 1989. Local versus global variations in ocean ridge basalt composition: a reply, *Journal of Geophysical Research,* **94,** 4241–4252.

—— & KARSTEN, J. L. 1995. Ocean-ridge basalts with convergent-margin geochemical affinities from the Chile ridge. *Nature,* **374,** 52–57.

——, LANGMUIR, C. H. & STAUDIGEL, H. 1991. Geochemistry of basalts from the Southeast Indian Ridge, 115°E–138°E. *Journal of Geophysical Research,* **96,** 2089–2108.

KLITGORD, K. D. & SCHOUTEN, H. 1986. Plate kinematics of the central Atlantic. *In:* VOGT, P. R. & TUCHOLKE, B. E. (eds) *The Western North Atlantic Region.* The Geology of North America, **M,** Geological Society of America, 351–378.

KUO, B.-Y. & FORSYTH, D. W. 1988. Gravity anomalies of the ridge-transform system in the south Atlantic between 31 and 34.5°S: upwelling centers and variations in crustal thickness. *Marine Geophysical Researches,* **10,** 205–232.

LANGMUIR, C. H., BENDER, J. F. & SHIREY, S. 1990. A trace-element enriched province on the East Pacific Rise north of the Orozco transform zone. *Transactions of the American Geophysical Union EOS,* **71,** 43, 1703.

——, —— & BATIZA, R. 1986. Petrological and

tectonic segmentation of the East Pacific Rise, 5°30'–14°30'N. *Nature, 322*, 422–429.

——, KLEIN, E. M. & PLANK, T. 1992. Petrological systematics of mid-ocean ridge basalts: constraints on melt generation beneath ocean ridges. *In*: PHIPPS MORGAN, J., BLACKMAN, D. K. & SINTON, J. M. (eds) *Mantle flow and melt generation at Mid-Ocean Ridges*. AGU Geophysical Monographs, **71**, 183–280.

LEDOUARAN, S. & FRANCHETEAU, J. 1981. Axial depth anomalies from 10 to 50° north along the Mid-Atlantic Ridge: correlation with other mantle properties. *Earth and Planetary Science Letters, 54*, 29–47.

LIN, J. & PHIPPS MORGAN, J. 1992. The spreading rate dependence of three-dimensional mid ocean ridge gravity structure. *Geophysical Research Letters, 19*, 13–16.

——, PURDY, G. M., SCHOUTEN, H., SEMPÉRÉ, J.-C. & ZERVAS, C. 1990. Evidence from gravity data for focused magmatic accretion along the Mid-Atlantic Ridge. *Nature, 344*, 627–632.

LONSDALE, P. 1991. Structural patterns of the Pacific floor offshore of Peninsular California. *In*: DAUPHIN, J. P. & SINONEIT, B. R. T. (eds) *The Gulf and Peninsular Province of the Californias*. American Association of Petroleum Geologists Memoirs, **47**, 87–125.

—— 1994. Geomorphology and structural segmentation of the crest of the southern (Pacific-Antarctic) East Pacific Rise. *Journal of Geophysical Research*, **99**, 4683–4702.

MACDONALD, K. C. 1986. The crest of the Mid-Atlantic Ridge: models for crustal generation processes and tectonics. *In*: VOGT, P. R. & TUCHOLKE, B. E. (eds) *The Western North Atlantic Region*. The Geology of North America, **M**, Geological Society of America, 51–68.

——, FOX, P. J., PERRAM, L. J., EISSEN, M. F. HAYMON, R. M., MILLER, S. P., CARBOTTE, S. M., CORMIER, M.-H. & SHOR, A. N. 1988. A new view of the mid-ocean ridge from the behaviour of ridge-axis discontinuities. *Nature, 335*, 217–225.

MACLEOD, C. J. & ROTHERY, D. A. 1992. Ridge axial segmentation in the Oman ophiolite: evidence from along-strike variations in the sheeted dyke complex. *In*: PARSON, L. M., MURTON, B. J. & BROWNING, P. (eds) *Ophiolites and their Modern Oceanic Analogues*. Geological Society, London, Special Publications, **60**, 39–63.

MCKENZIE, D. & BICKLE, M. J. 1988. The volume and composition of melt generated by extension of the lithosphere. *Journal of Petrology, 29*, 625–679.

MAHONEY, J. J., LE ROEX, A. P., PENG, Z., FISHER, R. L. & NATLAND, J. H. 1992. Western limits of Indian MORB mantle and the origin of low $^{206}Pb/^{204}Pb$ MORB: isotope systematics of the central Southwest Indian Ridge (17–50°E). *Journal of Geophysical Research, 97*, 19771–19790.

MAHONEY, J. J., SINTON, J. M., KURZ, M. D., MACDOUGALL, J. D., SPENCER, K. J. & LUGMAIR, G. W. 1994. Isotope and trace element characteristics of a super-fast spreading ridge: East Pacific Rise, 13°–23°S. *Earth and Planetary Science Letters, 121*, 173–193.

MENARD, H. W. 1978. Fragmentation of the Farallon plate by pivoting subduction. *Journal of Geology, 86*, 99–110.

MICHAEL, P. J., CHASE, R. L. & ALLAN, J. F. 1989. Petrologic and geological variations along the southern Explorer Ridge, northeast Pacific ocean. *Journal of Geophysical Research, 94*, 13 895–13 918.

——, FORSYTH, D. W., BLACKMAN, D. K., FOX, P. J., HANAN, B. B., HARDING, A. J., MACDONALD, K. C., NEUMANN, G. A., ORCUTT, J. A., TOLSTOY, M. & WEILAND, C. M. 1994. Mantle control of a dynamically evolving spreading center: Mid-Atlantic Ridge 31–34°S. *Earth and Planetary Science Letters, 121*, 451–468.

NATLAND, J. H. 1989. Partial melting lithologically heterogeneous mantle: inferences from crystallization histories of magnesian abyssal tholeiites from the Siqueiros fracture zone *In*: SAUNDERS, A. D. & NORRY, M. J. (eds) *Magmatism in the Ocean Basins*, Geological Society, London, Special Publications, **42**, 41–70.

NIU, Y. & BATIZA, R. 1991. An empirical method for calculating melt compositions produced beneath mid-ocean ridges: application for axis and off-axis (seamounts) melting. *Journal of Geophysical Research, 96*, 21 753–21 777.

—— & —— 1993. Chemical variation trends at fast and slow spreading mid-ocean ridges. *Journal of Geophysical Research, 98*, 7887–7902.

—— & —— 1994. Magmatic processes at a slow spreading ridge segment: 26°S Mid-Atlantic Ridge. *Journal of Geophysical Research, 99*, 19 719–19 740.

PARMENTIER, E. M. & PHIPPS MORGAN, J. 1990. Spreading rate dependence of three-dimensional structure in oceanic spreading centers. *Nature, 348*, 325–328.

PERFIT, M. R., FORNARI, D. J., SMITH, M. C., BENDER, J. F., LANGMUIR, C. H. & HAYMON, R. M. 1994. Small-scale spatial and temporal variations in mid-ocean ridge crest magmatic processes. *Geology, 22*, 375–379.

PHIPPS MORGAN, J. 1987. Melt migration beneath mid-ocean ridge spreading centers. *Geophysical Research Letters, 14*, 1238–1241.

—— & FORSYTH, D. W. 1988. Three-dimensional flow and temperature perturbations due to a transform offset: effects on oceanic crustal and upper mantle structure, *Journal of Geophysical Research, 93*, 2955–2966.

—— & CHEN, Y. J. 1993. The genesis of ocean crust: magma injection, hydrothermal circulation and crustal flow. *Journal of Geophysical Research, 98*, 6283–6297.

PRINZHOFER, A., LEWIN, E. & ALLEGRE, C. J. 1989. Stochastic melting of the marble cake mantle: evidence from local study of the East Pacific Rise at 12°50'N. *Earth and Planetary Science Letters, 92*, 189–206.

PYLE, D. G., CHRISTIE, D. M. & MAHONEY, J. J. 1992. Resolving an isotopic boundary within the Australia-Antarctic Discordance. *Earth and Planetary Science Letters, 112*, 161–178.

REYNOLDS, J. R., LANGMUIR, C. H., BENDER, J. F.,

KASTENS, K. A. & RYAN, W. B. F. 1992. Spatial and temporal variability in the geochemistry of basalts from the East Pacific Rise. *Nature*, **359**, 493–499.

ROYER, J. Y., SCLATER, J. G., SANDWELL, D. T., CANDE, S. C., SCHLICH, R., MUNSCHY, M., DYMENT, J., FISHER, R. L., MULLER, R. D., COFFIN, M. F., PATRIAT, P. & BERGH, H. W. 1992. Indian Ocean Reconstructions since the Late Jurassic. *In*: DUNCAN, R. A., REA, D. K., KIDD, R. B., VON RAD, U. & WEISSEL, J. K. (eds) *Synthesis of Results from Scientifc Drilling in the Indian Ocean*. AGU Geophysical Monographs **70**, 471–475.

RUBIN, A. M. & POLLARD, D. D. 1987. Origin of blade-like dikes in volcanic rift zones. *In*: DECKER, R. W., WRIGHT, T. L. & STAUFFER, P. H. (eds) *Volcanism in Hawaii*. US Geological Survey Professional Papers, **1350**, 1449–1470.

RUBIN, K. H. & MACDOUGALL, J. D. 1990. A radioactive decay technique for dating neovolcanic MORB. *Earth and Planetary Science Letters*, **101**, 313–322.

——, —— & PERFIT, M. R. 1994. ^{210}Po-^{210}Pb dating of recent volcanic eruptions on the sea floor. *Nature*, **368**, 841–844.

RYAN, M. P. 1987. Neutral buoyancy and the mechanical evolution of magmatic systems. *In*: MYSEN, B. O. (ed.) *Magmatic Processes: Physicochemical Principles*. The Geochemical Society, Special Publications, **1**, 259–287.

—— 1988. The mechanics and three-dimensional internal structure of active magmatic systems, Kilauea Volcano, Hawaii. *Journal of Geophysical Research*, **93**, 4213–4248.

—— 1993. Neutral bouyancy and the structure of mid-ocean ridge magma reservoirs. *Journal of Geophysical Research*, **98**, 22321–22338.

SCHEIRER, D. S. & MACDONALD, K. C. 1993. Variations in cross-sectional area of the axial ridge along the East Pacific Rise: evidence for the magmatic budget of a fast-spreading center. *Journal of Geophysical Research*, **98**, 7871–7885.

SCHILLING, J.-G. 1986. Geochemical and isotopic variation along the Mid-Atlantic Ridge axis from 79°N to 0°N. *In*: VOGT, P. R. & TUCHOLKE, B. E. (eds) *The Western North Atlantic Region*. The Geology of North America, **M**. Geological Society of America, 137–156.

—— 1991. Fluxes and excess temperatures of mantle plumes inferred from their interaction with migrating mid-ocean ridges. *Nature*, **352**, 397–403.

——, HANAN, B. B., MCCULLY, B. & KINGSLEY, R. H. 1994. Influence of the Sierra Leone mantle plume on the equatorial Mid-Atlantic Ridge: a Nd-Sr-Pb isotopic study. *Journal of Geophysical Research*, **99**, 12 005–12 028.

——, RUPPEL, C., DAVIS, A. N., MCCULLY, B., TIGHE, S. A., KINDSLEY, R. H. & LIN, J. 1995. Thermal structure of the mantle beneath Equatorial Mid-Atlantic Ridge: inferrences from spatial variations of dredged basalt glass compositions. *Journal of Geophysical Research*, **100**, 10 057–10 076.

SCHOUTEN, H. & KLITGORD, K. D. 1982. The memory of the accreting plate boundary and the continuity of fracture zones. *Earth and Planetary Science Letters*, **59**, 255–266.

—— & WHITE, R. S. 1980. Zero-offset fracture zones. *Geology*, **8**, 175–179.

——, KLITGORD, K. D. & WHITEHEAD, J. A. 1985. Segmentation of mid-ocean ridges. *Nature*, **317**, 225–229.

SHEN, Y. & FORSYTH, D. W. 1995. Geochemical constraints on initial and final depths of melting beneath mid-ocean ridges. *Journal of Geophysical Research*, **100**, 2211–2238.

SIGURDSSON, H. 1987. Dyke injection in Iceland: a review. *In*: HALLS, H. C. & FAHRIG, W. F. (eds) *Mafic Dyke Swarms*. Geological Society of Canada Special Papers, **34**, 55–64.

SINTON, J. M & DETRICK, R. S. 1992. Mid-ocean ridge magma chambers. *Journal of Geophysical Research*, **97**, 197–216.

——, WILSON, D. S., CHRISTIE, D. M., HEY, R. N. & DELANEY, J. T. 1983. Petrologic consequences of rift propagation on oceanic spreading centers. *Earth and Planetary Science Letters*, **62**, 193–207.

——, SMAGLIK, S. M., MAHONEY, J. J. & MACDONALD, K. C. 1991. Magmatic processes at superfast spreading oceanic ridges: Glass compositional variations along the East Pacific Rise, 13°S–23°S. *Journal of Geophysical Research*, **96**, 6133–6155.

SMITH, M. C., PERFIT, M. R. & JONASSON, I. R. 1994. Petrology and geochemistry of basalts from the southern Juan de Fuca ridge: controls on the spatial and temporal evolution of mid-ocean ridge basalt. *Journal of Geophysical Research*, **99**, 4787–4812.

SMITH, W. H. F., STAUDIGEL, H., WATTS, A. B. & PRINGLE, M. S. 1989. The Magellan seamounts: early Cretaceous record of the south Pacific isotopic and thermal anomaly. *Journal of Geophysical Research*, **94**, 10501–10523.

SOTIN, C. & PARMENTIER, E. M. 1989. Dynamical consequences of compositional and thermal density stratification beneath spreading centers. *Geophysical Research Letters*, **16**, 835–838.

SPARKS, D. W. & PARMENTIER, E. M. 1993. The structure of three-dimensional convection beneath oceanic spreading centers, *Geophysical Journal International*, **122**, 81–91.

——, & PHIPPS MORGAN, J. 1993. Three-dimensional mantle convection beneath a segmented spreading center: implications for along-axis variations in crustal thickness and gravity. *Journal of Geophysical Research*, **98**, 21977–21995.

STAKES, D. S., SHERVAIS, J. W. & HOPSON, C. H. 1984. The volcanic-tectonic cycle of the FAMOUS and AMAR valleys, Mid-Atlantic Ridge (36°47′N): evidence from basalt glass and phenocryst compositional variations for a steady state magma chamber beneath the valley mid-sections, AMAR 3. *Journal of Geophysical Research*, **89**, 6995–7028.

STAUDIGEL, H., GEE, J., TAUXE, L. & VARGA, R. J. 1992. Shallow intrusive directions in the Troodos ophiolite: Anisotropy of magnetic susceptibility and structural data. *Geology*, **20**, 841–844.

SU, W. & BUCK, R. 1993. Buoyancy effects on mantle

flow under mid-ocean ridges, *Journal of Geophysical Research*, **98**, 12 191–12 205.

TAYLOR, B., CROOK, K. & SINTON, J. 1994. Extensional transform zones and oblique spreading centres. *Journal of Geophysical Research*, **99**, 19 707–19 718.

THOMPSON, G., BRYAN, W. B., BALLARD, R., HAMURO, K. & MELSON, W. G. 1985. Axial processes along a segment of the East Pacific Rise, 10°–12°N. *Nature*, **318**, 429–433.

TOLSTOY, M., HARDING, A. J. & ORCUTT, J. A. 1993. Crustal thickness on the Mid-Atlantic Ridge: bull's-eye gravity anomalies and focused accretion. *Science*, **262**, 726–729.

TUCHOLKE, B. E. & LIN, J. 1994. A geologic model for the structure of ridge segments in slow spreading ocean crust. *Journal of Geophysical Research*, **99**, 11 937–11 958.

VOGT, P. R. 1971. Asthenosphere motion recorded by the seafloor south of Iceland, *Earth and Planetary Science Letters*, **13**, 153–160.

WANG, X. & COCHRAN, J. R. 1993. Gravity anomalies, isostasy and mantle flow at the East Pacific Rise crest. *Journal of Geophysical Research*, **98**, 19 505–19 531.

WANG, X. & COCHRAN, J. R. 1995. Along-axis gravity gradients at mid-ocean ridges: implications for mantle flow and axial morphology. *Geology*, **23**, 29–32.

WEIS, D., WHITE, W. M., FREY, F. A., DUNCAN, R. A., FISK, M. R., DEHN, J., LUDDEN, J., SAUNDERS, A. & STOREY, M. 1992. The influence of

mantle plumes in generation of Indian Ocean Crust. *In*: DUNCAN, R. A., REA, D. K., KIDD, R. B., VON RAD, U. & WEISSEL, J. K. (eds) *Synthesis of Results from Scientifc Drilling in the Indian Ocean*. American Geophysical Union, Geophysical Monographs, **70**, 57–89.

WEST, B. P., SEMPÉRÉ, J.-C., PYLE, D. G., PHIPPS MORGAN, J. & CHRISTIE, D. M. 1994. Evidence for variable upper mantle temperature and crustal thickness in and near the Australia-Antarctic Discordance. *Earth and Planetary Science Letters*, **128**, 135–153.

WHITE, W. M., HOFMANN, A. W. & PUCHELT, H. 1987. Isotope geochemistry of Pacific mid-ocean ridge basalt. *Journal of Geophysical Research*, **92**, 4881–4893.

WILSON, D. S. 1992. Focused mantle upwelling beneath mid-ocean ridges: evidence from seamount formation and isostatic compensation of topography. *Earth and Planetary Science Letters*, **113**, 41–55.

WRIGHT, D. J., HAYMON, R. M. & FORNARI, D. J. 1995. Crustal fissuring and its relationship to magmatic and hydrothermal processes on the East Pacific Rise crest (9°12' to 54'N). *Journal of Geophysical Research*, **100**, 6097–9120.

WRIGHT, T. L. & FISKE, R. S. 1971. Origin of the differentiated and hybrid lavas of Kilauea Volcano, Hawaii. *Journal of Petrology*, **12**, 1–65.

ZINDLER, A. & HART, S. R. 1986. Chemical Geodynamics, *Annual Reviews of Earth and Planetary Science*, **14**, 493–571.

Restricted melting under the very slow-spreading Southwest Indian ridge

C. J. ROBINSON, R.S. WHITE, M. J. BICKLE & T. A. MINSHULL

Bullard Laboratories, Madingley Rise, Madingley Road, Cambridge, CB3 0EZ, UK

Abstract: The major element chemistry of basaltic glass from two localities on the Southwest Indian Ridge is consistent with the occurrence of reduced melting under this very slow spreading ridge compared with that under 'normal' oceanic spreading centres. This agrees with theoretical models for mantle melting which include the influence of conductive heat loss from the mantle as it wells up beneath very slowly separating plates. There are distinct chemical differences between the two areas of study. Basaltic samples from the segment at 57°E have consistently lower Na_2O, Al_2O_3 and SiO_2 and higher FeO, CaO/Al_2O_3, K_2O and TiO_2 than those from 66°E. The 57°E lavas have an MgO range of 5.9–7.5 wt% and define approximately linear oxide trends versus MgO, while the 66°E lavas have too small a range of MgO values (6.8–7.8 wt%) to define any trends. The chemical data can be interpreted in terms of different extents of melting at different mean pressures.

The Southwest Indian Ridge is a very slow spreading ridge, with a full rate of only 13–16 mm a^{-1} (Patriat & Segoufin 1988; de Metz *et al.* 1990; Jestin *et al.* 1994). It is therefore an end-member case of mid-ocean ridge spreading. Theoretical models such as that by Bown & White (1994), using the mantle melting equations of McKenzie & Bickle (1988), predict that there is a reduction in crustal thickness and a change in the lava chemistry as the full spreading rate drops below 15–20 mm a^{-1}. This is because conductive heat loss from the upwelling asthenospheric mantle reduces the temperature in the melting region, resulting in smaller degrees of melting than are seen at 'normal' mid-ocean ridges; those spreading at rates above 20 mm a^{-1}. Figure 1 shows the predicted effect of the conductive cooling. The depth to the top of the melting zone increases and the degree of melting decreases as the spreading rate falls below 20 mm a^{-1}. The areas included in this study lie along the axis of the Southwest Indian Ridge and are far removed from any thermal anomalies caused by mantle plumes. They are therefore good locations for investigating the spreading rate dependent effects on lava chemistry.

Basaltic glass chips taken from dredge samples have been analysed for major elements. Analyses are interpreted in terms of extent and pressure of melting by correcting them to MgO = 8 wt% (Klein & Langmuir 1987), in order to reduce the effects of low pressure fractionation. Results are compared with the data compilation of Bown & White (1994) and with the qualitative

predictions of their model. These data are used to infer that conductive cooling affects the melting regime under the Southwest Indian Ridge. The chemical differences between the two areas of study are consistent with differences in the degree and mean pressure of melting.

Sampling

Two areas along the Southwest Indian Ridge were sampled in April–May 1994 during cruise 208 of the RRS *Discovery*. The dredge sites are shown in Fig. 2. The first area (57°E) is within a single segment which is adjacent to the Atlantis II Fracture Zone. The second is at 66°E, well away from any large fracture zones. Geophysical data indicate that the sampling here may span two small segments (Munschy 1987). There are no mantle plumes within 1000 km of either area.

Analytical method

Glass chips were used for analysis because they are considered to be the most representative samples of liquid composition (Byerly *et al.* 1977) and it is possible to avoid phenocrysts and to choose the least altered samples by examining the chips under a bifocal microscope. Analysis for major elements was done using the Department of Earth Sciences' electron microprobe at the University of Cambridge. Sodium was analysed using a wavelength dispersive spectrometer and counted on peak for 60 s and on background for 20 s. All other elements were analysed using an energy dispersive spectrometer. The analyses (Table 1) were made using an electron beam of 20 nA with a 30 μm diameter. They were corrected to the standard VG-2 (USNM 111240/91, Jarosewich *et al* 1980).

From: MacLeod, C. J., Tyler, P. A. & Walker, C. L. (eds) 1996, *Tectonic, Magmatic, Hydrothermal and Biological Segmentation of Mid-Ocean Ridges*, Geological Society Special Publication No. 118, pp. 131–141.

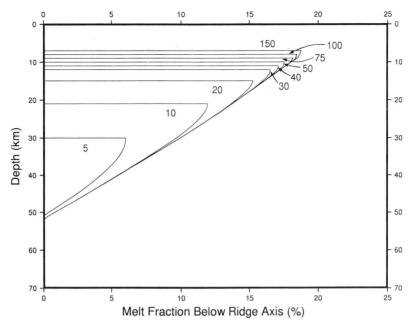

Fig. 1. The relationship between melt fraction and depth from the model of Bown & White (1994), annotated with full spreading rate in mm a^{-1}. Note the increase in depth of the top of the melting column and decrease in melt fraction as the spreading rate drops below 15–20 mm a^{-1}.

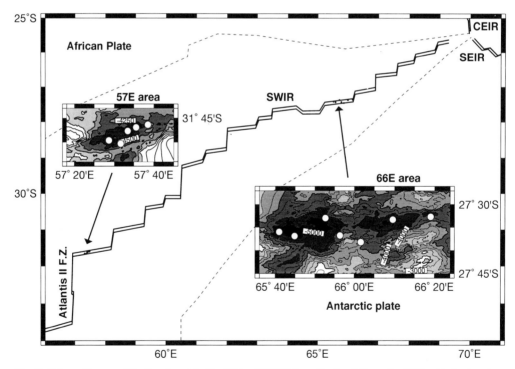

Fig. 2. Schematic map of the Southwest Indian Ridge (SWIR) from Patriat & Segoufin 1988, with dredge sites denoted by circles. The dotted lines show the trace of the triple junction with the Central Indian Ridge (CIR) and the Southeast Indian Ridge (SEIR). Insets show tectonic setting of the sample sites. The contour interval is 250 m, shading changes every 500 m and white dots mark the dredge sites.

Table 1. *Chemical analyses for samples*

	57°E													
	4/8a	4/8m(1)	4/8m(2)	4/9l	4/9m(1)	4/9m(2)	4/10a(1)	4/10a(2)	4/11d	4/11d(1)	4/11d(2)	4/11d(3)	4/11j(1)	4/11j(2)
SiO_2	50.97 (0.22)	50.76 (0.27)	50.11 (0.36)	50.59 (0.35)	50.34 (0.33)	50.16 (0.18)	50.49 (0.25)	49.83 (0.23)	50.15 (0.27)	49.70 (0.38)	50.54 (0.28)	49.82 (0.23)	50.85 (0.42)	50.84 (0.16)
TiO_2	1.88 (0.10)	1.91 (0.06)	1.86 (0.12)	2.28 (0.07)	2.26 (0.13)	2.29 (0.09)	2.21 (0.09)	2.30 (0.11)	2.15 (0.08)	2.19 (0.08)	2.24 (0.11)	2.20 (0.09)	1.85 (0.07)	1.89 (0.08)
Al_2O_3	15.34 (0.22)	15.34 (0.22)	15.09 (0.15)	14.86 (0.17)	14.74 (0.16)	14.64 (0.17)	14.80 (0.12)	14.51 (0.17)	15.15 (0.24)	14.96 (0.21)	15.25 (0.16)	15.01 (0.25)	15.33 (0.23)	15.27 (0.13)
FeO^T	9.95 (0.22)	9.98 (0.13)	9.85 (0.18)	11.02 (0.17)	11.08 (0.19)	11.10 (0.21)	10.71 (0.16)	11.06 (0.25)	10.64 (0.15)	10.99 (0.15)	11.00 (0.12)	10.95 (0.19)	9.67 (0.19)	9.99 (0.14)
MnO	0.19 (0.06)	0.18 (0.05)	0.18 (0.04)	0.20 (0.05)	0.20 (0.06)	0.20 (0.07)	0.19 (0.08)	0.20 (0.04)	0.18 (0.05)	0.21 (0.05)	0.19 (0.07)	0.19 (0.05)	0.18 (0.04)	0.18 (0.06)
MgO	7.39 (0.15)	7.14 (0.15)	7.32 (0.21)	6.75 (0.19)	6.64 (0.24)	6.66 (0.18)	6.86 (0.11)	6.60 (0.13)	6.70 (0.25)	6.61 (0.16)	6.78 (0.10)	6.56 (0.12)	7.51 (0.13)	7.44 (0.14)
CaO	10.16 (0.13)	10.21 (0.11)	10.07 (0.08)	9.63 (0.11)	9.61 (0.10)	9.63 (0.09)	9.42 (0.12)	9.61 (0.09)	9.51 (0.12)	9.72 (0.05)	9.68 (0.12)	9.69 (0.11)	9.97 (0.14)	10.16 (0.12)
Na_2O	3.34 (0.05)	3.43 (0.06)	3.35 (0.07)	3.61 (0.08)	3.63 (0.08)	3.62 (0.05)	3.56 (0.06)	3.59 (0.05)	3.52 (0.09)	3.58 (0.09)	3.57 (0.04)	3.56 (0.06)	3.37 (0.03)	3.35 (0.05)
K_2O	0.28 (0.03)	0.28 (0.05)	0.26 (0.03)	0.28 (0.03)	0.30 (0.04)	0.29 (0.04)	0.27 (0.07)	0.28 (0.04)	0.30 (0.04)	0.31 (0.04)	0.30 (0.03)	0.32 (0.05)	0.27 (0.03)	0.26 (0.03)
P_2O_5	0.19 (0.08)	0.21 (0.13)	0.20 (0.12)	0.29 (0.09)	0.32 (0.07)	0.31 (0.08)	0.28 (0.11)	0.28 (0.07)	0.27 (0.07)	0.25 (0.08)	0.24 (0.09)	0.26 (0.11)	0.20 (0.08)	0.18 (0.07)
Cr_2O_3	0.08 (0.05)	0.07 (0.06)	0.08 (0.07)	0.07 (0.06)	0.09 (0.02)	0.07 (0.05)	0.08 (0.05)	0.06 (0.06)	0.05 (0.06)	0.06 (0.06)	0.07 (0.05)	0.07 (0.06)	0.08 (0.05)	0.07 (0.09)
Total	99.02	98.74	98.73	98.88	98.52	99.36	98.16	98.73	97.90	98.95	99.14	99.03	98.51	98.90
Mg#	59.76	58.84	59.79	55.05	54.53	54.52	56.18	54.39	55.75	54.60	55.21	54.50	60.83	59.82

Each sample is taken from a separate piece of dredged rock. The first number refers to the site, the second to the sack and the letters are used to group samples that appear similar. Mg# is calculated assuming Fe^{2+} is 90% FeO^T (from analyses of Fe_2O_3 and FeO in Price *et al.* 1986). Each value in the table is an average of 10 measurements. Errors (2 s.d.) are in brackets beneath the measured concentrations.

Table 1. *Continued*

	57°E													
	4/12a(1)	412a(2)	4/13c(1)	4/13c(2)	4/13c(3)	5/15a	7/16a	7/17b	7/18p	7/18t	7/19a	7/20a	8/22d	8/23d
SiO_2	50.31	50.48	50.60	49.86	50.38	49.90	49.92	49.99	50.90	49.60	50.13	50.02	50.70	50.54
	(0.20)	(0.40)	(0.37)	(0.28)	(0.27)	(0.38)	(0.63)	(0.25)	(0.15)	(0.40)	(0.26)	(0.24)	(0.52)	(0.30)
TiO_2	1.90	1.90	1.92	1.88	1.90	1.70	2.03	2.02	1.95	2.04	2.06	2.41	2.01	2.03
	(0.07)	(0.08)	(0.06)	(0.08)	(0.07)	(0.06)	(0.11)	(0.08)	(0.06)	(0.13)	(0.11)	(0.08)	(0.05)	(0.09)
Al_2O_3	15.15	15.18	15.31	15.01	15.15	15.70	14.81	14.79	15.32	14.66	14.77	14.78	15.61	15.52
	(0.12)	(0.22)	(0.19)	(0.21)	(0.16)	(0.15)	(0.21)	(0.14)	(0.17)	(0.17)	(0.18)	(0.19)	(0.18)	(0.23)
FeO^T	10.03	10.01	10.01	9.78	10.02	9.78	10.48	10.41	9.97	10.41	10.52	11.52	10.00	9.96
	(0.16)	(0.11)	(0.09)	(0.21)	(0.13)	(0.18)	(0.13)	(0.16)	(0.15)	(0.21)	(0.20)	(0.16)	(0.16)	(0.18)
MnO	0.19	0.18	0.17	0.17	0.19	0.19	0.20	0.18	0.19	0.18	0.19	0.23	0.19	0.20
	(0.05)	(0.06)	(0.07)	(0.07)	(0.08)	(0.06)	(0.05)	(0.07)	(0.06)	(0.06)	(0.08)	(0.05)	(0.07)	(0.05)
MgO	7.30	7.24	7.13	7.12	7.25	7.51	7.02	6.93	7.24	6.82	7.05	6.13	6.52	6.43
	(0.16)	(0.40)	(0.30)	(0.24)	(0.20)	(0.14)	(0.13)	(0.15)	(0.20)	(0.21)	(0.15)	(0.14)	(0.12)	(0.13)
CaO	10.21	10.24	10.27	10.07	10.15	10.55	9.95	9.94	10.17	9.96	9.87	9.62	10.19	10.20
	(0.14)	(0.12)	(0.16)	(0.17)	(0.12)	(0.05)	(0.15)	(0.10)	(0.10)	(0.09)	(0.13)	(0.11)	(0.14)	(0.07)
Na_2O	3.41	3.36	3.38	3.42	3.32	3.17	3.44	3.42	3.42	3.61	3.51	3.70	3.45	3.61
	(0.07)	(0.04)	(0.15)	(0.06)	(0.09)	(0.07)	(0.12)	(0.07)	(0.05)	(0.05)	(0.07)	(0.08)	(0.07)	(0.09)
K_2O	0.27	0.26	0.28	0.27	0.28	0.22	0.27	0.28	0.30	0.28	0.28	0.34	0.27	0.28
	(0.04)	(0.03)	(0.06)	(0.05)	(0.04)	(0.06)	(0.03)	(0.03)	(0.07)	(0.03)	(0.03)	(0.02)	(0.14)	(0.02)
P_2O_5	0.20	0.20	0.21	0.22	0.19	0.17	0.22	0.20	0.22	0.23	0.19	0.25	0.23	0.23
	(0.07)	(0.10)	(0.06)	(0.09)	(0.06)	(0.11)	(0.08)	(0.07)	(0.06)	(0.07)	(0.08)	(0.14)	(0.08)	(0.05)
Cr_2O_3	0.05	0.08	0.07	0.08	0.08	0.07	0.07	0.07	0.05	0.08	0.06	0.05	0.08	0.10
	(0.04)	(0.03)	(0.04)	(0.07)	(0.05)	(0.07)	(0.07)	(0.05)	(0.09)	(0.05)	(0.06)	(0.08)	(0.05)	(0.06)
Total	99.38	99.51	99.71	98.23	99.28	99.29	98.80	98.62	98.97	98.25	99.01	99.43	98.45	98.35
Mg#	59.29	59.11	58.76	59.28	59.13	60.58	57.27	57.10	59.22	56.72	57.29	51.55	56.59	56.34

Table 1. *Continued*

	57°E									66°E				
	8/26f(1)	8/26f(2)	8/27a	9/28b	9/29a	9/29d	9/291	9/30a(1)	9/30a(2)	10/31a	10/32i(1)	10/32i(2)	10/33a(1)	10/33a(2)
SiO_2	49.87	50.60	50.09	50.10	50.44	50.31	50.85	50.20	49.96	51.33	50.29	50.68	50.49	50.80
	(0.38)	(0.36)	(0.26)	(0.27)	(0.20)	(0.32)	(0.26)	(0.39)	(0.31)	(0.34)	(0.30)	(0.46)	(0.30)	(0.22)
TiO_2	2.09	2.07	2.12	1.96	1.95	2.19	1.94	1.94	1.95	1.15	1.21	1.23	1.24	1.24
	(0.11)	(0.07)	(0.09)	(0.09)	(0.07)	(0.11)	(0.05)	(0.08)	(0.08)	(0.06)	(0.05)	(0.08)	(0.06)	(0.04)
Al_2O_3	15.28	15.43	15.35	15.17	15.35	15.46	15.52	15.26	15.20	17.04	17.49	17.42	17.49	17.26
	(0.19)	(0.19)	(0.12)	(0.12)	(0.24)	(0.17)	(0.10)	(0.23)	(0.14)	(0.16)	(0.48)	(0.25)	(0.20)	(0.14)
FeO^T	10.72	10.32	10.69	10.36	10.28	10.56	10.22	10.36	10.30	7.75	7.51	7.49	7.52	7.58
	(0.27)	(0.18)	(0.15)	(0.18)	(0.13)	(0.12)	(0.11)	(0.17)	(0.23)	(0.16)	(0.16)	(0.08)	(0.12)	(0.16)
MnO	0.20	0.18	0.20	0.18	0.21	0.20	0.19	0.18	0.19	0.14	0.14	0.13	0.13	0.15
	(0.08)	(0.11)	(0.06)	(0.08)	(0.07)	(0.07)	(0.07)	(0.08)	(0.04)	(0.07)	(0.07)	(0.07)	(0.07)	(0.08)
MgO	5.87	6.48	5.94	6.99	7.01	6.54	7.04	6.94	6.89	7.34	7.10	7.06	7.31	7.16
	(0.22)	(0.17)	(0.15)	(0.18)	(0.11)	(0.17)	(0.11)	(0.16)	(0.14)	(0.21)	(0.17)	(0.16)	(0.12)	(0.15)
CaO	9.77	10.11	9.73	10.08	10.15	9.84	10.16	10.15	10.16	10.33	10.45	10.32	10.43	10.33
	(0.07)	(0.15)	(0.10)	(0.12)	(0.11)	(0.13)	(0.09)	(0.13)	(0.13)	(0.08)	(0.21)	(0.10)	(0.10)	(0.12)
Na_2O	3.88	3.71	3.81	3.50	3.45	3.53	3.39	3.45	3.52	3.59	3.83	3.76	3.79	3.77
	(0.10)	(0.07)	(0.08)	(0.08)	(0.07)	(0.13)	(0.07)	(0.05)	(0.07)	(0.09)	(0.08)	(0.08)	(0.05)	(0.05)
K_2O	0.32	0.29	0.33	0.27	0.27	0.34	0.26	0.25	0.27	0.18	0.24	0.25	0.25	0.25
	(0.03)	(0.04)	(0.05)	(0.04)	(0.05)	(0.04)	(0.04)	(0.03)	(0.04)	(0.04)	(0.04)	(0.04)	(0.06)	(0.06)
P_2O_5	0.25	0.22	0.26	0.22	0.23	0.25	0.21	0.17	0.21	0.07	0.12	0.08	0.06	0.07
	(0.08)	(0.10)	(0.09)	(0.07)	(0.10)	(0.07)	(0.11)	(0.09)	(0.10)	(0.07)	(0.09)	(0.09)	(0.10)	(0.09)
Cr_2O_3	0.07	0.06	0.07	0.09	0.06	0.06	0.08	0.08	0.07	0.08	0.09	0.09	0.06	0.06
	(0.03)	(0.09)	(0.06)	(0.05)	(0.07)	(0.04)	(0.04)	(0.04)	(0.05)	(0.07)	(0.04)	(0.06)	(0.06)	(0.07)
Total	99.66	98.75	98.98	99.32	99.77	98.53	99.08	99.34	99.05	99.23	98.72	98.24	99.00	98.42
Mg#	52.26	55.69	52.65	57.43	57.67	55.33	57.94	57.27	57.22	65.44	65.41	65.35	66.03	65.39

Table 1. *Continued*

	66°E												
	11/34a	11/34d	11/34h	11/35a(2)	12/36b	12/37f(1)	12/37f(2)	12/37f(3)	12/37g(1)	12/37g(2)	12/37g(3)	12/37g(4)	13/40m(1)
SiO_2	51.64	51.08	51.30	51.30	50.32	50.17	50.18	50.27	50.63	50.69	50.59	50.51	49.92
	(0.44)	(0.39)	(0.94)	(0.28)	(0.28)	(0.31)	(0.15)	(0.23)	(0.24)	(0.39)	(0.25)	(0.43)	(0.35)
TiO_2	1.35	1.34	1.36	1.32	1.42	1.43	1.41	1.43	1.42	1.39	1.40	1.38	1.40
	(0.07)	(0.07)	(0.09)	(0.08)	(0.09)	(0.09)	(0.06)	(0.13)	(0.09)	(0.08)	(0.08)	(0.06)	(0.06)
Al_2O_3	17.28	17.33	17.32	17.24	16.92	16.96	16.96	16.94	17.08	16.99	17.01	16.96	16.91
	(0.14)	(0.17)	(0.43)	(0.11)	(0.14)	(0.21)	(0.26)	(0.17)	(0.26)	(0.20)	(0.14)	(0.21)	(0.21)
FeO^T	7.66	7.41	7.69	7.54	8.51	8.47	8.37	8.52	8.42	8.32	8.50	8.48	8.94
	(0.19)	(0.13)	(0.19)	(0.12)	(0.07)	(0.14)	(0.18)	(0.12)	(0.11)	(0.14)	(0.10)	(0.17)	(0.16)
MnO	0.15	0.15	0.15	0.16	0.16	0.15	0.16	0.16	0.16	0.16	0.15	0.14	0.16
	(0.07)	(0.06)	(0.06)	(0.07)	(0.05)	(0.08)	(0.07)	(0.08)	(0.05)	(0.06)	(0.04)	(0.06)	(0.05)
MgO	7.33	7.17	7.15	7.13	7.72	7.63	7.68	7.62	7.63	7.58	7.79	7.80	7.28
	(0.17)	(0.21)	(0.24)	(0.13)	(0.17)	(0.12)	(0.13)	(0.17)	(0.20)	(0.18)	(0.15)	(0.16)	(0.15)
CaO	10.33	10.29	10.47	10.29	10.05	10.12	10.04	10.11	10.11	10.00	10.00	9.92	9.91
	(0.13)	(0.10)	(0.16)	(0.04)	(0.11)	(0.15)	(0.12)	(0.10)	(0.12)	(0.13)	(0.08)	(0.09)	(0.07)
Na_2O	3.73	3.83	3.83	3.75	3.76	3.84	3.81	3.76	3.71	3.64	3.76	3.75	3.85
	(0.07)	(0.07)	(0.07)	(0.07)	(0.06)	(0.07)	(0.05)	(0.06)	(0.11)	(0.04)	(0.10)	(0.08)	(0.07)
K_2O	0.26	0.24	0.24	0.26	0.18	0.17	0.18	0.19	0.17	0.19	0.18	0.16	0.19
	(0.03)	(0.04)	(0.03)	(0.03)	(0.03)	(0.03)	(0.05)	(0.02)	(0.03)	(0.04)	(0.03)	(0.03)	(0.05)
P_2O_5	0.09	0.07	0.10	0.11	0.13	0.14	0.09	0.10	0.10	0.10	0.08	0.09	0.11
	(0.09)	(0.09)	(0.07)	(0.09)	(0.08)	(0.09)	(0.07)	(0.07)	(0.09)	(0.09)	(0.10)	(0.05)	(0.12)
Cr_2O_3	0.08	0.08	0.08	0.07	0.06	0.08	0.09	0.08	0.09	0.08	0.08	0.10	0.09
	(0.06)	(0.07)	(0.05)	(0.04)	(0.07)	(0.06)	(0.08)	(0.06)	(0.04)	(0.04)	(0.08)	(0.05)	(0.07)
Total	98.99	98.11	99.95	98.90	99.53	99.48	99.25	99.45	98.66	98.92	98.71	98.44	99.02
Mg#	65.67	65.92	65.03	65.43	64.45	64.32	64.71	64.14	64.44	64.58	64.72	64.80	61.95

Table 1. *Continued*

66°E

	13/40m(2)	13/40m(3)	13/40m(4)	14/41c	14/41d	14/41d(2)	15/42a(1)	15/42a(2)	15/43c	16/46a	16/46g(1)	16/47g(2)	VG-2	Reported values
SiO_2	49.84 (0.34)	50.34 (0.31)	50.06 (0.21)	50.34 (0.21)	50.76 (0.31)	51.48 (0.31)	50.76 (0.36)	50.85 (0.43)	50.22 (0.28)	50.93 (0.24)	51.14 (0.54)	51.16 (0.26)	50.58 (0.72)	50.81
TiO_2	1.39 (0.05)	1.38 (0.07)	1.38 (0.06)	1.35 (0.08)	1.35 (0.08)	1.36 (0.07)	1.46 (0.08)	1.46 (0.05)	1.45 (0.10)	1.51 (0.10)	1.51 (0.04)	1.49 (0.06)	1.97 (0.04)	1.85
Al_2O_3	16.95 (0.20)	16.95 (0.18)	16.87 (0.13)	16.92 (0.27)	16.99 (0.21)	17.24 (0.17)	16.97 (0.17)	17.01 (0.17)	16.89 (0.22)	16.53 (0.26)	16.43 (0.25)	16.42 (0.28)	13.71 (0.20)	14.06
FeO^T	8.91 (0.17)	8.98 (0.08)	8.92 (0.15)	8.16 (0.15)	8.22 (0.09)	8.24 (0.17)	8.12 (0.15)	8.12 (0.13)	8.77 (0.10)	8.38 (0.15)	8.32 (0.21)	8.53 (0.17)	11.91 (0.13)	11.84
MnO	0.17 (0.08)	0.16 (0.05)	0.16 (0.04)	0.14 (0.06)	0.14 (0.07)	0.14 (0.06)	0.16 (0.07)	0.16 (0.05)	0.17 (0.07)	0.15 (0.09)	0.16 (0.07)	0.18 (0.06)	0.23 (0.04)	0.22
MgO	7.26 (0.14)	7.36 (0.11)	7.22 (0.19)	7.33 (0.33)	7.34 (0.13)	7.40 (0.21)	6.86 (0.15)	6.85 (0.23)	7.02 (0.16)	7.33 (0.07)	7.54 (0.17)	7.57 (0.21)	6.86 (0.19)	6.71
CaO	9.92 (0.12)	9.86 (0.12)	9.88 (0.05)	10.17 (0.06)	10.24 (0.10)	10.21 (0.08)	10.08 (0.11)	10.12 (0.28)	10.33 (0.14)	9.77 (0.13)	9.62 (0.16)	9.64 (0.10)	11.12 (0.26)	11.12
Na_2O	3.82 (0.05)	3.79 (0.06)	3.71 (0.16)	3.76 (0.09)	3.64 (0.07)	3.74 (0.06)	3.96 (0.04)	3.93 (0.06)	3.67 (0.06)	3.95 (0.07)	3.76 (0.15)	4.04 (0.07)	2.72 (0.06)	2.62
K_2O	0.20 (0.03)	0.20 (0.04)	0.20 (0.04)	0.21 (0.03)	0.22 (0.04)	0.23 (0.04)	0.26 (0.04)	0.24 (0.08)	0.16 (0.02)	0.26 (0.03)	0.28 (0.03)	0.28 (0.06)	0.20 (0.03)	0.19
P_2O_5	0.09 (0.07)	0.08 (0.09)	0.08 (0.11)	0.10 (0.08)	0.12 (0.09)	0.12 (0.11)	0.12 (0.06)	0.13 (0.12)	0.14 (0.06)	0.15 (0.09)	0.12 (0.09)	0.16 (0.05)	0.18 (0.03)	na
Cr_2O_3	0.06 (0.03)	0.07 (0.06)	0.06 (0.05)	0.08 (0.05)	0.09 (0.05)	0.06 (0.07)	0.05 (0.05)	0.07 (0.06)	0.07 (0.05)	0.09 (0.11)	0.08 (0.05)	0.07 (0.06)	0.06 (0.04)	na
Total	98.87	98.31	98.34	98.84	99.36	99.33	99.06	99.21	99.16	99.38	98.80	98.79	99.48	99.42
Mg#	61.96	62.12	61.81	64.24	64.10	64.24	62.84	62.78	61.55	63.62	64.44	63.99		

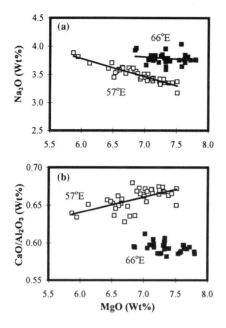

Fig. 3. Averaged glass analyses of (**a**) Na_2O and (**b**) CaO/Al_2O_3 versus wt% MgO. Solid squares are from the 66°E lavas and open squares are from the 57°E lavas. Linear regressions used to correct Na_2O to Na_8 are superimposed on the data from each area. The equations used are $Na_8 = Na_2O - 0.3221(8\text{-MgO})$ and $Na_8 = Na_2O - 0.0763(8\text{-MgO})$ for 57°E and 66°E respectively.

Fractionation effects

The composition of erupted basalt is affected by low pressure fractionation. The Mg numbers of these Southwest Indian Ridge samples range from 52 to 61 for the 57°E lavas (with 5.9–7.5 wt% MgO) and from 62 to 66 for the 66°E lavas (6.8–7.8 wt% MgO). If it is assumed that a primary MORB is in equilibrium with mantle olivine (approximately Fo_{91}) and therefore has an Mg# of approximately 70 (Wilson 1989), it is evident that these glasses have all been fractionated to some degree and that the degree of fractionation is lower in the 66°E lavas than in the 57°E lavas. Since it is necessary to use primary melt compositions in order to infer melting conditions such as mantle temperature and pressure (Klein & Langmuir 1987; McKenzie & Bickle 1988), data are corrected to MgO = 8 wt% in an attempt to see through the low-pressure fractionation effects (Klein & Langmuir 1987).

The data from the 57°E area define approximately linear trends in MgO variation diagrams for Na_2O, FeO, TiO_2, SiO_2, CaO and Al_2O_3. The 66°E data form clusters with slight elongation opposing the 57°E trends for all elements except Na_2O where there is a weak trend approximately parallel to that of the 57°E data (Fig. 3a).

The correction to MgO = 8 wt% for Na2O is done by linear regression through the data from each dredge area (Fig. 3a). However, for CaO, FeO, SiO_2 and Al_2O_3, data from the individual dredge sites define trends which are roughly parallel to each other, but not necessarily to the trend of the whole area. If these trends are compared with some predicted liquid lines of descent from the model of Weaver & Langmuir (1990) seen in Langmuir *et al.* (1992), it appears likely that the individual site trends are controlled by fractionation but in some cases the 'area' trends may be affected by magma mixing. Fractionation and mixing form parallel trends in the case of Na_2O. This could explain why considerably less scatter is seen there than for the other oxides. The correction to MgO = 8 wt% for CaO/Al_2O_3 and SiO_2/FeO is made parallel to the 'area' trends of the 57°E lavas because they form definite trends and the 'site' trends are approximately parallel to the 'area' trend. This simplistic correction is performed purely to enable a comparison between the two areas in terms of the relative degrees and pressures of melting.

Interpretation

Na_2O behaves moderately incompatibly and is therefore expected to enter the melt preferentially at an early stage, becoming diluted as the degree of melting increases. The CaO/Al_2O_3 ratio increases with increasing extents of melting as long as clinopyroxene remains in the solid residue (Niu & Batiza 1991; Langmuir *et al.* 1992), because Al_2O_3 behaves incompatibly during melting. This is assumed to hold for the Southwest Indian Ridge glasses.

Values for Na_8 obtained from the regression shown in Fig. 3a and both corrected and uncorrected CaO/Al_2O_3 values (Fig. 3b) are used as indicators of the extent of melting (Klein & Langmuir 1987; Niu & Batiza 1991). Uncorrected CaO/Al_2O_3 values are used to make a comparison with the data compilation of Bown & White (1994). In contrast to Na_2O, SiO_2 and FeO are sensitive to the mean pressure of melting. As the pressure increases, the SiO_2 concentration in the melt decreases and FeO increases (Niu & Batiza 1991). They also exhibit

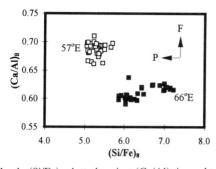

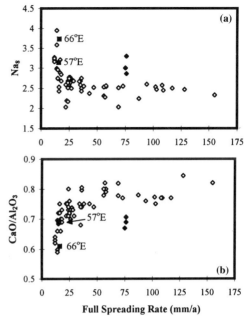

Fig. 4. $(Si/Fe)_8$ plotted against $(Ca/Al)_8$ is used to imply the relative extent and mean pressure of melting for the two study areas. Solid squares represent the 66°E lavas and open squares are the 57°E lavas. Arrows labelled F and P show the directions of trends produced by increased degree of melting and increased pressure of melting respectively (Niu & Batiza 1991). The trend seen in the 66°E data is approximately parallel to that from the Mid-Atlantic Ridge at 26°S from Niu & Batiza (1991) and is a representation of the 'local' trend of Klein & Langmuir (1987). It is likely that no 'local' trend is seen in the 57°E data because the sample sites are too closely spaced and therefore do not span enough of the segment. The equations used for both data sets were $(Ca/Al)_8 = CaO/Al_2O_3 + 0.0257(8-MgO)$ and $(Si/Fe)_8 = SiO_2/FeO + 0.3952(8-MgO)$.

Fig. 5. Variation of (**a**) Na_8 and (**b**) CaO/Al_2O_3 with spreading rate, compared with global data compiled by Bown & White (1994). Solid diamonds denote data points from the unusually deep Australia–Antarctic Discordance on the Southeast Indian Ridge which shows evidence for restricted melting. Values from this study are shown as filled squares. Data are averaged over the whole of each study area and are labelled accordingly. Na_8 values are obtained using the correction listed in the caption of Fig. 3. The average for 57°E is 3.14 ± 0.2 (2σ) and for 66°E is 3.73 ± 0.2 (2σ). Averaged values for CaO/Al_2O_3 are 0.69 ± 0.02 (2σ) for 57°E and 0.61 ± 0.02 (2σ) for 66°E.

some dependence on the degree of melting, but it is possible to infer the relative mean pressures of melting by plotting $(Si/Fe)_8$ against $(Ca/Al)_8$ since $(Ca/Al)_8$ depends only on the extent of melting (Fig. 4).

Corrected Na_8 values for basaltic glass samples from 57°E and 66°E (Fig. 5a) are consistently higher than those of 'normal' mid-ocean ridges and values for CaO/Al_2O_3 (Fig. 5b) are lower. This is consistent with the occurrence of low degree melting under the Southwest Indian Ridge relative to 'normal' mid-ocean ridges. Na_8 values can be used to estimate the crustal thickness, for example, using fig. 59 of Langmuir et al (1992). Na_8 values of 3.14 and 3.73 imply crustal thicknesses of approximately 4–4.5 km and 1.5–2.5 km for 57°E and 66°E, respectively. The estimate for 57°E is similar to the measurement of 4.5 km obtained by Muller et al. (1995) from seismic profiles in the region around ODP hole 735B.

The chemical differences between the two areas are clear, not only in Na_2O and CaO/Al_2O_3 (Figs 3a & b), but also in FeO, TiO_2, K_2O and SiO_2 concentrations (Table 1). At the same MgO content, the 57°E lavas contain less Na_2O,

Al_2O_3 and SiO_2 and more FeO, CaO/Al_2O_3, K_2O and TiO_2 than the 66°E lavas. The difference in CaO/Al_2O_3 between the two areas is controlled entirely by Al_2O_3 variation, CaO concentrations being indistinguishable between the two. All the major element distinctions between the 57°E and 66°E lavas, except for the K_2O and TiO_2 data, are consistent with the 57°E lavas being formed by higher degrees of melting at higher mean pressure than the 66°E lavas (Fig. 4).

Conclusions

The major element compositions from both areas are consistent with a reduction in melt fraction and an increase in melting depth compared with 'normal' mid-ocean ridges, as predicted by the Bown & White (1994) melting model. Comparison with their data compilation of CaO/Al_2O_3 and Na_8 for many mid-ocean ridges shows that these Southwest Indian Ridge samples fall on the low spreading rate trend, along with sparse previous data from the Southwest Indian Ridge (Price *et al.* 1986; Johnson & Dick 1992) and other slow spreading ridges such as the Cayman Trough (Thompson *et al.* 1980) (Fig. 5). Qualitatively, the predictions of Bown & White (1994) indicate that this trend is probably due to conductive cooling of the upwelling mantle. This conclusion is supported by seismic crustal thickness measurements. Muller *et al.* (1995) find that the crustal thickness near the 57°E area is only 4.5 km, considerably less than the 'normal' oceanic crustal thickness of about 7 km (White *et al.* 1992). Minshull & White (in press) and Francis & Raitt (1967) report a similar crustal thickness (approximately 5 km) along the Lusiad 33 profile (Fig. 2), on crust created at about 34 Ma at a full spreading rate of 13–15 mm a^{-1} (Patriat & Segoufin 1988). Rare earth element inversions (based on McKenzie & O'Nions 1991) from preliminary analyses of some of the samples reported here also indicate that mantle melting is restricted beneath the Southwest Indian Ridge (Robinson *et al.* 1995).

The Na_8, $(Ca/Al)_8$ and $(Si/Fe)_8$ values indicate that the 57°E lavas were produced by higher degrees of melting at higher mean pressures than the lavas sampled at 66°E. This is perhaps surprising because there is only a 0.5 mm a^{-1} difference in the full spreading rate (using Jestin *et al.* 1994) and the 57°E lavas were collected near the large offset Atlantis II fracture zone. If the fracture zone has some kind of cooling affect on the melt regime of the 57°E area, it may explain why these lavas appear to have been formed at higher pressures than the 66°E lavas, but it does not explain the inferred higher melt fraction. There are other differences between the two areas, for example, in geophysical setting and topography, although it is not clear why these should relate to differences in magma composition. The ridge at 66°E has only been spreading for the past 10 Ma whereas the ridge at 57°E has been actively spreading for at least 50 Ma (Patriat & Segoufin 1988; Royer *et al.* 1988). The style of spreading is reported to be different in the two areas (Sclater *et al.* 1981) and the 66°E area is within a region of considerably rougher topography than the 57°E area.

Magma may be transported substantial distances along the spreading axis within the ocean crust and the lavas sampled at 57°E, which are 25–50 km from the fracture zone, may have been produced several tens of kilometres further away, but this would not necessarily account for their production by a relatively high degree of melting. The major element data indicate that both areas are affected by magma mixing to some degree. Magmas may be less completely mixed at very slow spreading ridges and this is likely to complicate the chemical signatures sampled at the surface. The significance of the compositional variation seen in these basalts needs further investigation.

We are grateful to the officers and crew of the RRS *Discovery* for their help at sea dredging the samples and to the Natural Environment Research Council for financial support. We thank D. Sauter (Ecole et Observatoire de Physique du Globe, Strasbourg) for providing us with unpublished swath bathymetric data. T.A.M. is supported by a Royal Society University Research Fellowship. Department of Earth Sciences, University of Cambridge, contribution no. 4694.

References

Bown, J. W. & White, R. S. 1994. Variation with spreading rate of oceanic crustal thickness and geochemistry. *Earth and Planetary Science Letters*, **121**, 435–449.

Byerly, G. R., Melson, W. G., Nelen, J. A. & Jarosewich, E. 1977. Abyssal basaltic glasses as indicators of magma compositions. *Smithsonian Contributions to the Earth Sciences*, **19**, 22–30.

de Metz, C., Gercher, R. G., Argus, D. F. & Stein, S. 1990. Current plate motions. *Geophysical Journal International*, **101**, 425–478.

Francis, T. J. G. & Raitt, R. W. 1967. Seismic refraction measurements in the Southern Indian Ocean. *Journal of Geophysical Research*, **72**, 3013–3041.

Jarosewich, E., Nelen, J. A. & Norberg, J. A. 1980. Reference samples for electron microprobe analysis. *Geostandards Newsletter*, **4**, 43–47.

Jestin, F., Huchon, P. & Gaulier, J. M. 1994. The Somalia plate and the East African Rift System: present day kinematics. *Geophysical Journal International*, **116**, 637–654.

Johnson, K. T. & Dick, H. J. B. 1992. Open system melting and temporal and spatial variation of peridotite and basalt at the Atlantis II Fracture Zone. *Journal of Geophysical Research*, **97**, 9219–9241.

Klein, E. & Langmuir, C. H. 1987. Global correlations of ocean ridge basalt chemistry with axial depth and crustal thickness. *Journal of Geophysical Research*, **92**, 8089–8115.

Langmuir, C. H., Klein, E. & Plank, T. 1992. Petrological systematics of mid-ocean ridge basalts: constraints on melt generation beneath ocean ridges. *In*: Morgan, J. P., Blackman, D. K. & Sinton, J. M. (eds) *Mantle Flow and Melt Generation at Mid Ocean Ridges*. AGU Geophysical Monograph, **71**, 183–280.

McKenzie, D. P. & Bickle, M. J. 1988. The volume and composition of melt generated by extension of the lithosphere. *Journal of Petrology*, 625–679.

—— & O'Nions, R. K. 1991. Partial melt distributions from inversion of Rare Earth Element concentrations. *Journal of Petrology*, 1021–1091.

Minshull, T. A. & White, R. S. Thin crust on the flanks on the slow-spreading Southwest Indian Ridge, *Geophysical Journal International*, **125**, 139–148.

Muller, M. R., Minshull, T. A. & White, R. S. 1995. Crustal structure at the very slow-spreading Southwest Indian Ridge. *InterRidge News*, **4**, 3–6.

Munschy, M. 1987. *Etude Géophysique du Point Triple de Rodriguez et la Zone Axiale des Trois Dorsales Associées (Océan Indien)*. PhD thesis, Université Louis Pasteur de Strasbourg, France.

Niu, Y. & Batiza, R. 1991. An empirical method for calculating melt compositions produced beneath mid-ocean ridges: application for axis and off-axis (seamounts) melting. *Journal of Geophysical Research*, **96**, 21753–21777.

Patriat, P. & Segoufin, J. 1988. Reconstruction of the Central Indian Ocean. *Tectonophysics*, **155**, 211–234.

Price, R. C., Kennedy, A. K., Riggs-Sneeringer, M. & Frey, F. A. 1986. Geochemistry of basalts from the Indian Ocean triple junction: implications for the generation and evolution of Indian Ocean ridge basalts. *Earth and Planetary Science Letters*, **78**, 379–396.

Robinson, C. J., White, R. S., Bickle, M. J. & Minshull, T. A. 1995. Melting beneath the very slow spreading South West Indian Ridge: from analysis of basaltic glass. *TERRA abstracts*, **7**, 150.

Royer, J.-Y., Patriat, P., Bergh, H. W. & Scotese, C. R. 1988. Evolution of the Southwest Indian Ridge from the Late Cretaceous (anomaly 34) to the Middle Eocene (anomaly 20). *Tectonophysics*, **155**, 235–260.

Sclater, J. G., Fisher, R. L., Patriat, P., Tapscott, C. & Parsons, B. 1981. Eocene to recent development of the Southwest Indian Ridge, a consequence of the evolution of the Indian Ocean Triple Junction. *Geophysical Journal of the Royal Astronomical Society*, **64**, 587–604.

Thompson, G., Bryan, W. B. & Melson, W. G. 1980. Geological and geophysical investigation of the Mid-Cayman Rise spreading center; geochemical variation and petrogenesis of basalt glasses. *Journal of Geology*, **88**, 41–55.

Weaver, J. & Langmuir, C. H. 1990. Calculation of phase equilibrium in mineral–melt systems. *Computers and Geosciences*, **16**, 1–19.

White, R. S., McKenzie, D. P. & O'Nions, R. K. 1992. Oceanic crustal thickness from seismic measurements and rare earth element inversions. *Journal of Geophysical Research*, **97**, 19 683–19 715.

Wilson, M. 1989. *Igneous Petrogenesis*. Unwin Hyman.

A review of the petrology of harzburgites at Hess Deep and Garrett Deep: implications for mantle processes beneath segments of the East Pacific Rise

STEPHEN J. EDWARDS[1], TREVOR J. FALLOON[2], JOHN MALPAS[3] & ROLF B. PEDERSEN[4]

[1]*Department of Environmental and Geographical Sciences, Manchester Metropolitan University, Chester Street, Manchester M1 5GD, UK*
Present address: School of Earth Sciences, University of Greenwich, Medway Campus, Pembroke, Chatham Maritime, Kent ME4 4AW, UK
[2]*Department of Geology, University of Tasmania, GPO Box 252C, Hobart, Tasmania 7001, Australia*
[3]*Department of Earth Sciences, University of Hong Kong, Pokfulam Road, Hong Kong*
[4]*Geologisk Institutt, Universitetet i Bergen, Allegaten 41, N-5007 Bergen, Norway*

Abstract: In recent years a unique set of samples of uppermost mantle at the mantle–crust transition zone have been collected from two different environments along the fast-spreading East Pacific Rise (EPR): a 'normal' spreading segment (represented by samples from Hess Deep) and the end of a spreading segment where the EPR meets the Garrett transform fault (represented by samples from Garrett Deep). A review of the petrology of harzburgites from the two sites demonstrates that these rocks were produced by partial melting of adiabatically upwelling mantle and, subsequently, at the top of the mantle, they were impregnated by reactive and crystallizing mid-ocean ridge basaltic (MORB) melts. Despite this similar history, non-impregnated harzburgites at Garrett Deep have a more fertile spinel chemistry than those at Hess Deep, which is consistent with reduced partial melting of shallow mantle as a transform fault is approached – the 'transform fault effect'.

The extent of reaction between melt and harzburgite during the impregnation event suggests that melt arrived in the uppermost mantle in a highly reactive state because along the adiabatic path it had been highly channelled in spatially restricted conduits. This implies that mantle upwelling below the EPR was, and presumably still is, dominantly two dimensional (sheet-like). Within this framework, the chemical evolution of MORB melt below fast-spreading ridges will be significantly affected by melt–peridotite reaction only when melt reaches the uppermost mantle and mantle–crust transition zone, where along-axis transport of melt may also be important.

Although the harzburgites from Hess Deep and Garrett Deep formed and evolved beneath different parts of different first-order segments of the EPR, the petrology of these rocks presents the best analogue available for defining real variations in mantle processes along a single first-order ridge segment in a fast-spreading environment.

Understanding tectonic and magmatic segmentation of mid-ocean ridges (MORs) is fundamental for constraining the chemical and physical evolution and behaviour of the mantle underlying the ocean basins. Batiza (this volume) has eloquently reviewed segmentation along MORs and several relevant points from this review are summarized here. MORs are tectonically segmented at various scales along their length by triple junctions, transform and non-transform fracture zones, and a variety of smaller features (e.g. Macdonald *et al.* 1988; Lonsdale 1994). Macdonald *et al.* (1988) defined a first-order segment as the region between transform or major non-transform fracture zones (scale 10^2 km). Superimposed on a first-order segment are progressively smaller scale segments (down to a scale of 10 km), often referred to as second-, third- and fourth-order segments (Macdonald *et al.* 1988).

There exists good qualitative evidence for a hierarchical system linking tectonic and magmatic segmentation along MORs, which is most applicable to fast-spreading ridges (Macdonald *et al.* 1988; Sinton *et al.* 1991). Macdonald *et al.* (1988) argue that first-order segments reflect

From: MacLeod, C. J., Tyler, P. A. & Walker, C. L. (eds) 1996, *Tectonic, Magmatic, Hydrothermal and Biological Segmentation of Mid-Ocean Ridges*, Geological Society Special Publication No. 118, pp. 143–156.

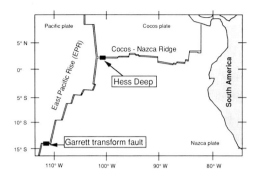

Fig. 1. Location of Hess Deep and the Garrett transform fault in the eastern Pacific Ocean.

cells of upwelling mantle, and this explains why large petrological and geochemical provinces in the oceanic crust are bounded by major tectonic discontinuities. Higher order segmentation relates to a complex interplay of melt segregation, dynamics of melt lenses in the shallow mantle and the crust, and behaviour of melt and mush zones in sub-axial crustal magma chambers (Langmuir *et al.* 1986; Sinton *et al.* 1991; Sinton & Detrick 1992).

An important aspect of MOR segmentation, which requires a great deal of study, is the variation of mantle composition within and between segments because it has a fundamental bearing on magmatic and tectonic segmentation. This composition reflects the degree of partial melting and efficiency of melt extraction (e.g. Dick & Bullen 1984; Dick & Fisher 1984; Dick *et al.* 1984; Michael & Bonatti 1985; Elthon 1990; Johnson *et al.* 1990), and the extent of subsequent melt–peridotite reaction (e.g. Dick 1989; Kelemen *et al.* 1990, 1992; Edwards & Malpas 1996). Closely linked with mantle composition is the mode and rate of melt extraction, which may reflect whether mantle upwelling and melt transport beneath MORs should be considered in terms of two-dimensional (sheet-like) or three-dimensional (diapiric) flow (e.g. Parmentier & Phipps Morgan 1990). Studies investigating these aspects of ridge segmentation may be possible in the mantle sections of ophiolites (e.g. Nicolas 1989; Ceuleneer 1991), but are very limited in the ocean basins because of the difficulties associated with sampling and mapping. In recent years, however, the East Pacific Rise (EPR) has been extensively studied and there now exists a very significant collection of mantle and crustal rocks formed along this fast-spreading MOR system. Samples have been collected of uppermost mantle harzburgite at the

mantle–crust transition zone, and these are representative of two different environments along the EPR: a 'normal' spreading segment (represented by samples from Hess Deep) and the end of a segment where the EPR meets the Garrett transform fault (represented by samples from Garrett Deep) (Fig. 1). The Hess Deep samples are unique because there are only a few places where tectonic windows through fast-spreading crust give access to the underlying mantle, and all but these samples are from first-order segment boundaries (e.g. Garrett Deep samples), where the segmentation process itself may have influenced the lithosphere-forming processes.

The present study provides a brief review of the texture, mineralogy and mineral chemistry of samples of uppermost mantle harzburgite at the mantle–crust transition zone preserved at Hess Deep and Garrett Deep. As such, it permits the first integrated examination of mantle flow, mantle composition, melt extraction and melt–peridotite reaction within and between regions of mantle underlying different segments of a fast-spreading MOR.

Site descriptions

Hess Deep (2°14′N, 101°33′W) lies to the east of the EPR (Fig. 1) and forms the deepest part of an amagmatic oceanic rift valley that is propagating westward into the eastern flank of the EPR ahead of the Cocos–Nazca spreading centre (Lonsdale 1988). The rift valley is bounded by normal fault escarpments, along which oceanic lithosphere (including mantle harzburgite) that originally formed at a 'normal' spreading segment of the EPR is exposed (Francheteau *et al.* 1990; MacLeod *et al.* 1996).

The Garrett transform fault lies several first-order ridge segments to the south of Hess Deep, at a latitude of about 13°20′S–13°30′S (Fig. 1). The fault offsets two fast-spreading segments of the EPR by some 130 km. Garrett Deep is situated in a deep trough in the ocean floor produced by the fault. In contrast to Hess Deep, harzburgites recovered from Garrett Deep are believed to have originally formed beneath the EPR where it meets the Garrett transform fault, before being transported laterally and uplifted along the transform fault (Hébert *et al.* 1983).

Despite oceanic lithosphere at Hess Deep and Garrett Deep originating far from and close to a transform fault respectively, both sites exhibit many similarities. Both segments of lithosphere formed at very fast spreading rates: 130–135 mm

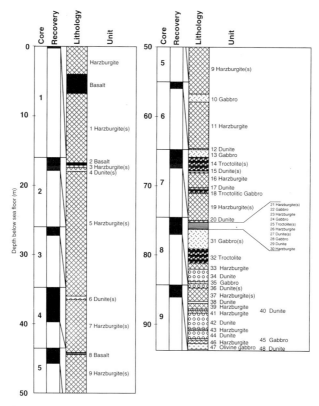

Fig. 2. Schematic lithostratigraphic column for Hole 895D (redrawn from Gillis *et al.* 1993) as an example of the interbanding of mafic and ultramafic rock types in the mantle–crust transition zone preserved at Hess Deep. Units studied by Edwards & Malpas (1996) are denoted by (s).

a^{-1} for Hess Deep (Lonsdale 1988; Francheteau *et al.* 1990) and a strike-slip motion of 145 mm a^{-1} for the Garrett transform fault (Naar & Hey 1989). The difference in age of the lithosphere at the extremities of the 130 km-long Garrett transform fault is estimated at 1.8 Ma (Naar & Hey 1989), which implies that the rocks at Garrett Deep are significantly younger than 1.8 Ma and may compare with the age of 1 Ma predicted by Francheteau *et al.* (1990) for the EPR lithosphere exposed by rifting at Hess Deep.

The rock types at both locations are very similar in terms of their mineral assemblage, and comprise harzburgites, dunites, wehrlites, troctolites, gabbros, dolerite dykes and basaltic volcanic rocks (Hébert *et al.* 1983; Cannat *et al.* 1990; Francheteau *et al.* 1990; Hekinian *et al.* 1992, 1993; Girardeau & Francheteau 1993; Gillis *et al.* 1993; Dick *et al.* 1994; Edwards & Malpas 1996). With the exception of dolerite dykes and basaltic volcanic rocks, these authors agree that the ultramafic and mafic rocks at both

sites are part of the uppermost mantle and lower crust, including the mantle–crust transition zone, of the ocean floor. This is particularly well exemplified by the interbanding of rock types in Hess Deep (Fig. 2), which bears many similarities with the mantle–crust transition zone preserved in ophiolites (e.g. Benn *et al.* 1988; Nicolas 1989). Similar crystallization sequences are documented in the crustal cumulates at both sites, with olivine, spinel and plagioclase as early phases and clinopyroxene as a late phase (Hébert *et al.* 1983; Gillis *et al.* 1993). This sequence is expected for low-pressure crystallization of mid-ocean ridge basaltic (MORB) melt (Shido *et al.* 1971; Shibata 1976; Bender *et al.* 1978; Walker *et al.* 1979).

Harzburgite

Sampling

Information on Garrett Deep harzburgite comes primarily from the four samples examined by

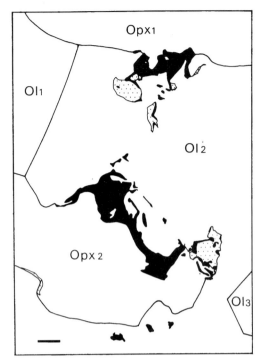

Fig. 3. Line drawing of a backscattered electron image of an olivine–orthopyroxene–spinel–clinopyroxene assemblage in impregnated harzburgite from Hole 895D at Hess Deep (from Edwards & Malpas 1996). Note the replacement of orthopyroxene porphyroclasts by undeformed grains of olivine, spinel (black) and clinopyroxene (stipple), and the intergrowth of spinel and clinopyroxene. Scale bar represents 100 μm.

Cannat *et al.* (1990). These samples were dredged from 13°28.98′S, 111°29.61′W in the valley floor of the Garrett transform fault zone during dredge haul SO12-91D of the R.V. *Sonne* as part of the 1980 GEOMETEP (Geothermal Metallogenesis in the East Pacific) cruise (Hébert *et al.* 1983). Information on harzburgite from around 2°17′N, 101°27′W on the southern slope of an intra-rift ridge to the northeast of Hess Deep itself comes mainly from two sources: eight samples studied by Hekinian *et al.* (1993) were obtained on the 1988 NAZCOPAC diving cruise during Dive 17 of the French submersible *Nautile* (Francheteau *et al.* 1990); 13 samples examined by Edwards & Malpas (1996) were retrieved from Hole 895D during Ocean Drilling Program (ODP) Leg 147 at Site 895 (Gillis *et al.* 1993).

Texture and mineralogy

In agreement with studies of the transition zone preserved in ophiolites, harzburgites from Hess Deep and Garrett Deep record a two-stage evolution (Cannat *et al.* 1990; Hekinian *et al.* 1993; Allen & Dick 1996; Arai & Matsukage 1996; Dick & Natland 1996; Edwards & Malpas 1996): (i) production of a plastically deformed refractory harzburgite residue (olivine + orthopyroxene + Cr-spinel ± clinopyroxene) by partial melting of adiabatically upwelling mantle peridotite and effective melt extraction beneath the spreading ridge; (ii) migration of basaltic melt through this plastically deformed residue, whereby melt reacted with harzburgite and/or crystallized liquidus phases, which resulted in dissolution of orthopyroxene, textural and chemical re-equilibration of residual olivine and spinel, and formation of undeformed or weakly deformed grains of olivine, spinel, clinopyroxene and plagioclase (Figs 3, 4 & 5). Those samples exhibiting the latter stage will be referred to as 'impregnated' from here on.

The period of mantle upwelling is recorded by the porphyroclastic texture of residual minerals. This texture, and associated fabrics, developed at temperatures in excess of 1000°C and are somewhat similar in Hess Deep and Garrett Deep harzburgites (Boudier *et al.* 1996). Notably, olivine fabrics in Hess Deep (Boudier *et al.* 1996) and Garrett Deep (Cannat *et al.* 1990) harzburgites are inferred by Boudier *et al.* (1996) to record flow at the top of a zone of mantle upwelling, where plastic deformation developed in the presence of melt (cf. Rabinowicz *et al.* 1987). A residual clinopyroxene component is rare. Where present in Hess Deep harzburgites, it occurs as exsolution lamellae in orthopyroxene porphyroclasts (Hekinian *et al.* 1993; Boudier *et al.* 1996), and in Garrett Deep harzburgites it occurs as rare exsolution lamellae in orthopyroxene porphyroclasts and, possibly, as rare interstitial grains associated with orthopyroxene (Cannat *et al.* 1990).

All harzburgite samples examined here from Hole 895D are impregnated, whereas only some of those examined from Dive 17 (samples 17-1, 17-2, 17-8 and 17-9 of Hekinian *et al.* 1993) and Garrett Deep (samples SO12-91D-13 and SO12-91D-B2 of Cannat *et al.* 1990) show evidence of melt impregnation. Impregnation is obvious as undeformed grains of clinopyroxene that are interstitial with respect to the deformed mineral matrix of harzburgite (Boudier *et al.* 1996; Edwards & Malpas 1996), and this texture is seen in all impregnated samples. A particular characteristic of Hole 895D harzburgites is the

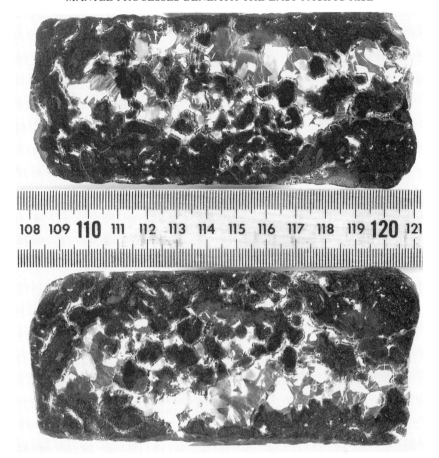

Fig. 4. Photograph of the interval 04R-02 (108–121 cm) from Hole 895C at Hess Deep showing dunite grading into olivine gabbro (plagioclase is white; clinopyroxene is grey; olivine is black). Hole 895C is only a few metres from Hole 895D.

Fig. 5. Dykelet of plagioclase (coarse stipple) and clinopyroxene (black) in a matrix of olivine (unpatterned) and orthopyroxene (fine stipple) in impregnated harzburgite sample SO12-91D-B2 from Garrett Deep (redrawn from Cannat *et al.* 1990). Scale bar represents 1 mm.

occurrence of undeformed clinopyroxene–spinel intergrowths (Fig. 3). Plagioclase has not been observed in the matrix of Hole 895D harzburgites, but these harzburgites are inter-

banded with well-defined bodies and irregular patches of troctolite and gabbro that often have dunitic margins (Figs 2 & 4). In Hess Deep sample 17-9 plagioclase occurs in wehrlite in contact with harzburgite and spinel occurs only in the harzburgitic part of this sample (Hekinian *et al.* 1993). All mineral phases in this wehrlite are considered by Girardeau & Francheteau (1993) to be magmatic in origin. Plagioclase occurs in both impregnated harzburgites from Garrett Deep, where it exists in millimetre-wide gabbroic dykelets which intersect the plastic foliation of harzburgite at shallow and steep angles (Cannat *et al.* 1990; Fig. 5).

Mineral chemistry

The mineral compositions obtained from samples from Garrett Deep (Hébert *et al.* 1983; Cannat *et al.* 1990) and from Dive 17 at Hess

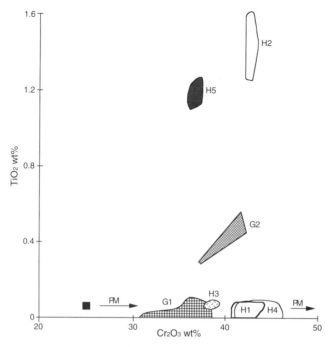

Fig. 6. TiO$_2$ v. Cr$_2$O$_3$ of spinel. Hess Deep: melt impregnated harzburgites of Hole 895D (H1) except sample 895D-02R-02 (8–12 cm) (H2) and sample 895D-03R-01 (38–43 cm) (H3) (from Edwards & Malpas 1996); harzburgites and melt impregnated harzburgites from Dive 17 (H4) except wehrlite-bearing sample 17–9 (H5) (from Hekinian *et al.* 1993). Garrett Deep: harzburgites and melt impregnated harzburgites (G1) and melt impregnated harzburgite sample SO12-91D-B2 (G2) (from Cannat *et al.* 1990). Average abyssal spinel peridotite (black square) from Dick & Fisher (1984); PM is the partial melting trend.

Deep (Hekinian *et al.* 1993) were determined by the same technique on a CAMEBAX MBX electron microprobe at Microsonde de l'Ouest, Brest, France. Mineral compositions from samples from Hole 895D at Hess Deep (Edwards & Malpas 1996) were obtained using a similar technique on a CAMEBAX SX50 electron microprobe at Memorial University of Newfoundland, Canada. The reader is also referred to further, independent mineralchemical studies of ODP Cores from Hess Deep by Allan & Dick (1996), Arai & Matsukage (1996) and Dick & Natland (1996). Precise details of the mineral compositions and analytical conditions can be found in the respective studies. The similarity of mineral compositions for most Hess Deep harzburgites suggests that data obtained using the two techniques are comparable.

All samples of harzburgite are highly serpentinised and have experienced late-stage subsolidus re-equilibration (Cannat *et al.* 1990; Hekinian *et al.* 1993; Edwards & Malpas 1996). The original mineralogy of the harzburgites and the extent of alteration of this mineralogy are

broadly similar and, consequently, mineral compositions do not have to be corrected to enable definition of major compositional differences within and between samples from Hess Deep and Garrett Deep. The composition of spinel is worthy of special mention. The original Mg# (for spinel Mg# is 100Mg/(Mg + Fe^{2+}), where Fe^{2+} is calculated assuming charge balance and perfect stoichiometry) of this mineral in mafic and ultramafic rocks is normally reduced by sub-solidus Fe–Mg exchange with silicate minerals (e.g. Irvine 1967). Because the original mineralogy and degree of alteration of harzburgites are similar, the broad compositional trends exhibited by spinel are expected to reflect original chemical variation prior to late-stage sub-solidus chemical exchange and alteration, as will be demonstrated.

Spinel in all harzburgites is moderately to highly enriched in Cr (Fig. 6). With the exception of Ti-rich spinel in impregnated harzburgite samples 895D-02R-02 (8–12 cm), SO12-91D-B2 and 17-9, spinel in impregnated and non-impregnated harzburgites has TiO$_2$ <0.10 wt%,

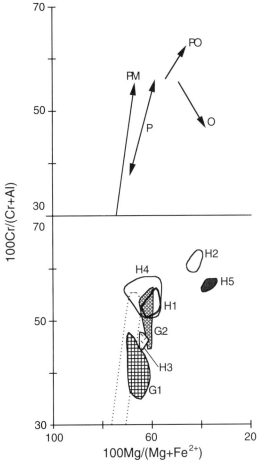

Fig. 7. Cr# v. Mg# of spinel. Symbols as in Fig. 6, except for the field of abyssal spinel peridotites (dotted outline) from Dick & Fisher (1984). Trends shown for fractional crystallisation of olivine (O) and plagioclase and olivine (PO) are from Dick & Bullen (1984). The partial melting trend (PM) is the vector defined by the field of abyssal spinel peridotites, and the plagioclase control line (P) is inferred from the Al trend defined in melt impregnated harzburgites in Hole 895D (see text).

and plots along a partial melting trend (Fig. 6). Similarly, spinel in these impregnated and non-impregnated harzburgites plots along a trend of increasing Cr# [100Cr/(Cr + Al)] and decreasing Mg# that is expected for spinel in abyssal peridotites that have experienced progressively higher degrees of partial melting (Fig. 7). This result confirms that found independently by Allan & Dick (1996) in Site 895 samples.

Spinel and silicate mineral compositions in several samples exhibit significant deviation from the majority of compositions found within a particular site. Spinel is the best indicator of this deviation. Sample SO12-91D-B2 from Garrett Deep is cut by a gabbroic dykelet (Fig. 5) and spinel in the harzburgitic portion of this sample has higher Cr, Ti and Cr# relative to non-impregnated and other impregnated harzburgites from Garrett Deep (Figs 6 & 7); this deviation will be referred to as the Ti trend from here on. Two trends are defined with respect to the majority of spinel compositions in Hess Deep harzburgites. Spinel in sample 895D-03R-01 (38–43 cm) has a low Cr# (Fig. 7) resulting from a low Cr content (Fig. 6) and high Al content (Edwards & Malpas 1996), and defines what will be referred to as the Al trend. Apart from having a slightly higher than normal abundance of spinel (2 modal %), this sample is mineralogically and texturally very similar to the other Hole 895D harzburgites. Sample 895D-02R-02 (8–12 cm) is also compositionally distinct and preserves a sharp contact between dunite and harzburgite. Mineral compositions across this contact are very similar (Edwards & Malpas 1996), but relative to the majority of Hole 895D harzburgites, spinel shows an increase in Cr# for no change in Cr content (Figs 6 & 7). This spinel composition reflects large decreases in Al (Al_2O_3 decreases from 23.4–26.7 wt% to 17.2–18.8 wt%) and Mg (MgO decreases from 12.5–13.9 wt% to 8.6–9.5 wt%), accompanied by a significant increase in Fe (total FeO increases from 16.1–19.6 wt% to 26.6–28.2 wt%) with a low Fe^{2+}/Fe^{3+} ratio (from 5.8–12.8 to 3.9–4.6) (Edwards & Malpas 1996). These spinels are highly enriched in Ti (Fig. 6) and define what will be referred to as the Fe–Ti trend. The Fe–Ti trend is also observed in spinel in harzburgite adjacent to wehrlite in Hess Deep sample 17-9 (Figs 6 & 7).

Olivine compositions are plotted in Fig. 8. With the exception of Hess Deep sample 895D-02R-02 (8–12 cm), in which olivine has a low Mg# (for silicates Mg# is 100Mg/(Mg + Fe), where Fe is total Fe as Fe^{2+}) consistent with Fe–Ti enrichment in associated spinel, olivines in impregnated and non-impregnated harzburgites from Hess Deep and Garrett Deep have broadly similar refractory compositions. Olivine compositions in these harzburgites are typical of those found in residual mantle peridotites and plot in the region of a partial melting trend (Fig. 8).

Pyroxenes in most Hess Deep and Garrett Deep samples have similar major element compositions, with orthopyroxene being intermediate between enstatite and bronzite and clinopyroxene between diopside and endiop-

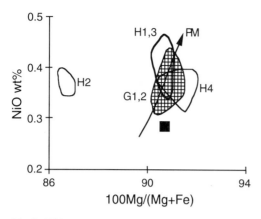

Fig. 8. NiO v. Mg# of olivine. Symbols as in Fig. 6 and the progressive partial melting trend (PM) is from Girardeau & Francheteau (1993).

Discussion

Within the framework of MOR segmentation, low-order and high-order variation in the composition of mantle peridotites probably relates primarily to a dynamic interplay of the degree and scale of partial melting and melt extraction, and the distribution and reaction/crystallization of melt. This section discusses how mineralogical and mineral chemical variation arises in harzburgites from Hess Deep and Garrett Deep, and the implications of this variation for mantle flow and melt extraction beneath the EPR.

Mantle melting

Partial melting of peridotitic mantle below MORs occurs during adiabatic upwelling of this mantle (e.g. McKenzie & Bickle 1988). A polybaric, near-fractional or incremental melting process can account for the observed compositions of MORBs and residual abyssal peridotites, with the typical oceanic basalt evolving from aggregated fractional melts (e.g. Johnson *et al.* 1990; Johnson & Dick 1992). Disequilibrium between shallow residual abyssal peridotites and aggregate MORB melts results because the residue was in equilibrium with the last, most depleted melt fraction to be extracted during the melting process. Non-impregnated harzburgites from Hess Deep and Garrett Deep are expected to represent residues of shallow melting because textures and fabrics in these residual rocks are inferred to record mantle flow at the top of upwelling zones (Boudier *et al.* 1996). In these harzburgites the composition of spinel and the presence of residual clinopyroxene or exsolution lamellae of clinopyroxene in residual orthopyroxene suggests melting did not proceed beyond the clinopyroxene-in–clinopyroxene-out phase boundary. This control has been documented for mantle beneath other MOR segments and is expected for the global MOR system (Dick & Bullen 1984; Dick & Fisher 1984). The non-impregnated harzburgites from Hess Deep are more refractory than those from Garrett Deep (Figs 6 & 7), and are moderately to highly refractory with respect to the majority of abyssal peridotites examined by Dick & Bullen (1984), Dick & Fisher (1984) and Dick (1989). Therefore, relative to the global MOR system and non-impregnated harzburgites from Garrett Deep, these Hess Deep harzburgites are products of a very high degree of total partial melting, a conclusion also reached by Gillis *et al.* (1993) and Dick & Natland (1996). Compared to spinel in non-impregnated Hess Deep harzburgites,

side. Exceptionally, pyroxenes in Hess Deep sample 895D-02R-02 (8–12 cm) exhibit slight, but obvious, Fe–Ti enrichment (Table 1) which correlates with the composition of olivine and spinel in this sample; orthopyroxene in the harzburgitic portion of Hess Deep sample 17-9 also exhibits Fe enrichment (Hekinian *et al.* 1993). The majority of harzburgites have pyroxenes with compositions intermediate between those of residual abyssal peridotites and highly refractory peridotites from ophiolites, with enstatite tending toward undersaturation with respect to diopside (Table 1).

Comparison of the composition of olivine (Fig. 8) and spinel (Figs 6 & 7) with that of pyroxenes in the same sample (Table 1) reinforces the compositional trends defined by spinel: (i) the Ti trend defined by spinel in Garrett Deep sample SO12-91D-B2 is also seen as higher Ti and lower Al and Cr concentrations in orthopyroxene, little or no change in the Mg# of silicates or the Ni content of olivine, and clinopyroxene rich in Ti and Na; (ii) in Hess Deep sample 895D-03R-01 (38–43 cm) the Al trend exhibited by spinel is associated with Ti-poor, Na-rich clinopyroxene, orthopyroxene showing slight enrichment in Al and Cr (which is also seen in clinopyroxene), a slightly lowered Ni content in olivine (Edwards & Malpas 1996), and unchanged Mg#s and Ti concentrations in silicates; (iii) the Fe–Ti trend defined by spinel in Hess Deep sample 895D-02R-02 (8–12 cm) is also seen as a decrease in Mg# of all silicates, a slightly lowered Ni content in olivine (Fig. 8), Ti enrichment and Al and Cr depletion in orthopyroxene, and clinopyroxene rich in Ti and Na.

Table 1. *Representative compositions of pyroxene in harzburgite*

	HD1	HD2	HD3	GD1	GD2	GD3	AASP	BIO
Orthopyroxene								
SiO_2	56.1	55.7	55.5	54.47	55.19	55.25	54.7	57.58
TiO_2	bd	bd	0.25	0.01	0.05	0.15	0.05	0.00
Al_2O_3	2.1	2.7	1.8	2.62	2.46	1.87	4.20	0.92
Cr_2O_3	0.66	0.87	0.56	0.80	0.83	0.69	0.43	0.52
FeO	5.8	5.8	8.2	5.84	5.64	5.77	5.61	5.29
MnO	0.13	0.10	0.17	0.20	0.15	0.12	0.13	0.11
NiO	0.12	0.07	0.09	0.10	0.09	0.09	bd	0.08
MgO	33.1	33.1	31.6	34.69	32.67	34.67	32.3	34.93
CaO	2.13	1.56	1.82	1.52	2.35	0.96	2.12	1.03
Na_2O	0.02	0.03	0.05	0.03	0.02	0.00	0.43	0.04
Total	100.16	99.93	100.04	100.28	99.45	99.57	99.97	100.50
Mg#	91.0	91.0	87.3	91.1	91.2	91.5	91.1	92.2
En	87.4	88.3	84.2	88.5	87.1	89.8	87.4	90.4
Fs	8.6	8.7	12.3	8.6	8.4	8.4	8.5	7.7
Wo	4.0	3.0	3.5	2.9	4.5	1.8	4.1	1.9
Clinopyroxene								
SiO_2	52.4	52.5	52.6	52.03	52.10	51.73	51.1	55.95
TiO_2	0.07	0.05	0.49	0.05	0.36	0.38	0.15	0.02
Al_2O_3	2.7	3.0	2.5	3.41	2.89	3.07	5.64	1.14
Cr_2O_3	1.02	1.38	1.08	1.31	1.19	1.24	0.88	0.97
FeO	2.8	2.6	3.4	2.88	2.67	2.66	3.10	1.77
MnO	0.06	0.07	bd	0.06	0.06	0.05	0.08	0.09
NiO	0.11	0.04	0.06	0.03	0.04	0.05	bd	0.04
MgO	18.3	17.2	16.7	17.37	17.74	18.32	18.2	17.94
CaO	22.0	22.8	22.6	21.17	21.86	21.04	19.8	22.58
Na_2O	0.02	0.40	0.74	0.56	0.37	0.32	0.38	0.50
Total	99.48	100.04	100.17	98.87	99.28	98.86	99.33	101.00
Mg#	92.1	92.2	89.7	91.5	92.2	92.4	91.3	94.8
En	51.3	49.1	47.9	50.8	50.7	52.4	53.3	51.0
Fs	4.4	4.1	5.5	4.7	4.3	4.3	5.1	2.8
Wo	44.3	46.8	46.6	44.5	45.0	43.3	41.6	46.2

Impregnated samples from Hole 895D at Hess Deep are massive harzburgites 895D-02R-01 038-041 cm #06A (HD1) and 895D-03R-01 038-043 cm #06D (HD2) and harzburgite directly adjacent to dunite in 895D-02R-02 008-012 cm #01B (HD3) (Edwards & Malpas 1996); samples from Garrett Deep are harzburgites (GD1) SO12-91D-7-3 (orthopyroxene) and SO12-91D-B3 (clinopyroxene) and impregnated harzburgite SO12-91D-B2 away from a gabbroic dykelet (GD2) and associated with this dykelet (GD3) (Cannat *et al.* 1990); average abyssal spinel peridotite (AASP) from Dick & Fisher (1984); refractory residual harzburgite TM1245 from Table Mountain of the Bay of Islands ophiolite (BIO) (Suhr & Robinson 1994).

Mg# = 100Mg/(Mg + Fe); En = 100Mg/(Mg + Fe + Ca); Fs = 100Fe/(Mg + Fe + Ca); Wo = 100Ca/(Mg + Fe + Ca); bd = below detection.

the relatively fertile composition of spinel in non-impregnated harzburgites from Garrett Deep can be explained by the 'transform fault effect' of lower total partial melting of mantle as a transform fault is approached (Langmuir & Bender 1984; Bender *et al.* 1984).

Although non-impregnated harzburgites from Hess Deep and Garrett Deep formed beneath different parts of different first-order segments of the EPR (Fig. 1), the compositional differences between these two sites are probably the best analogue available for the compositional variation expected in shallow mantle below a

first-order ridge segment in a fast-spreading environment. Obviously, sampling of mantle along the entire length of a single ridge segment is necessary to determine this variation more accurately, especially in terms of the extent of high-order mantle heterogeneity on the scale of a first-order ridge segment.

Melt-peridotite reaction

Melt-peridotite reaction is recognized as an important process affecting the texture, mineralogy and composition of mantle peridotites (e.g. Dick 1977; Quick 1981; Nicolas 1989; Kelemen 1990) and the composition of melt involved in reaction (e.g. Fisk 1986; Kelemen 1990; Kelemen *et al.* 1990; Niu & Batiza 1993). The compositional changes experienced by melt and peridotite during reaction are strongly influenced by the melt/peridotite ratio and the connectivity of melt. This ratio is low where residual pyroxene remains (Edwards & Malpas 1996), but high where dunites form by reaction (Kelemen *et al.* 1995*a*) and where gabbros form by fractional crystallization of melt which may also react (Edwards & Malpas 1996). The texture, mineralogy and mineral chemistry of impregnated harzburgites from Hess Deep and Garrett Deep reflect melt activity in residual uppermost harzburgitic mantle during or after plastic deformation (Cannat *et al.* 1990; Boudier *et al.* 1996; Edwards & Malpas 1996). Melt impregnation resulted in the Al, Ti and Fe–Ti trends. These chemical trends, discussed below, provide insight into some mechanisms for chemical transfer between melt and harzburgite which may produce heterogeneity in mantle beneath segments of the EPR and other MORs.

The Ti and Fe–Ti trends are observed adjacent to zones of high melt/harzburgite ratio, i.e. gabbro and dunite, respectively. In samples exhibiting the latter trend, all silicates are enriched in Fe and silicates and spinel plot off partial melting trends and, in some cases, along crystallization trends (Figs 6, 7 & 8). As shown earlier, the Ti trend observed in harzburgite sample SO12-91D-B2 from Garrett Deep is seen as enrichment of spinel and orthopyroxene in Ti, increase in the Cr content and Cr# of spinel, and Al depletion in orthopyroxene. These variations increase with proximity to dykelets containing plagioclase and clinopyroxene, which suggests the effects of chemical transfer across the melt-solid interface decrease away from the melt-rich zone. The composition of spinel in sample 5012-91D-B2 may be explained by the peritectic reaction described by Thayer (1956) and Irvine (1976), in which melt extracts the $MgAl_2O_4$ component from spinel during formation of plagioclase and olivine. Hence, the composition of spinel in harzburgite may have been controlled by the formation of plagioclase in gabbro dykelets (Fig. 7).

The Al trend correlates with low melt/harzburgite ratio and in many ways the mineral chemical changes of this trend are the opposite of those of the Ti trend. This is particularly obvious from the composition of spinel (Fig. 7) and Al enrichment in orthopyroxene (Table 1). Based on the studies of Kelemen (1990), Edwards (1991) and Sühr & Robinson (1994),

Edwards & Malpas (1996) explained the Al trend in terms of complete reaction of interstitial plagioclase or liquidus plagioclase with the residual harzburgite mineral assemblage, which resulted in formation of clinopyroxene ± spinel and re-equilibration of residual spinel. This process leaves the Mg and Fe content of olivine and orthopyroxene unchanged. With the exception of sample 895D-02R-02 (8–12 cm), mineral compositions in all impregnated Hole 895D harzburgites are expected to reflect plagioclase enrichment (Edwards & Malpas 1996). In these harzburgites the original Cr# of residual spinel is expected to have been higher, but no greater than 60 because of the control of the clinopyroxene-in–clinopyroxene-out melting boundary on the composition of spinel (Dick & Bullen 1984).

There is no mineralogical or mineral chemical evidence that basaltic melt completely crystallised in any of the impregnated harzburgites. However, mineral compositions in these harzburgites give some idea of melt composition, although at low melt/harzburgite ratio the composition of melt is strongly buffered by that of harzburgite (Kelemen 1986, 1990). The melt responsible for the Fe–Ti trend was clearly evolved in terms of major element composition and high Fe^{3+}/Fe^{2+} ratio (see earlier). In contrast, the development of the Al and Ti trends appears to have involved relatively primitive melts. The low concentration of Ti in clinopyroxene and spinel in harzburgites exhibiting the Al trend is consistent with reaction of a plagioclase component from a Ti-poor melt. Cannat *et al.* (1990) explain the Ti trend by melt controlling Ti enrichment and being in equilibrium with harzburgite for Fe, Mg and Ni. In the model of Hekinian *et al.* (1995) this trend should correlate with enriched-MORB magmatism along the Garrett transform fault. Although the mineralogy of the gabbro dykelets linked with this trend may be consistent with off-axis magmatism (see later), further comment at this stage is not possible with the information available.

Mantle flow and melt extraction

Currently there exist two debates on mantle processes beneath MORs which have a profound bearing on models of segmentation of MORs. One is the style of mantle flow in upwelling zones (e.g. Batiza this volume), and the other concerns the mode of extraction of melt from the mantle (e.g. Kelemen *et al.* 1995*b*). Together, these processes have a major influence, at small and large scales, on the extent of melt-peridotite reaction in deep and shallow mantle, the composition of oceanic crust and mantle, and the morphology and structure of MORs.

Melt extraction. Once melt is produced in the mantle it may migrate either by hydrofracture or by focused porous flow (see review by Kelemen *et al.* 1995*b*). In both cases melt channels may be walled by dunite formed by melt-peridotite reaction (Nicolas 1986; Kelemen *et al.* 1995*a, b*) and melt is extracted rapidly with little chance to re-equilibrate during ascent. However, in the uppermost mantle at the mantle–crust transition zone, residual peridotites are pervasively soaked by reactive, crystallizing melt. This arises because melt transport by hydrofracture is difficult to sustain at the mantle–crust transition zone and melt is dispersed into the peridotite matrix (e.g. Nicolas 1989) or, as melt cools and partially crystallizes, melt channels become blocked and porous flow becomes more diffuse (Tait & Jaupart 1991; Tait *et al.* 1992; Bédard *et al.* 1992).

During melt extraction the thermal regime in the mantle controls the phases formed by melt-peridotite reaction and melt crystallisation. Studies by Ceuleneer (1991) and Kelemen *et al.* (1995*a*) of harzburgitic upper mantle preserved in the Oman ophiolite demonstrate that below the axis of a spreading ridge mantle flow is near-vertical and adiabatic, whereas off-axis flow is near-horizontal and a conductive geotherm dominates. Adiabatic conditions below the ridge may be maintained to near the base of the mantle–crust transition zone (Kelemen *et al.* 1995*a, b*). Under adiabatic conditions, basaltic melt will react with peridotite to form dunite (Kelemen *et al.* 1995*a*), but clinopyroxene is unlikely to form because adiabatically ascending basaltic melts are not saturated with clinopyroxene and clinopyroxene is not a product of most reactions involving adiabatically ascending mantle peridotite and basaltic melt (Kelemen 1990; Kelemen *et al.* 1992, 1995*b*). Hence, clinopyroxene is expected to form in a conductive thermal regime, such as in gabbroic dykes in the zone of divergent mantle flow (Ceuleneer 1991; Kelemen *et al.* 1995*a*), or as interstitial grains and gabbroic patches and rare dykes in peridotites in the crust-mantle transition zone at the top of the zone of vertical mantle upwelling (Ceuleneer 1991).

Within this framework, the adiabatic history of mantle upwelling, melting, melt extraction and melt-peridotite reaction preserved in harzburgitic mantle at Hess Deep and Garrett Deep has been overprinted by melt reaction and melt crystallization in a conductive geotherm. A dynamic situation of melting, melt extraction, melt–peridotite reaction and, finally, melt crystallization is, therefore, envisaged for mantle as it upwells below the EPR axis and then diverges away from it. Evidence for this is preserved at Site 895, where Boudier *et al.* (1996) have documented melt activity associated with a near-vertical mantle flow trajectory which shallows toward the top of the mantle section. At Hess Deep and Garrett Deep the distribution of undeformed, interstitial grains of clinopyroxene in harzburgite adjacent to dunitic and gabbroic zones suggests a change from focused to diffuse porous flow as melt channels became clogged by solid products of reaction and crystallization.

Mantle flow. Two contrasting types of mantle flow are presumed to exist beneath MORs: buoyancy-driven, diapiric upwelling (three dimensional) and passive, plate-driven upwelling (two dimensional or sheet-like) (e.g. Parmentier & Phipps Morgan 1990; Lin & Phipps Morgan 1992). Whereas three-dimensional upwelling is favoured at slow-spreading ridges, the dominance of three-dimensional or two-dimensional upwelling below fast-spreading ridges is uncertain (e.g. Batiza 1989; Wilson 1992; Wang & Cochran 1993), although two-dimensional upwelling is favoured by fast spreading and low mantle viscosity (Lin & Phipps Morgan 1992). Highly channelized extraction of melt in spatially restricted conduits would be expected with two-dimensional mantle upwelling. In this situation, melts are unable to re-equilibrate during ascent, and when they reach the shallow mantle and are dispersed throughout the peridotite matrix they react extensively with orthopyroxene. This is precisely what is seen in the mantle-crust transition zone preserved at Hess Deep, suggesting that two-dimensional upwelling may dominate below 'normal' spreading segments of the EPR.

Melt–peridotite reaction at fast-spreading and slow-spreading MORs

The previous discussion supports two-dimensional mantle upwelling and channelled extraction of melt in spatially restricted conduits below fast-spreading ridge segments. This may be in keeping with the observation of Niu & Batiza (1993), reviewed by Batiza (this volume), that MORBs erupted at fast-spreading MOR segments exhibit very different chemical systematics from those at slow-spreading segments because the systematics at slow-spreading segments may be a function of melt dissolving pyroxene from peridotite during melting and ascent of mantle and melt. Extensive reaction over most of the melt ascent path is consistent with three-dimensional upwelling in which a high ratio of mineral surface area to melt volume exists throughout the peridotite matrix; this in turn should favour buoyancy-driven flow in a relatively large volume of mantle.

Whereas polybaric melt-peridotite reaction may exert a significant control on the composition of MORB beneath slow-spreading ridge segments, Dick *et al.* (1994) and Dick & Natland (1996) suggest the chemical evolution of MORB beneath the EPR may be strongly coupled with mineralogical and chemical heterogeneity produced by melt–harzburgite reaction in the uppermost mantle. The present study supports this conclusion, especially as significant modification of melt composition by melt–peridotite reaction is not expected at depth in the mantle below the EPR. In this situation, the further reactive melt travels through the shallow mantle, the greater will be the integrated effect of melt–harzburgite reaction on the composition of melt. The gravity data of Wang & Cochran (1993, 1995) support along-axis transport of melt in the upper mantle, which is most efficient at fast-spreading ridges. Thus, melt extraction from mantle beneath fast-spreading ridges may be dominated initially by highly channelized flow throughout the main adiabatic ascent path, and later by along-axis flow in mantle directly below the ridge axis.

More studies are required on the compositional evolution of MORB melt below fast- and slow-spreading ridge segments in terms of mantle and melt flow and melt–peridotite reaction. The effect of these on MORB composition may be as important as the compositional changes produced by intracrustal melt differentiation (Sinton & Detrick 1992) and reaction of melt with ultramafic and mafic cumulates (Bédard 1993). Evidently, parameters of mantle and melt flow and melt–peridotite reaction must continue to be upgraded and integrated into models of segmentation and MORB evolution along MORs.

The authors would like to take this opportunity to thank all members of the ODP Leg 147 Scientific Party for their assistance and discussions during and after the cruise, especially members involved in studies of peridotites. The thoughtful, constructive reviews of Ian Parkinson, an anonymous reviewer and, in particular, C. MacLeod contributed significantly to improving an earlier version of the manuscript. K. Hudson-Edwards and D. Groom are thanked for their assistance with drafting diagrams. Correspondence to S. J. Edwards (e-mail: s.j.edwards@greenwich.ac.uk).

References

ALLAN, J. F. & DICK, H. J. B. 1996. Cr-rich spinel as a tracer for melt migration and melt–wall rock interaction in the mantle: Hess Deep, Leg 147. *In*: MÉVEL, C., GILLIS, K. M., ALLAN, J. F. & MEYER, P. S. (eds) *Proceedings of the Ocean Drilling Program, Scientific Results, 147*. College Station, Texas (Ocean Drilling Program), 157–172.

ARAI, S. & MATSUKAGE, K. 1996. Petrology of gabbro-troctolite-peridotite complex from Hess Deep, equatorial Pacific: implications for mantle-melt interaction within the oceanic lithosphere. *In*: MÉVEL, C., GILLIS, K. M., ALLAN, J. F. & MEYER, P. S. (eds) *Proceedings of the Ocean Drilling Program, Scientific Results, 147*. College Station, Texas (Ocean Drilling Program), 135–155.

BATIZA, R. 1996. Magmatic segmentation of mid-ocean ridges: a review. *This volume*.

—— 1989. Petrology and geochemistry of eastern Pacific spreading centres. *In*: WINTERER, E. L., HUSSONG, D. M. & DECKER, R. W. (eds) *The Geology of North America (volume N): The Eastern Pacific Ocean and Hawaii*. Geological Society of America, Boulder, Colorado, 145–160.

BÉDARD, J. H. 1993. Oceanic crust as a reactive filter: Synkinematic intrusion, hybridization, and assimilation in an ophiolitic magma chamber, western Newfoundland. *Geology*, **21**, 77–80.

——, KERR, R. C. & HALLWORTH, M. A. 1992. Porous sidewall and sloping floor crystallization experiments using a reactive mush: implications for the self-channelization of residual melts in cumulates. *Earth and Planetary Science Letters*, **111**, 319–329.

BENDER, J. F., HODGES, F. N. & BENCE, A. E. 1978. Petrogenesis of basalts from the project FAMOUS area: experimental study from 0 to 15 kbars. *Earth and Planetary Science Letters*, **41**, 277–302.

——, LANGMUIR, C. H. & HANSON, G. N. 1984. Petrogenesis of basalt glasses from the Tamayo region, East Pacific Rise. *Journal of Petrology*, **25**, 213–254.

BENN, K., NICOLAS, A. & REUBER, I. 1988. Mantle-crust transition zone and origin of wehrlitic magmas: evidence from the Oman ophiolite. *Tectonophysics*, **151**, 75–85.

BOUDIER, F., MACLEOD, C. J. & BOLOU, L. 1996 Structures in peridotites from Site 895, Hess Deep: implications for the geometry of mantle flow beneath the East Pacific Rise. *In*: MÉVEL, C., GILLIS, K. M., ALLAN, J. F. & MEYER, P. S. (eds) *Proceedings of the Ocean Drilling Program, Scientific Results, 147*. College Station, Texas (Ocean Drilling Program), 347–356.

CANNAT, M., BIDEAU, D. & HÉBERT, R. 1990. Plastic deformation and magmatic impregnation in serpentinized ultramafic rocks from the Garrett transform fault (East Pacific Rise). *Earth and Planetary Science Letters*, **101**, 216–232.

CEULENEER, G. 1991. Evidence for a paleo-spreading center in the Oman ophiolite: mantle structures in the Maqsad area. *In*: PETERS, T., NICOLAS, A. & COLEMAN, R. G. (eds) *Ophiolite Genesis and Evolution of Oceanic Lithosphere*. Kluwer Academic Publishers, Dordrecht, 147–173.

DICK, H. J. B. 1977. Evidence of partial melting in the Josephine peridotite. *In*: DICK, H. J. B. (ed.) *Magma Genesis*. Oregon Department of Geology and Mineral Industries Bulletin, **96**, 59–62.

—— 1989. Abyssal peridotites, very slow spreading ridges and ocean ridge magmatism. *In*: SAUNDERS, A. D. & NORRY, M. J. (eds) *Magmatism in the Ocean Basins*. Geological Society of London, Special Publications, **42**, 71–105.

—— & BULLEN, T. 1984. Chromian spinel as a petrogenetic indicator in abyssal and alpine-type peridotites and spatially associated lavas. *Contributions to Mineralogy and Petrology*, **86**, 54–76.

—— & FISHER, R. L. 1984. Mineralogic studies of residues of mantle melting: abyssal and alpine-type peridotites. *In*: KORNPROBST, J. (ed.) *Kimberlites II: the Mantle and Crust-Mantle Relationships*. Elsevier, Amsterdam, 295–308.

——, —— & BRYAN, W. B. 1984. Mineralogical variability of the uppermost mantle along mid-ocean ridges. *Earth and Planetary Science Letters*, **69**, 88–106.

DICK, H. J. B. & NATLAND, J. H. 1996. Late-stage melt evolution and transport in the shallow mantle beneath the East Pacific Rise. *In*: MÉVEL, C., GILLIS, K. M., ALLAN, J. F. & MEYER, P. S. (eds) *Proceedings of the Ocean Drilling Program, Scientific Results, 147*. College Station, Texas (Ocean Drilling Program), 103–134.

——, NATLAND, J. & THE LEG 147 SCIENTIFIC PARTY 1994. Melt transport and evolution in the shallow mantle beneath the East Pacific Rise: preliminary results from ODP Site 895. *Mineralogical Magazine*, **58A**, 229–230.

EDWARDS, S. J. 1991. *Magmatic and fluid processes in the upper mantle: a study of the Bay of Islands Ophiolite Complex, Newfoundland*. PhD thesis, Memorial University of Newfoundland.

—— & MALPAS, J. 1996. Melt-peridotite interactions in shallow mantle at the East Pacific Rise: evidence from ODP Site 895 (Hess Deep). *Mineralogical Magazine*, **60**, 191–206.

ELTHON, D. 1990. The petrogenesis of primary mid-ocean ridge basalts. *Reviews in Aquatic Science*, **2**, 27–53.

FISK, M. R. 1986. Basalt-magma interactions with harzburgite and the formation of high magnesium andesites. *Geophysical Research Letters*, **13**, 467–470.

FRANCHETEAU, J., ARMIJO, R., CHEMINÉE, J. L., HEKINIAN, R., LONSDALE, P. & BLUM, N. 1990. 1 Ma East Pacific Rise oceanic crust and uppermost mantle exposed by rifting in Hess Deep (Equatorial Pacific Ocean). *Earth and Planetary Science Letters*, **101**, 281–295.

GILLIS, K., MÉVEL, C., ALLAN, J., ET AL. 1993. *Proceedings of the Ocean Drilling Program, Initial Reports, 147*. College Station, Texas (Ocean Drilling Program).

GIRARDEAU, J. & FRANCHETEAU, J. 1993. Plagioclase-wehrlites and peridotites on the East Pacific Rise (Hess Deep) and the Mid-Atlantic Ridge (DSDP Site 334): evidence for magma percolation in the oceanic upper mantle. *Earth and Planetary Science Letters*, **115**, 137–149.

HÉBERT, R., BIDEAU, D. & HEKINIAN, R. 1983. Ultramafic and mafic rocks from the Garrett transform fault near 13°30'S on the East Pacific Rise: igneous petrology. *Earth and Planetary Science Letters*, **65**, 107–125.

HÉKINIAN, R., BIDEAU, D., HÉBERT, R. & NIU, Y. 1995. Magmatism in the Garrett transform fault (East Pacific Rise near 13°27'S). *Journal of Geophysical Research*, **100**, 10163–10185.

——, ——, CANNAT, M., FRANCHETEAU, J. & HÉBERT, R. 1992. Volcanic activity and crust-mantle exposure in the ultrafast Garrett transform fault near 13°28'S in the Pacific. *Earth and Planetary Science Letters*, **108**, 259–275.

——, ——, FRANCHETEAU, J., CHEMINÉE, J. L., ARMIJO, R., LONSDALE, P. & BLUM, N. 1993. Petrology of the East Pacific Rise crust and upper mantle exposed in Hess Deep (eastern Equatorial Pacific). *Journal of Geophysical Research*, **98**, 8069–8094.

IRVINE, T. N. 1967. Chromian spinel as a petrogenetic indicator. Part 2, petrologic applications. *Canadian Journal of Earth Sciences*, **4**, 71–103.

—— 1976. Chromite crystallization in the join Mg_2SiO_4–$CaMgSi_2O_6$–$CaAl_2Si_2O_8$–$MgCr_2O_4$–SiO_2. *Carnegie Institution of Washington Year Book*, **76**, 465–472.

JOHNSON, K. T. M. & DICK, H. J. B. 1992. Open system melting and the temporal and spatial variation of peridotite and basalt compositions at the Atlantis II Fracture Zone. *Journal of Geophysical Research*, **97**, 9219–9241.

——, —— & SHIMIZU, N. 1990. Melting in the oceanic upper mantle: an ion microprobe study of diopsides in abyssal peridotites. *Journal of Geophysical Research*, **95**, 2661–2678.

KELEMEN, P. B. 1986. Assimilation of ultramafic rock in subduction-related magmatic arcs. *Journal of Geology*, **94**, 829–843.

—— 1990. Reaction between ultramafic rock and fractionating basaltic magma I. Phase relations, the origin of calc-alkaline magma series, and the formation of discordant dunite. *Journal of Petrology*, **31**, 51–98.

——, DICK, H. J. B. & QUICK, J. E. 1992. Formation of harzburgite by pervasive melt/rock reaction in the upper mantle. *Nature*, **358**, 635–641.

——, SHIMIZU, N. & SALTERS, V. J. M. 1995a. Extraction of mid-ocean-ridge basalt from the upwelling mantle by focused flow of melt in dunite channels. *Nature*, **375**, 747–753.

——, JOYCE, D. B., WEBSTER, J. D. & HOLLOWAY, J. R. 1990. Reaction between ultramafic rock and fractionating basaltic magma II. Experimental investigation of reaction between olivine tholeiite and harzburgite at 1150–1050°C and 5 kb. *Journal of Petrology*, **31**, 99–134.

——, WHITEHEAD, J. A., AHARONOV, E. & JORDAHL, K. A. 1995b. Experiments on flow focusing in soluble porous media, with applications to melt extraction from the mantle. *Journal of Geophysical Research*, **100**, 475–496.

LANGMUIR, C. H. & BENDER, J. F. 1984. The

geochemistry of oceanic basalts in the vicinity of transform faults: observations and implications. *Earth and Planetary Science Letters,* **69**, 107–127.

——, —— & BATIZA, R. 1986. Petrological and tectonic segmentation of the East Pacific Rise, 5°30′–14°30′N. *Nature,* **322**, 422–429.

LIN, J. & PHIPPS MORGAN, J. 1992. The spreading rate dependence of three-dimensional mid-ocean ridge gravity structure. *Geophysical Research Letters,* **19**, 13–16.

LONSDALE, P. 1988. Structural pattern of the Galapagos microplate and evolution of the Galapagos triple junction. *Journal of Geophysical Research,* **93**, 13 551–13 574.

—— 1994. Geomorphology and structural segmentation of the crest of the southern (Pacific–Antarctic) East Pacific Rise. *Journal of Geophysical Research,* **99**, 4683–4702.

MACDONALD, K. C., FOX, P. J., PERRAM, L. J., EISSEN, M. F., HAYMON, R. M., MILLER, S. P., CARBOTTE, S. M., CORMIER, M.-H. & SHOR, A. N. 1988. A new view of the mid-ocean ridge from the behaviour of ridge-axis discontinuities. *Nature,* **335**, 217–225.

MACLEOD, C. J., CÉLÉRIER, B., FRÜH-GREEN, G. L. & MANNING, C. E. 1996. Tectonics of Hess Deep: a synthesis of drilling results from Leg 147. *In*: MÉVEL, C., GILLIS, K. M., ALLAN, J. F. & MEYER, P. S. (eds) *Proceedings of the Ocean Drilling Program, Scientific Results, 147*. College Station, Texas (Ocean Drilling Program), 461–475.

MCKENZIE, D. P. & BICKLE, M. J. 1988. The volume and composition of melt generated by extension of the lithosphere. *Journal of Petrology,* **29**, 625–679.

MICHAEL, P. J. & BONATTI, E. 1985. Peridotite composition from the North Atlantic: regional and tectonic variations and implications for partial melting. *Earth and Planetary Science Letters,* **73**, 91–104.

NAAR, D. F. & HEY, R. M. 1989. Recent Pacific-Nazca plate motions. *In*: SINTON, J. M. (ed.) *The Evolution of Mid-Ocean Ridges*. American Geophysical Union Geophysical Monographs, **57**, 9–30.

NICOLAS, A. 1986. A melt extraction model based on structural studies in mantle peridotites. *Journal of Petrology,* **27**, 999–1022.

—— 1989. *Structures of Ophiolites and Dynamics of Oceanic Lithosphere*. Kluwer Academic Publishers, Dordrecht.

NIU, Y. & BATIZA, R. 1993. Chemical variation trends at fast and slow spreading mid-ocean ridges. *Journal of Geophysical Research,* **98**, 7887–7902.

PARMENTIER, E. M. & PHIPPS MORGAN, J. 1990. Spreading rate dependence of three-dimensional structure in oceanic spreading centers. *Nature,* **348**, 325–328.

QUICK, J. E. 1981. The origin and significance of large, tabular dunite bodies in the Trinity peridotite, northern California. *Contributions to Mineralogy and Petrology,* **78**, 413–422.

RABINOWICZ, M., CEULENEER, G. & NICOLAS, A. 1987. Melt segregation and flow in mantle diapirs below spreading centers: evidence from the Oman ophiolite. *Journal of Geophysical Research,* **92**, 3475–3486.

SHIBATA, T. 1976. Phenocryst-bulk rock composition relations of abyssal tholeiites and their petrogenetic significance. *Geochimica et Cosmochimica Acta,* **40**, 1407–1417.

SHIDO, F., MIYASHIRO, A. & EWING, M. 1971. Crystallization of abyssal tholeiites. *Contributions to Mineralogy and Petrology,* **31**, 251–266.

SINTON, J. M. & DETRICK, R. S. 1992. Mid-ocean ridge magma chambers. *Journal of Geophysical Research,* **97**, 197–216.

——, SMAGLIK, S. M., MAHONEY, J. J. & MACDONALD, K. C. 1991. Magmatic processes at superfast spreading oceanic ridges: glass compositional variations along the East Pacific Rise, 13°S–23°S. *Journal of Geophysical Research,* **96**, 6133–6155.

SUHR, G. & ROBINSON, P. T. 1994. Origin of mineral chemical stratification in the mantle section of the Table Mountain massif (Bay of Islands ophiolite, Newfoundland, Canada). *Lithos,* **31**, 81–102.

TAIT, S. & JAUPART, C. 1991. Compositional convection in a reactive crystalline mush and melt differentiation. *Journal of Geophysical Research,* **97**, 6735–6756.

——, JAHRLING, K. & JAUPART, C. 1992. The planform of compositional convection in a mushy layer. *Nature,* **359**, 406–408.

THAYER, T. P. 1956. Mineralogy and geology of chromium. In UDY, M. J. (ed.) *Chromium. American Chemical Society Monograph,* **132**, 14.

WALKER, D., SHIBATA, T. & DELONG, S. E. 1979. Abyssal tholeiites from the Oceanographer Fracture Zone. II. Phase equilibria and mixing. *Contributions to Mineralogy and Petrology,* **70**, 111–126.

WANG, X. & COCHRAN, J. R. 1993. Gravity anomalies, isostasy and mantle flow at the East Pacific Rise crest. *Journal of Geophysical Research,* **98**, 19505–19531.

—— & —— 1995. Along-axis gravity gradients at mid-ocean ridges: implications for mantle flow and axial morphology. *Geology,* **23**, 29–32.

WILSON, D. S. 1992. Focused mantle upwelling beneath mid-ocean ridges: evidence from seamount formation and isostatic compensation of topography. *Earth and Planetary Science Letters,* **113**, 41–55.

The response of ridge-crest hydrothermal systems to segmented, episodic magma supply

RACHEL M. HAYMON

Department of Geological Sciences and Marine Science Institute, University of California, Santa Barbara, Santa Barbara, CA 93106, USA

Abstract: Segmented, episodic magma supply and cracking models, advanced to explain ridge-axis discontinuities and along-strike variability, have provoked a recent shift in mid-ocean ridge crest hydrothermal models. Segment-scale and time-series observations of hydrothermal vents and plumes along fast and intermediate spreading centres reveal a close coupling between magma supply and hydrothermal fluxes. Based on the premise that magma supply exerts a primary control on hydrothermal processes, a new paradigm of segmented hydrothermal circulation has arisen featuring independent temporal evolution of hydrothermal systems along individual ridge segments. This paradigm explains observed segment-to-segment variations in the chemical composition of vent fluids and plumes ('chemical segmentation'). Hydrothermally active segments can be classified, independently of spreading rate, into two types that represent end-members for the coupling of hydrothermal and magmatic processes: magma-rich segments and magma-starved segments. Along magma-rich segments, magma intrudes to shallow depths where water:rock ratios are high. Seafloor vent distribution tends to be magmatically controlled along dyke-induced fissures. Along magma-starved segments, intrusion is at greater depth, and deep cracks (usually faults) are required to transport water to and from the heat source. Water:rock ratios are low and vent locations are influenced by crustal permeability structure, as well as heat source location. Although both segment types can occur at any spreading rate, magma-rich segments are more common at faster spreading rates, and magma-starved segments are more typical at slow spreading rates.

'No man's knowledge here can go beyond his experience' (Locke 1690).
'Theory helps us to bear our ignorance of fact' (Santayana 1896).

Conceptual paradigms for mid-ocean ridge (MOR) hydrothermal systems over the past twenty years have been based on scant empirical data about MOR thermal structure and permeability. Since 1982, mounting evidence for magmatic segmentation of the MOR (see review by Macdonald *et al.* 1993, and references therein) has called attention to along-axis and temporal variability in ridge crest thermal structure and permeability, and to the potential consequences for the distribution of hydrothermal vents. Recent mapping of ridge crest hydrothermal vent and plume distributions at different scales of segmentation (Haymon *et al.* 1991; Baker & Hammond 1992; Delaney *et al.* 1992; Charlou & Donval 1993; Baker *et al.* 1994, 1995*a*; Murton *et al.* 1994; Embley *et al.* 1995; German *et al.* 1995; Urabe *et al.* 1995; Wright *et al.* 1995; Wilson *et al.* in press), and studies of temporal evolution in hydrothermal processes (Haymon *et al.* 1991, 1993; Butterfield &

Massoth 1994; Embley *et al.* 1995; Von Damm *et al.* 1995; Wright *et al.* £995), now indicate that there is a close coupling between magma supply and hydrothermal processes on individual ridge segments. A new paradigm of hydrothermal processes on magma-rich vs. magma-starved MOR segments is presented here to incorporate the growing body of evidence for spatial-temporal response of ridge crest hydrothermal systems to a segmented, episodic magma supply.

Pre-segmentation hydrothermal paradigm

Before 1982, wide-beam echo sounding maps of the MOR depicted a highly linear ridge interrupted at widely-spaced intervals (50–500 km along strike) by major transform faults (e.g. Heezen & Tharp 1977). The ridge axis was envisioned at that time as a line source of magmatic heat with large thermal gradients perpendicular to the ridge axis (e.g. Sleep 1975), and localized 'edge effects' near transform faults (Ballard & Francheteau 1982). Heat-flow measurements collected in off-axis

From: MacLeod, C. J., Tyler, P. A. & Walker, C. L. (eds) 1996, *Tectonic, Magmatic, Hydrothermal and Biological Segmentation of Mid-Ocean Ridges*, Geological Society Special Publication No. 118, pp. 157–168.

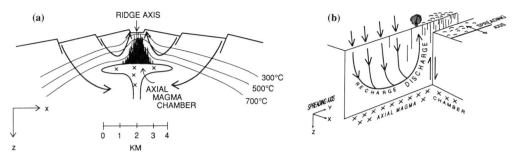

Fig. 1. (**a**) Schematic diagram of pre-segmentation paradigm, showing a cross-section perpendicular to the ridge axis. Driven by cross-strike thermal gradients, seawater circulates perpendicular to the ridge axis along faults, reaching depths adjacent to the axial magma chamber (Bonatti 1975; Wolery & Sleep 1976; Lister 1977). (**b**) Schematic diagram of circulation along ridge axis-parallel fractures and faults on the ridge crest; modified after Sleep (1984), and Strens & Cann (1986). In Figs 1a, b and 2, z = depth, x = perpendicular to ridge axis, y = parallel to ridge axis.

sediment ponds along transects perpendicular to the ridge axis (e.g. Talwani *et al.* 1971; Williams *et al.* 1974) provided the first data about the thermal structure of the MOR. These measurements revealed anomalously low and oscillatory heat flow perpendicular to the ridge axis, and were interpreted as evidence for large-scale heat removal from new oceanic crust by vigorous hydrothermal circulation within 15 km of the axis.

The bathymetric continuity of the ridge crest between transform faults in early maps of the MOR suggested that thermal gradients along the ridge axis were minimal, and presumably much less than ridge-perpendicular thermal gradients. These observations led directly to a paradigm of two-dimensional, ridge-perpendicular convection driven by steep thermal gradients normal to the ridge axis (Bonatti 1975; Wolery & Sleep 1976; Lister 1977). Realizing that fracture permeability could be important, and taking into account some off-axis observations of increased heat flow near the margins of fault-bounded abyssal hills, Bonatti (1975) proposed as part of his paradigm that seawater flowed downward along off-axis faults, migrated at depth toward the ridge axis along unspecified pathways, and travelled back to the seafloor along faults close to the ridge axis.

Studies of hydrothermal alteration in ophiolites were consistent with deep penetration of hydrothermal fluids to regions adjacent to and just above the axial magma chamber (e.g. Spooner & Fyfe 1973; Gregory & Taylor 1981; MacLeod & Rothery 1992). Axial magma chambers were thought to be rather wide

(15–20 km) based on the ophiolite model of Pallister & Hopson (1981), but subsequent seismic data from modern MORs revealed narrower (<2–4 km wide) melt lenses beneath the ridge axis (McClain *et al.* 1985; Toomey *et al.* 1990; Harding *et al.* 1993).

Figure 1a depicts the prevailing paradigm for MOR hydrothermal circulation that existed prior to 1982, and incorporates the following features: (1) two-dimensional, ridge-perpendicular convection driven by steep thermal gradients; (2) ridge-perpendicular flow facilitated by inward-facing normal faults; and, (3) deep penetration of water to the proximity of the magma chamber. A modification of the axial magma chamber width to 4 km has been made to make this figure consistent with current seismic data.

The pre-segmentation paradigm is not necessarily incorrect; however, it is incomplete. Cann (1979) pointed out that the permeability of young seafloor is dominated by ridge-parallel cracking, and that ridge-parallel hydrothermal flow should result (Fig. 1b). In 1982, high-resolution multibeam sonar surveys of the East Pacific Rise (EPR) revealed morphological evidence of ridge segmentation between major transform faults (Macdonald *et al.* 1984; Schouten *et al.* 1985; Lonsdale 1985). A segmented magma supply model was subsequently developed to explain: overlapping spreading centres and along-strike variations in axial depth and morphology of the EPR (Macdonald *et al.* 1984, 1988); along-strike variations in axial lava flow compositions (Langmuir *et al.* 1986); and along-strike discontinuities in a sub-axial seismic reflector beneath the EPR that

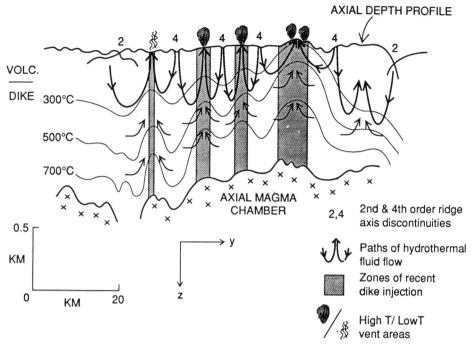

Fig. 2. Schematic diagram of segmented hydrothermal circulation, showing a cross-section parallel to the ridge axis. Episodes of dyke injection (stippled) along magmatically activated segments create thermal gradients along-strike that drive circulation (arrows) parallel to the ridge axis. Heavier arrows at shallow depths represent greater fluid flux in the upper crust compared to the lower crust, due to higher permeability in the volcanic section and volcanic/dyke transition zone relative to the sheeted dyke complex and gabbro section. Hydrothermal flux is greatest above more recent magma injection zones that are still transferring significant amounts of heat to fluids circulating in the upper crust. Variations in crustal permeability along strike, related to along-strike variations in magma supply and the length of time elapsed since the last eruptive episode, also influence locations of hydrothermal discharge and recharge (Rosenberg *et al.* 1993).

appeared to be the top of a segmented axial magma chamber (Detrick *et al.* 1987, 1993).

Segmented, episodic magma supply to the ridge crest introduced the likely possibility of along-strike thermal gradients and variations in permeability that could result in vigorous axis-parallel ridge crest hydrothermal convection (Fig. 2). Thus the segmentation paradigm for MOR magmatic and tectonic processes that was developed in the mid-1980s inspired a revised 1990s paradigm of MOR hydrothermal processes, with some testable predictions.

Hydrothermal segmentation paradigm: some initial predictions

(1) The fundamental premise of the segmentation paradigm for MOR hydrothermal processes (Fig. 2) is that hydrothermal vent distribution and flux are ultimately controlled by the location, volume, and timing of magma injection along the MOR. This leads directly to two predictions: (a) at long-wavelength, long-lived, first- and second-order scales of segmentation, abundant vents and large hydrothermal flux are expected to occur along inflated, magma-rich segments; (b) at short-wavelength, short-duration fourth-order scales of segmentation, vent locations and depth of circulation are controlled by dyke injections.

(2) Due to time-dependent hydrothermal and conductive heat loss from magmatic intrusions, individual fourth-order ridge segments created or inflated by episodes of magma injection are expected to evolve independently in time from a state of high temperature/large hydrothermal

flux to a state of lower temperature/decreased hydrothermal flux.

(3) Ridge-parallel hydrothermal circulation should exist within the sheeted dyke complex at MORs, driven energetically by along-strike thermal gradients in the dyke complex, and facilitated dynamically by along-strike anisotropy in permeability and along-strike variations in crack abundance (Rosenberg et al. 1993).

Testing and development of hydrothermal segmentation paradigm

Early work

Prediction 1a (above) was initially argued by Crane et al. (1985) on the basis of apparent temperature elevation of bottom waters above axial highs along the Juan de Fuca Ridge and northern EPR. Field evidence for hydrothermal discharge above shallow dyke intrusion zones (prediction 1b) first came from geologic mapping around the Bayda massive sulphide deposit in the northern Oman ophiolite (Haymon et al. 1989, 1990). Prediction 2 derived support from the detection of short-lived 'megaplumes' along the Juan de Fuca Ridge in 1986–93 (Baker et al. 1989, 1995b). From studies of hydrothermal alteration and veins in the sheeted dykes of ophiolites in Oman and Cyprus, a good case also was made for prediction 3 (Richardson et al. 1987; Nehlig & Juteau 1988; Haymon et al. 1990).

The first seafloor study specifically designed to test the predictions of the new paradigm was a 1989 near-bottom survey of the fast-spreading ($11 \, \text{cm a}^{-1}$) EPR 9–10°N ridge segment (Haymon et al. 1991; Wright et al. 1995b). The results of this survey are discussed below.

Segment-scale observations of hydrothermal vents at EPR 9–10°N

Using the deep-towed *Argo I* optical/acoustic system, a dense and continuous survey along an 83 km long segment of the EPR axial zone (9°09'–54'N) established the abundance, distribution and characteristics of hydrothermal vents and other fine-scale volcanic and tectonic features of a fast-spreading MOR at second- and fourth-order scales of ridge segmentation. Spatial correlations observed between hydrothermal, magmatic, and tectonic features were

quite coherent at both scales of segmentation (Fig. 3). Listed below are some critical observations that emerge from the data shown in Fig. 3.

(1) Vent abundance (and flux; Baker et al. 1994) correlate extremely well with ridge inflation at the second-order scale (Haymon et al. 1991; Wright et al. 1995a). Note that axial cross-sectional area is a particularly good index of ridge inflation, much better than axial depth (Scheirer & Macdonald 1993). There is excellent along-strike co-variation of axial cross-sectional area and vent abundance in Fig. 3. This co-variation is entirely consistent with the prediction that hydrothermal vent abundance and distribution are primarily controlled at the 1st and 2nd order scale by magmatic injection (prediction 1a above).

(2) Inflated segments of the ridge exhibit younger lavas cut by sparse, 'hot cracks' (wide, presumably deep, fissures; see Wright et al. (1995b) for discussion about the relationship between fissure width and depth). Submersible observations have shown that these wide cracks are often eruptive, presumably because they are deep enough to tap melt, and most high-temperature venting in the *Argo I* survey area is located along them (Haymon et al. 1993; Wright et al. 1995a, b) (Fig. 3). Deflated, hydrothermally inactive segments of the ridge exhibit older flows that are densely fissured by narrow, presumably shallow, 'cold cracks' that have accumulated in the crust over time, probably in response to far-field tensile stresses (Wright et al. 1995a, b). At first glance, the association of wide cracks with younger lavas, and narrower cracks with older lavas, seems counter-intuitive, as we might expect axial cracks to widen with time. This apparent paradox is resolved if the wide cracks in younger lavas are deep, eruptive cracks that have poured out new lava flows and buried older, narrower fissures.

The observed spatial association of hydrothermal vents with wide, eruptive cracks at EPR 9–10°N is consistent with predictions 1b (above) that hydrothermal vent distribution at the fourth order scale is controlled by dyke injection. This prediction has been verified a second time on the Co-Axial Segment of the Juan de Fuca Ridge, where migration of earthquakes over a period of two weeks in June–July, 1993, documented along-axis propagation of a dyke northward from Axial Seamount (Dziak et al. 1995). Camera surveys and submersible studies conducted subsequently in this area confirmed eruption of fresh volcanic flows

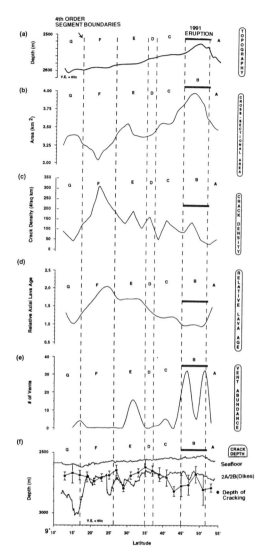

Fig. 3. Correlated axial profiles along the EPR crest 9°12′–54′N, from Wright *et al.* (1995*b*), based on near-bottom *Argo I* and multibeam SeaBeam surveys. Vertical dashed lines mark the latitudes of fourth-order segment boundaries as determined by Haymon *et al.* (1991). Fourth-order segments are lettered A–G after Haymon *et al.* (1991). Horizontal bars mark bounds of the 1991 eruption site (Haymon *et al.* 1993). (**a**) Seafloor topography from the SeaBeam bathymetry of Macdonald *et al.* (1992); VE = × 93. (**b**) Smoothed along-strike variation in ridge cross-sectional area (an index for magmatic inflation) from Scheirer & Macdonald (1993); (**c**) Along-strike variation in crack density (= number of fissures per sq km of seafloor imaged with the *Argo I* video camera) from Wright *et al.* (1995*a*). (**d**) Along-strike variation in age of axial lava flows from Wright *et al.* (1995*a*), where 1.0 = youngest and 2.0 = oldest as described in Haymon *et al.* (1991). (**e**) Along-strike variation in the total abundance of high-T + low-T hydrothermal vents (an index of hydrothermal flux) observed to be actively discharging fluids in December, 1989 (data from Haymon *et al.* (1991), and Wright *et al.* (1995*a*). (**f**) Along-strike variation in crack depth calculated from observed crack widths (Wright *et al.* 1995*b*; Layer 2A/2B boundary converted by Wright *et al.* (1995*b*) from two-way travel times of Harding *et al.* (1993), Kent *et al.* (1993), and Vera & Diebold (1994); VE= × 60.

1993, dyking/eruption event (Embley *et al.* 1995).

Temporal evolution of individual segments

From the results of the *Argo I* survey at EPR 9–10°N, Haymon *et al.* (1991) concluded that individual fourth-order segments along this fast spreading centre are created by individual dyking episodes at different times and locations along the axial zone (Fig. 4), and that the individual fourth-order segments are in different phases of a prescribed volcanic-hydrothermal-tectonic cycle. The cycle begins with an episode of dyke intrusion and volcanic eruption from discontinuous fissures along the length of a segment. This magmatic activity is accompanied and immediately followed by hydrothermal activity, along with magma drainage and consequent gravitational collapse leading to formation of an axial summit caldera (ASC; Haymon *et al.* 1991; Fornari *et al.* 1990). The cycle continues with waxing of hydrothermal activity and onset of amagmatic tectonic cracking, and concludes with waning of hydrothermal activity, continued crustal cracking, and

and venting of hydrothermal fluids from hot fissures above the dyke (Embley *et al.* 1995).

(3) Fourth-order segments of different relative ages exhibit a decrease in total vent abundance (and flux; Baker *et al.* 1994) with increasing age of axial lavas (Haymon *et al.* 1991; Wright *et al.* 1995*a, b*) (Fig. 3). This is consistent with prediction 2 (above) that individual segments evolve independently over time from hot to cold following magma injection along the segment. Similarly, vents along the Co-Axial Segment of the Juan de Fuca Ridge cooled significantly and reduced flow within three months of the June–July,

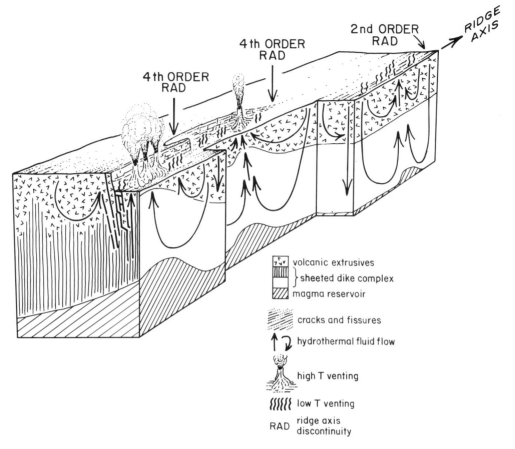

Fig. 4. Schematic sketch of hydrothermal geometry for a fast-spreading, magma-rich segment of the MOR crest; from Haymon *et al.* (1991). Cross-section parallel to ridge axis follows one side of the ASC until it disappears, then shifts to the ridge axis and follows a highly fissured segment of the axial zone. Notice that vent locations, vent abundance (i.e. hydrothermal flux), and fourth-order segmentation of hydrothermal circulation are primarily controlled by the spatial-temporal distribution of magma injection along the ridge crest, (represented as bumps on the tops of the magma reservoir and sheeted dykes). Shallow intrusion effectively transfers heat to permeable, water-rich levels of the crust, and drives fluid circulating at depth toward the seafloor. This diagram depicts a two-layered model of flow, with 3D hydrothermal circulation in a layer of highly permeable volcanic flows superimposed onto 2D, axis-parallel circulation of fluids along zones of enhanced permeability (deep axial fissures and cracked ASC margins) in the sheeted dyke layer.

widening of the ASC by mass wasting along its margins. An ASC may evolve over time into a fault-bounded axial summit graben (ASG); however this stage of the cycle has not (yet?) occurred on the EPR 9–10°N segment. Further investigation of axial summit troughs on the MOR are needed to assess whether ASCs evolve routinely into ASGs.

The duration of the entire cycle at EPR 9–10°N was proposed to be on the order of hundreds of years (Haymon *et al.* 1991; Wright *et*

al. 1995*a*), although an active segment may experience frequent eruptions every few years to decades, which restart the cycle before the tectonic phase is well developed. This resetting of the cycle to time-zero was dramatically illustrated by the eruption of young Segment B (9°52′–45′N) during the AdVenture '91 dive program only 15 months after the *Argo I* survey was completed. Observations made on Segment B during AdVenture '91 (Haymon *et al.* 1993) were followed by time-series monitoring during

Table 1. *Evolution of hydrothermal fluid on the East Pacific Rise, 9–10°N*

Component or parameter	Change in fluids from 'Aa' vent (9°46.5'N) 1991–1994		
Temperature*	Decreased 17%		
H_2S*	Decreased 85%		
Si*	Increased 85%		
(Cl^-)*	Increased 90%		
CO_2†	Decreased		
CH_4†	Decreased		
H_2†	Decreased		
1991 Interfield Variation in $(Cl^-)_{vent}/(Cl^-)_{seawater}$	9°46.5'N (new vent)	9°33.5'N (mature vent)	9°29.7'N (oldest vent)
	0.08*	c. 1.30‡	c. 1.00‡

* Calculated from data in Von Damm *et al.* (1995).
† From Lilley *et al.* (1994).
‡ From Von Damm *et al.* (1991).

four subsequent dive programs conducted 11 months (Haymon *et al.* 1992), 32 months (Lutz & Haymon 1994; Lutz *et al.* 1994; Shank *et al.* in press), 36 months, and 42 months (Von Damm *et al.* 1994, 1995) after the 1991 eruption. These time-series observations have provided valuable information about the short-term nature and sequence of changes in features of the ridge crest and hydrothermal system at the beginning of the volcanic–hydrothermal–tectonic cycle.

The 1991–1994 submersible studies at EPR 9–10°N made the following important observations about temporal change in the hydrothermal system.

(1) Thermal evolution of segments is accompanied by chemical evolution of hydrothermal vent fluids (Von Damm *et al.* 1994, 1995; Lilley *et al.* 1994) (Table 1). This results in both 'chemical clocks' and 'chemical segmentation'. Initially following an event of shallow intrusion and eruption, volatile-rich, low-salinity vapor, created by subseafloor phase separation, escape the seabed. The residual volatile-depleted, conjugate brine phase left within the seafloor is pumped out later in time. When this brine reservoir is exhausted, fluids of seawater salinity, that have not undergone phase separation, are vented. The maturity, or relative age, of a given fourth-order segment is therefore reflected in the salinity and gas composition of its vent fluids and plumes, and gas:chloride or volatile:non-volatile species ratios can be used as chemical 'clocks' to distinguish youthful segments from older segments. Chemical segmentation (segment-to-segment variation in hydrothermal fluid and plume compositions

along-strike) is thus the inevitable result of variations in segment maturity (i.e. time elapsed since last magmatic episode).

(2) Thermal and chemical evolution of hydrothermal systems on individual fourth-order segments appears to be quite rapid during the first three years following shallow dyke injection and eruption, as the dyke intrusion chills and freezes downward (Von Damm *et al.* 1995). The time-scale on which this occurs will depend generally on the volume of the intrusion and on seafloor permeability. At EPR 9°16.8'N, a time scale of 3–5 years was observed between onset of vapour discharge and onset of brine discharge (Von Damm *et al.* 1994, 1995). Similarly, rapid hydrothermal cooling over a period of only 3 months was observed along the eruptive fissure monitored at Co-Axial segment, Juan de Fuca Ridge (Dziak *et al.* 1995; Embley *et al.* 1995).

(3) Faunal recovery and succession at hydrothermal vents following an eruption are also very rapid. In only three years, fauna at new vents evolved from rampant bacterial mats (<30 days after the eruption; Haymon *et al.* 1993) to mature-looking, diverse hydrothermal vent communities with clumps of giant vestimentiferans over one meter in length (Lutz & Haymon 1994; Lutz *et al.* 1994; Shank *et al.* in press).

Multi-segment-scale observations of hydrothermal plumes

Near-bottom ROV surveys and on-bottom submersible studies are necessary for the detailed fourth-order-scale observations and time-series studies required to determine interrelationships

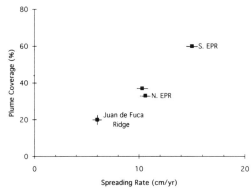

Fig. 5. Plot showing spreading rate v. percentage of ridge axis covered by hydrothermal plumes in three areas where strike-continuous multisegment plume surveys have been conducted; modified from Baker *et al.* (1994), with an additional data point for the southern EPR from Urabe *et al.* (1995). This plot indicates the direct proportionality of plume coverage (a proxy for hydrothermal flux) to spreading rate (a reasonable proxy for time-integrated magma flux), and strongly suggests that magma supply is the primary control on hydrothermal flux at a multisegment spatial scale.

of magmatic, hydrothermal and tectonic processes on MORs. However, plume studies conducted from surface ships are the most efficient and cost-effective means for determining integrated hydrothermal fluxes at a multisegment-scale. These studies have been done thus far along the fast-spreading northern EPR at 8.6–11.8°N (Lupton *et al.* 1993; Baker *et al.* 1994; Feely *et al.* 1994), along the ultrafast-spreading southern EPR at 13.5–19°S (Baker *et al.* 1995*a*; Urabe *et al.* 1995), along the medium-rate Juan de Fuca Ridge (Baker & Hammond 1992; Baker *et al.* 1995*b*; Lupton *et al.* 1995), and along the slow-spreading Mid-Atlantic Ridge at 11–40°N (Klinkhammer *et al.* 1985; Charlou & Donval 1993; German *et al.* 1994; Murton *et al.* 1994; Wilson *et al.* in press).

The plume data from fast, ultrafast, and intermediate-rate spreading centres show consistent positive correlations along strike between extent of hydrothermal flux, as inferred from percentage of the ridge axis covered by hydrothermal plumes, and magma budget, as inferred from ridge inflation, from depth/continuity of the axial magma reservoir, and from spreading rate (Fig. 5) (Baker & Hammond 1992; Baker *et al.* 1994, 1995*a*). The correlation between high hydrothermal and magmatic fluxes is particularly evident in comparative along-axis plots of plume data and axial inflation; (see plots in Baker & Hammond 1992 and Baker *et al.* 1994).

Figure 5 shows a steady increase in observed hydrothermal plume coverage with increasing spreading rate (a time-averaged measure of magma input to the ridge). New data from the northern Mid-Atlantic Ridge, not shown in Fig. 5, suggest that the covariation between hydrothermal flux and spreading rate (time-averaged magmatic flux) for the global MOR is quite linear. This is very compelling evidence that there is direct linkage between magmatic and hydrothermal fluxes along the MOR.

Vent-fluid and plume studies also show along-strike variations in compositions of hydrothermal plumes (chemical segmentation) that correlate with relative recency of magmatic injection along different fourth-order segments (see discussion in previous section and in Baker *et al.* 1994). Ratios of volatile:non-volatile species, such as hydrogen sulphide:iron and methane:manganese, seem to be particularly useful plume indicators of chemical segmentation and segment maturity, as these ratios are much higher above young, volcanically active segments (Baker *et al.* 1994; Feely *et al.* 1994). Concentrations of bacteria are also higher in young plumes (Feely *et al.* 1994; Urabe *et al.* 1995), consistent with seafloor observations of prolific bacterial growth during volcanic eruptions at EPR, 9°45′N–52′N (Haymon *et al.* 1993) and along the Coaxial segment of the Juan de Fuca Ridge (Juniper *et al.* 1995).

Incorporation of magma supply elements into the hydrothermal segmentation paradigm: magma-rich v. magma-starved segments

The excellent degree of correlation between magma budget and hydrothermal flux observed in the segment-scale and global scale datasets shown in Figs 3 and 5, respectively, leave little doubt that *magma supply exerts primary control on hydrothermal flux magnitude and vent distribution along the MOR*. At any spreading rate, the decoupled, episodic supply of magma to individual segments causes some segments to experience excessive magma supply, relative to spreading rate, while other segments experience deficient magma supply relative to spreading rate. Though faster spreading centres have greater rates of magma supply and younger axial lavas on average, even along the slow-spreading Mid-Atlantic Ridge, some segments are magma-rich and exhibit a very shallow rift valley floored with relatively young lavas; conversely, even at the fastest spreading rates along the southern EPR, some magma-starved segments exhibit deep axial summit grabens floored with relatively old lavas. Regardless of spreading rate, magma-rich vs. magma-starved segments represent two end-members for the coupling of hydrothermal processes to magmatic processes.

Along magma-rich segments, magma intrudes into shallow levels of the oceanic crust, where permeability and water:rock ratios are high. This occurs by dyke intrusion, which can follow existing cracks, or open cracks all the way to the seafloor, and literally bring the magma up to meet the water in the system (for example, as observed at EPR 9–10°N and Co-Axial segment, Juan de Fuca Ridge). On the EPR at 17.5°S, the seismic reflector interpreted as the top of the magma chamber rises in a narrow spike to within 800 m of the seafloor (Detrick *et al.* 1993). On magma-rich segments, the location of hydrothermal venting is controlled directly by magmatic injection, and the bulk of hydrothermal circulation in the axial zone tends to be relatively shallow, in the volcanic and dyke sections, with only small volumes of fluid reaching and reacting with gabbro above the magma chamber.

On magma-starved segments, magma resides at depth, and water descends to the magma along deep cracks (typically along faults). Discharge is also crack- or fault-controlled, and location of hydrothermal vents is strongly influenced by tectonic permeability as well as location of the heat source at depth. Where 'transfer faults' act as relay zones linking offset fourth-order segments, intersections of two or more crack sets create high permeability zones that are particularly good conduits for ascending vent fluids to follow: for example, at the Endeavour site, Juan de Fuca Ridge (Delaney *et al.* 1992), in the TAG area on the Mid-Atlantic Ridge (Karson & Rona 1990), and at the Broken Spur site on the Mid-Atlantic Ridge (Murton *et al.* 1995). Fluid circulation close to the axis is typically deeper on magma-starved segments than it is on magma-rich segments, so that larger volumes of fluid penetrate into and react with gabbro above the magma chamber.

In effect, behavior of the magma-starved segments can be viewed as similar to the pre-segmentation hydrothermal paradigm for MORs (Fig. 1), with the addition of a ridge-parallel component of hydrothermal flow. Magma-rich segments may behave more like the systems depicted in Figs 2 and 4.

In the absence of magmatic replenishment, magma-rich hydrothermal segments tend to evolve in character toward magma-starved hydrothermal segments. Although both types of systems can develop at any spreading rate, magma-rich systems are the most common type at fast-spreading centres, whereas magma-starved systems dominate at slow-spreading centres.

Concluding remarks

'When there is much desire to learn, there of necessity will be much arguing, much writing, and many opinions; for opinions in good men is knowledge in the making' (Milton 1644).
'It is the customary fate of new truths to begin as heresies and to end as superstitions' (Huxley 1880).

Paradigms are evolving conceptual constructs, destined for modification as more observations are made. The current 4D, spatial-temporal view of segmented, episodic, magmatically-driven MOR hydrothermal systems is certainly more complex and more complete than the pre-segmentation hydrothermal paradigm of 10–20 years ago; yet much is still left out. For example, what is the nature of axial hydrothermal systems at depth, and how do we visualize the relationship between transient circulation due to shallow intrusion/dyking and longer-lived, deep circulation above the magma reservoir? What are the detailed interactions between magma injection, cracking, structure of the ridge crest, and fluid flow? Does cracking trigger dyke injection, or does dyke injection trigger crack propagation? What is the temporal frequency of episodic magma injection along different types of segments, and at different spreading rates? How long does the hydrothermal part of the magmatic-hydrothermal-tectonic cycle last? Can enough data be gathered to constrain statistical estimates of the relative contributions to global hydrothermal fluxes of: megaplumes; other short-lived events; very young ridge crest systems; older, more sustained ridge crest systems; low-temperature ridge flank systems?

Like intellectual milestones, paradigms mark the distance travelled from a state of ignorance toward a state of comprehension. Guided by unanswered questions, aided by new technology, and illuminated by exploratory flashes of discovery, we collectively continue to make exciting progress toward understanding the largest and most dynamic geological feature on Earth, the mid-ocean ridge.

I would like to thank C. Walker and J. Cann for the opportunity to address the Geological Society of London in April 1995, and C. MacLeod for his patience and encouragement with regard to this manuscript. The manuscript was improved by the comments of C. Manning, C. MacLeod, K. Macdonald, and an anonymous reviewer. Haymon's EPR

9–10°N studies have been supported by grants from the National Science Foundation Ocean Science Division (OCE-88-17587; OCE-89-11396; OCE-90-20111; OCE-92-17746; OCE-92-96158 (to K. L. Von Damm)).

References

BAKER, E. T. & HAMMOND, S. R. 1992. Hydrothermal venting and the apparent magmatic budget of the Juan de Fuca Ridge. *Journal of Geophysical Research*, **97**, 3443–3456.

——, FEELY, R. A., MOTTL, M. J., SANSONE, F. J., WHEAT, C. G., RESING, J. A. & LUPTON, J. E. 1994. Hydrothermal plumes along the East Pacific Rise, 8°40′ to 11°50′N: plume distribution and relationship to the apparent magmatic budget. *Earth and Planetary Science Letters*, **128**, 1–17.

——, GERMAN, C. R. & ELDERFIELD, H. 1995a. Hydrothermal plumes over spreading-center axes: global distributions and geological inferences. *In: Seafloor hydrothermal systems: physical, chemical, biological, and geological interactions*. American Geophysical Union, Geophysical Monographs, **91**, 47–71.

——, LAVELLE, J. W., FEELY, R.A., MASSOTH, G. J., WALKER, S. L. & LUPTON, J. E. 1989. Episodic venting of hydrothermal fluids from the Juan de Fuca ridge. *Journal of Geophysical Research*, **94**, 9237–9250.

——, MASSOTH, G. J., FEELY, R. A., EMBLEY, R. W., THOMSON, R. E. & BURD, B. J. 1995b. Hydrothermal event plumes from the CoAxial seafloor eruption site, Juan de Fuca ridge. *Geophysical Research Letters*, **22**, 147–150.

BALLARD, R. D. & FRANCHETEAU, J. 1982. The relationship between active sulfide deposition and the axial processes of the mid-ocean ridge. *Marine Technical Society Journal*, **16**, 8–20.

BONATTI, E. 1975. Metallogenesis at oceanic spreading centers. *Annual Reviews of Earth and Planetary Sciences*, **3**, 401–431.

BUTTERFIELD, D. A. & MASSOTH, G. J. 1994. Geochemistry of north Cleft segment vent fluids: temporal changes in chlorinity and their possible relation to recent volcanism. *Journal of Geophysical Research*, **99**, 4951–4968.

CANN, J. R. 1979. Metamorphism in the ocean crust. *In:* TALWANI, M., HARRISON, C. G. & HAYES, D. E. (eds) *Deep drillling in the Atlantic ocean: ocean crust*. American Geophysical Union, Maurice Ewing Series, **2**, 230–238.

CHARLOU, J. L. & DONVAL, J. P. 1993. Hydrothermal methane venting between 12°N and 26°N along the Mid-Atlantic Ridge. *Journal of Geophysical Research*, **98**, 9625–9642.

CRANE, K., AIKMAN, F., III, EMBLEY, R., HAMMOND, S., MALAHOFF, A. & LUPTON, J. 1985. The distribution of geothermal fields on the Juan de Fuca Ridge. *Journal of Geophysical Research*, **90**, 727–744.

DELANEY, J. R., ROBIGOU, V., McDUFF, R. E. & TIVEY, M. K. 1992. Geology of a vigorous hydrothermal system on the Endeavor segment, Juan de Fuca ridge. *Journal of Geophysical Research*, **97**, 19 663–19 682.

DETRICK, R. S., BUHL, P., VERA, E., ORCUTT, J., MADSEN, J. & BROCHER, T. 1987. Multi-channel seismic imaging of a crustal magma chamber along the East Pacific Rise. *Nature*, **326**, 35–41.

——, HARDING, A. J., KENT, G. M., ORCUTT, J. A., MUTTER, J. C. & BUHL, P. 1993. Seismic structure of the southern East Pacific Rise. *Science*, **259**, 499–503.

DZIAK, R. P., FOX, C. G. & SCHREINER, A. E. 1995. The June–July 1993 seismo-acoustic event at CoAxial segment, Juan de Fuca ridge: evidence for a lateral dike injection. *Geophysical Research Letters*, **22**, 135–138.

EMBLEY, R. W., CHADWICK, W. W., JR., JONASSON, I. R., BUTTERFIELD, D. A. & BAKER, E. T. 1995. Initial results of the rapid response to the 1993 CoAxial event: relationships between hydrothermal and volcanic processes. *Geophysical Research Letters*, **22**, 143–146.

FEELY, R. A., GENDRON, J. F., BAKER, E. T. & LEBON, G. T. 1994. Hydrothermal plumes along the East Pacific Rise, 8°40′ to 11°50′N: particle distribution and composition. *Earth and Planetary Science Letters*, **128**, 19–36.

FORNARI, D. J., HAYMON, R. M., EDWARDS, M. H. & MACDONALD, K. C. 1990. Volcanic and tectonic characteristics of the East Pacific Rise crest, 9°09′N to 9°54′N: implications for fine-scale segmentation of the plate boundary. *EOS, Transaction of American Geophysical Union*, **71**, 625.

GERMAN, C. R., BAKER, E. T. & KLINKHAMMER, G. 1995. The regional setting of hydrothermal activity. *In:* PARSON, L. M., WALKER, C. L. & DIXON, D. R. (eds) *Hydrothermal Vents and Processes*. Geological Society, London, Special Publications, **87**, 3–16.

——, BRIEM, J., CHIN, C., DANIELSON, M., HOLLAND, S., JAMES, R., JONSDOTTIR, A., LUDFORD, E., MOSER, C., OLAFSSON, J., PALMER, M. R. & RUDNICKI, M. D. 1994. The Steinaholl vent field at 63°06′N. *Earth and Planetary Science Letters*, **121**, 647–654.

GREGORY, R. T. & TAYLOR, H. P. 1981. An oxygen isotope profile in a section of Cretaceous oceanic crust, Samail ophiolite, Oman: evidence of $\delta^{18}O$ buffering of the oceans by deep (>5 km) seawater-hydrothermal circulation at mid-ocean ridges. *Journal of Geophysical Research*, **86**, 2737–2755.

HARDING, A. J., KENT, G. M. & ORCUTT, J. A. 1993. A multichannel seismic investigation of upper crustal structure at 9°N and on the East Pacific Rise: implications for crustal accretion. *Journal of Geophysical Research*, **98**, 13 925–13 944.

HAYMON, R. M., FORNARI, D. J., EDWARDS, M., CARBOTTE, S. M., WRIGHT, D. & MACDONALD, K. C. 1991. Hydrothermal vent distribution along the East Pacific crest (9°09′–54′N) and its relationship to magmatic and tectonic processes on fast spreading mid-ocean ridges, *Earth Planetary Science Letters*, **104**, 513–534.

——, ——, LUTZ, R., VON DAMM, K., PERFIT, M., LILLEY, M., SHANKS, W. C., III, MACDONALD, K.,

EDWARDS, M., NELSON, D., COLODNER, D., KAPPUS, M., WRIGHT, D., BLACK, M., SCHEIRER, D., EDMONDS, H., OLSON, E. & GEISELMAN, T. 1992. 1991 eruption site on the East Pacific Rise at 9°45'–52'N is evolving rapidly: results of AdVenture '92 dive series. *RIDGE Events*, **3**, 1–2.

——, ——, VON DAMM, K. L., LILLEY, M. D., PERFIT, M. R., EDMOND, J. M., SHANKS, W. C., III, LUTZ, R. A., GREBMEIER, J. M., CARBOTTE, S., WRIGHT, D., MCLAUGHLIN, E., SMITH, M., BEEDLE, N. & OLSON, E. 1993. Volcanic eruption of the mid-ocean ridge along the East Pacific Rise crest at 9°45–52'N: direct submersible observations of seafloor phenomena associated with an eruption event in April, 1991. *Earth and Planetary Science Letters*, **119**, 85–101.

——, KOSKI, R. & ABRAMS, M. 1989. Hydrothermal discharge zones beneath massive sulfide deposits mapped in the Oman ophiolite. *Geology*, **17**, 531–535.

——, ——, & STAKES, D. 1990. *Field guide to massive sulfide deposits in N. Oman*. Ministry of Petroleum and Minerals, Sultanate of Oman.

HEEZEN, B. & THARP, M. 1977. *World Ocean Floor Panorama*, US Office of Naval Research, Washington, DC.

HUXLEY, T. H. 1880. *The Coming of Age of the Origin of Species*.

JUNIPER, S. K., MARTINEU, P., SARRAZIN, J. & GELINAS, G. 1995. Microbial-mineral floc associated with nascent hydrothermal activity on CoAxial segment, Juan de Fuca ridge. *Geophysical Research Letters*, **22**, 179–181.

KARSON, J. A. & RONA, P. A. 1990. Block-tilting, transfer faults, and structural control of magmatic and hydrothermal processes in the TAG area, Mid-Atlantic Ridge 26°N. *Geological Society of America Bulletin*, **102**, 1635–1645.

KENT, G. M., HARDING, A. J. & ORCUTT, J. 1993. Distribution of magma beneath the East Pacific Rise near the 9°03'N overlapping spreading center from forward modelling of CDP data. *Journal of Geophysical Research*, **98**, 13 971–13 996.

KLINKHAMMER, G., RONA, P., GREAVES, M. & ELDERFIELD, H. 1985. Hydrothermal manganese plumes in the Mid-Atlantic Ridge rift valley. *Nature*, **314**, 727–731.

LANGMUIR, C. M., BENDER, J. F. & BATIZA, R. 1986. Petrological and tectonic segmentation of the East Pacific Rise, 5°30'–14°30'N. Nature, **322**, 422–429.

LILLEY, M. D., OLSON, E. & LUPTON, J. 1994. The behavior of carbon dioxide, hydrogen, and methane in nascent hydrothermal systems. *EOS, Transactions of the American Geophysical Union*, **75**, 618.

LISTER, C. R. B. 1977. Qualitative models of spreading center processes, including hydrothermal penetration. *Tectonophysics*, **37**, 203–218.

LOCKE, J. 1690. *Essay concerning human understanding*. Chapter 1, Section 19.

LONSDALE, P. 1985. Non-transform offsets of the Pacific-Cocos plate and their traces on the rise flank. *Geological Society of American Bulletin*, **96**, 313–327.

LUPTON, J. E., BAKER, E. T., MASSOTH, G. J.,

THOMSON, R. E., BURD, B. J., BUTTERFIELD, D. A., EMBLEY, R. W. & CANNON, G. A. 1995. Variations in water column helium-3/heat ratios associated with the 1993 CoAxial event, Juan de Fuca ridge. *Geophysical Research Letters*, **22**, 155–158.

——, ——, MOTTL, M. J., SANSONE, F. J., WHEAT, C. G., RESING, J. A., MASSOTH, G. J., MEASURES, C. I. & FEELY, R. A. 1993. Chemical and physical diversity of hydrothermal plumes along the East Pacific Rise, 8°45'N to 11°50'N. *Geophysical Research Letters*, **24**, 2913–2916.

LUTZ, R. & HAYMON, R. 1994. Rebirth of a deep sea vent. *National Geographic Magazine*, **186**, 115–126.

——, SHANK, T. M., FORNARI, D. J., HAYMON, R. M., LILLEY, M. D., VON DAMM, K. L. & DESBRUYERES, D. 1994. Rapid growth at deep-sea vents. *Nature*, **371**, 663–664.

MACDONALD, K. C. *et al.* 1992. The East Pacific Rise and its flanks 8–18°N: history of segmentation, propagation, and spreading direction based on SEAMARC II and Sea Beam studies. *Marine Geophysical Researches*, **14**, 299–344.

——, FOX, P. J., PERRAM, L. J., EISEN, M. F., HAYMON, R. M., MILLER, S. P., CARBOTTE, S. M., CORMIER, M.-H. & SHOR, A. N. 1988. A new view of the mid-ocean ridge from the behaviour of ridge-axis discontinuities. *Nature*, **335**, 217–225.

——, SCHEIRER, D. S., CARBOTTE, S. & FOX, P. J. 1993. It's only topography. *GSA Today*, **3**, 24–25, 29–30, 34–35.

——, SEMPERE, J.-C. & FOX, P. J. 1984. East Pacific Rise from Siquieros to Orozco fracture zones: along-strike continuity of axial neovolcanic zone and structure and evolution of overlapping spreading centers. *Journal of Geophysical Research*, **89**, 6049–6069.

MACLEOD, C. J. & ROTHERY, D. A. 1992. Ridge axial segmentation in the Oman ophiolite: evidence from along-strike variations in the sheeted dyke complex. *In*: PARSON, L. M., MURTON, B. J. & BROWNING, P. (eds) *Ophiolites and their Modern Ocean Analogues*. Geological Society, London, Special Publications, **60**, 39–63.

MCCLAIN, J. S., ORCUTT, J. A. & BURNETT, M. 1985. The East Pacific Rise in Cross Section: A Seismic Model. *Journal of Geophysical Research*, **90**, 8627–8640.

MILTON, J. 1644. *Areopagitica*.

MURTON, B. J., KLINKHAMMER, G., BECKER, K., BRIAIS, A., EDGE, D., HAYWARD, N., MILLARD, N., MITCHELL, I., ROUSE, I., RUDNICKI, M., SAYANAGI, K., SLOAN, H. & PARSON, L. 1994. Direct evidence for the distribution and occurrence of hydrothermal activity between 27°N–30 degs N on the Mid Atlantic Ridge. *Earth and Planetary Science Letters*, **125**, 119–128.

——, VAN DOVER, C. & SOUTHWARD, E. 1995. Geological setting and ecology of Broken Spur hydrothermal vent field: 29.1°N on the Mid-Atlantic Ridge. *In*: PARSON, L. M., WALKER, C. L. & DIXON, D. R. (eds) *Hydrothermal Vents and Processes*. Geological Society, London, Special Publications, **87**, 33–41.

NEHLIG, P. & JUTEAU, T. 1988. Flow porosities, permeabilities, and preliminary data on fluid inclusions and thermal gradients in the crustal sequence of the Sumail ophiolite (Oman). *Tectonophysics*, **151**, 199–221.

PALLISTER, J. S. & HOPSON, C.A. 1981. Semail ophiolite plutonic suite: field relations, phase variations, cryptic variation and layering, and a model of a spreading ridge magma chamber. *Journal of Geophysical Research*, **86**, 2593–2644.

RICHARDSON, C. J., CANN, J. R., RICHARDS, H. G. & COWAN, J. G. 1987. Metal depleted root-zones of the Troodos ore-forming hydrothermal systems, Cyprus. *Earth and Planetary Science Letters*, **87**, 243–254.

ROSENBERG, N. D., SPERA, F. J. & HAYMON, R. M. 1993. The relationship between flow and permeability field in seafloor hydrothermal systems. *Earth and Planetary Science Letters*, **116**, 135–153.

SANTAYANA, G. 1896. *The Sense of Beauty, Part III: Form.*

SCHEIRER, D. S. & MACDONALD, K. C. 1993. Variation in cross-sectional area of the axial ridge along the East Pacific Rise: evidence for the magmatic budget of a fast spreading center. *Journal of Geophysical Research*, **98**, 7871–7885.

SCHOUTEN, H., KLITGORD, K. D. & WHITEHEAD, J. A. 1985. Segmentation of mid-ocean ridges. *Nature*, **317**, 225–229.

SHANK, T. M., FORNARI, D. J., LUTZ, R. A., HAYMON, R. M., VON DAMM, K. L., LILLEY, M. D., PERFIT, M. R., EDWARDS, M. H., SHANKS, W. C., III, NELSON, D., VAN DOVER, C., BLACK, M. & DESBRUYERES, D. Establishment of long-term observatory at nascent hydrothermal vents on the East Pacific Rise crest at 9°49'–50'N: baseline observations. *Journal of Geophysical Research*, in press.

SLEEP, N. H. 1975. Formation of oceanic crust: Some thermal constraints. *Journal of Geophysical Research*, **80**, 4037–4042.

—— 1984. A hydrothermal convection at ridge axes, *In*: RONA, P. A., BOSTROM, K., LAUBIER, L. & SMITH, K., JR. (eds) *Hydrothermal Processes at Seafloor Spreading Centers*. Plenum Press, 71–82.

SPOONER, E. T. C. & FYFE, W. S. 1973. Sub-seafloor metamorphism, heat, and mass transfer. *Contributions to Mineralogy and Petrology*, **42**, 287–304.

STRENS, M. R. & CANN, J. R. 1986. A fracture-loop thermal balance model of black smoker circulation. *Tectonophysics*, **122**, 307–324.

TALWANI, M., WINDISCH, C. C. & LANGSETH, M. G. 1971. Reykjanes ridge crest: a detailed geophysical study. *Journal of Geophysical Research*, **76**, 473–517.

TOOMEY, D. R., PURDY, G. M., SOLOMON, S. C. & WILCOCK, W. S. D. 1990. The three-dimensional seismic velocity structure of the East Pacific Rise near latitude 9°30'N. *Nature*, **347**, 639–645.

URABE, T., BAKER, E. T. *et al.* 1995. The effect of magmatic activity on hydrothermal venting along the superfast-spreading East Pacific Rise. *Science*, **269**, 1092–1095.

VERA, E. E. & DIEBOLD, J. B. 1994. Seismic imaging of oceanic layer 2A between 9°30'N on the East Pacific Rise from two-ship wide-aperture profiles. *Journal Geophysical Research*, **99**, 3031–3042.

VON DAMM, K. L., OOSTING, S. E. & BUTTERMORE, L. G. 1994. Chemical evolution of hydrothermal fluids at 9–10 degs.N EPR since the 1991 eruption: 1994 results. *EOS, Transaction of the American Geophysical Union*, **75**, 618.

——, OOSTING, S. E., KOZLOWSKI, R., BUTTERMORE, L. G., COLODNER, D. C., EDMONDS, H. N., EDMOND, J. M. & GREBMEIER, J. M. 1995. Evolution of East Pacific Rise hydrothermal vent fluids following a volcanic eruption. *Nature*, **375**, 47–50.

WILLIAMS, D. L., VON HERZEN, R. P., SCLATER, J. & ANDERSON, R. N. 1974. Galapagos spreading center: Lithospheric cooling and hydrothermal circulation. *Geophysical Journal of Royal Astronomical Society*, **38**, 587–608.

WILSON, C., SPEER, K., CHARLOU, J. L., BOUGAULT, H. & KLINKHAMMER, G. Mid-Atlantic Ridge hydrography and anomalies at Lucky Strike. *Journal of Geophysical Research*, in press.

WOLERY, T. J. & SLEEP, N. H. 1976. Hydrothermal circulation and geothermal flux at mid-ocean ridges. *Journal of Geophyscial Research*, **84**, 249–275.

WRIGHT, D. J., HAYMON, R. M. & FORNARI, D. J. 1995a. Crustal fissuring and its relationship to magmatic and hydrothermal processes on the East Pacific Rise crest (9 degs.12' to 54°N). *Journal of Geophysical Research*, **100**, 6097–6120.

——, —— & MACDONALD, K. C. 1995b. Breaking new ground: estimates of crack depth along the axial zone of the East Pacific Rise (9 degs.12'–54'N). *Earth and Planetary Science Letters*, **134**, 441–457.

Hydrothermal activity and ridge segmentation on the Mid-Atlantic Ridge: a tale of two hot-spots?

C. R. GERMAN[1], L. M. PARSON[1], B. J. MURTON[1] & H. D. NEEDHAM[2]

[1]*Southampton Oceanography Centre, Southampton SO14 3ZH, UK*
[2]*IFREMER Centre de Brest, BP70, 29263 Plouzané, France*

Abstract: In the past five years we have completed systematic seabed and water column surveys of two key sections of the Reykjanes Ridge and the Mid-Atlantic Ridge, between 36° and 38° North, close to the Iceland and Azores hot-spots, respectively. These surveys have provided a data set which provides a unique opportunity to address the relationship between the incidence and location of hydrothermal venting relative to volcanism/tectonism and ridge segmentation. Along the Reykjanes Ridge, running SW from Iceland, deep-tow (TOBI) sidescan sonar and swath bathymetry data have indicated a pattern dominated by volcanism in the form of axial volcanic ridges with only minor tectonic fracturing. A water column survey along this section of ridge crest, using conventional profiling techniques, indicated no evidence for high-temperature hydrothermal actvity except at 63°N where the Reykjanes Ridge intercepts the bathymetric platform ($\leq$500 m water depth) which surrounds Iceland. Along the Mid-Atlantic Ridge between 36–38°N, running SW from the Azores archipelago, TOBI and swath bathymetry data have imaged a ridge with spreading rate and obliquity of spreading motion similar to the Reykjanes Ridge. But this ridge-section is dominated by tectonism rather than volcanism, the axis being partitioned into a series of short (20–55 km length) segments of spreading axis, each systematically displaced from its neighbours by right-lateral non-transform offsets of 15–50 km. Hydrothermal activity (as detected by a combination of deep-tow instruments and conventional profiling) is much more common along this latter section of spreading centre: evidence for discrete sources of hydrothermal discharge was obtained on fourteen different occasions, in seven separate segments of ridge crest. Sources of hydrothermal activity, including the major Rainbow hydrothermal field, were observed to coincide consistently with ridge-offset intersections suggesting a strong link between tectonic segmentation and the occurrence of high temperature venting.

High temperature hydrothermal activity was first discovered at 21°N on the East Pacific Rise in 1979 as part of a joint US-French deep-diving submersible expedition to that section of the fast-spreading ridge-crest (Spiess *et al.* 1980). Since that time, a wealth of further hydrothermal vent sites have been discovered associated with both fast and slow-spreading sections of the global mid-ocean ridge system (e.g. German *et al.* 1995a; Baker *et al.* 1995). A key advance in attempts to discover and locate new hydrothermal vent fields has been the recognition that high-temperature venting gives rise to pronounced chemical and physical enrichments in the neutrally buoyant hydrothermal plumes which are generated in the overlying water column, 100–500 m above the seabed, and that these plume signals can be detected several miles distant (Baker *et al.* 1995). Typical plume enrichments include high concentrations of suspended particulate matter and high concentrations of the dissolved tracers ^{3}He, Mn and CH_4 when compared to background open-ocean

seawater. Consequently, use of in situ optical sensors (nephelometer, transmissometer) and shipboard measurements of dissolved tracers (Mn, Si, CH_4) have been adopted widely as a method of prospecting for hydrothermal plumes in the water column overlying a section of ridge crest (Baker *et al.* 1995). This approach has proved much more efficient for locating new sites of hydrothermal activity than simple visual inspection of the entire seafloor, whether through submersible operations or use of deep-tow camera systems. What has remained elusive, to date, has been the capacity to determine where hydrothermal activity might be expected to occur with respect to volcanic and tectonic processes at mid-ocean ridge crests.

Swath sonar and multibeam bathymetry mapping over the past two decades has enabled mid-ocean ridge researchers to determine the detailed geometry and spatial organization of slow spreading plate boundaries. At a finer scale than the large-offset transform fault zones which interrupt spreading axes with major displace-

From: MacLeod, C. J., Tyler, P. A. & Walker, C. L. (eds) 1996, *Tectonic, Magmatic, Hydrothermal and Biological Segmentation of Mid-Ocean Ridges*, Geological Society Special Publication No. 118, pp. 169–184.

ments, it is widely accepted that most ridges opening at less that $40 \, \text{mm} \, \text{a}^{-1}$ (full rate) comprise series of numerous, discrete sections of active spreading axis, ranging in length from a few kilometres to tens of kilometres, which are offset or separated from one another by lineaments, or ridge discontinuities. These sections are referred to as 'second-order' ridge segments, after their classification, largely in terms of their scale, into a hierarchy or 'order' of segmentation (Macdonald *et al.* 1988, 1991). Some of the factors which control the degree to which a 'typical' ridge is partitioned in this way are believed to include variations in spreading rate, spreading direction and spreading symmetry, as well as spatial and temporal variations in magmatic supply. Debate continues on the relative significance of each of these parameters (see e.g. Macdonald *et al.* 1988, 1991; Semperé *et al.* 1990, 1993).

For the purposes of this paper we can summarize current knowledge on second order segmentation by generalizing that the central portion of the second-order segments is considered to be supplied with a relatively continuous magmatic source, which decreases towards segment ends. It is assumed that there is a progressive and concomitant increase in extensional processes dominated by tectonic processes such as faulting and crustal attenuation, towards the segment terminations. Within the offsets, there is normally a complete absence of fabrics associated with strike-slip (transform) motion which leads to them being referred to as 'non-transform discontinuities' (NTDs), but they nonetheless accommodate transverse motion between adjacent plates. Off-axis, the NTD traces frequently display rapidly varying orientations, which indicate that fluctuations in magmatism within the bounding segments can lead to periodic propagation/retreat of segment ends. The NTDs commonly occupy a wider offset zone than might be expected to be generated by their relatively small displacement, which we consider to be an indication of instability in the ridge spreading processes (Parson *et al.* 1994). It is also recognised that regional variations in thermal structure of the ridge, such as its proximity to a hotspot, can affect the segmentation style. Many workers have discussed the lack of NTDs along the Reykjanes Ridge southwest of Iceland (Talwani *et al.* 1971; Appelgate & Shor 1994), but point out that a form of segmentation, probably derived from focussed magmatism along the axis, persists (Searle & Laughton 1981; Parson *et al.* 1993). Here, volcanic constructional centres form as an array of en echelon ridges at intervals

of a few kilometres along the axis. The spacing and dimensions of these axial volcanic ridges (AVRs) is similar to the more 'typical' segments described above, but their geometry along this oblique spreading centre is quite different (Parson *et al.* 1993; Murton & Parson 1993).

It would be an important advance if some coherent relationship between the incidence of hydrothermal activity and styles of tectonic segmentation could be established because remote sensing by satellite can provide, at least, a base map of ridge geometry and, hence, some indications of the volcanic and tectonic processes active around the entire global ridge-crest system, including many remote areas inaccessible by research vessels. This rapid and comprehensive geophysical surveying cannot be directly matched by present methods of water column investigations and/or sampling.

The present study focuses upon the relationships between hydrothermal activity, tectonism and magmatism, in the context of hot-spot activity, along the slow-spreading Mid-Atlantic Ridge (MAR) southwest from the Azores Triple Junction and along the Reykjanes Ridge southwest from Iceland. In both areas we have completed combined seabed investigations using swath bathymetry and IOSDL's deep-towed sidescan sonar instrument TOBI, complemented by optical and chemical investigations of the overlying water column to prospect for neutrally buoyant hydrothermal plume signals. Following this strategy, our rationale has been to identify new sites of hydrothermal activity and simultaneously determine their volcano-tectonic setting on the underlying seabed. Our objective has been to provide better constraints upon the location of high-temperature venting on slow-spreading mid-ocean ridges and, in so doing, to develop improved predictive models to aid in future mid-ocean ridge exploration.

Reykjanes Ridge v. Mid-Atlantic Ridge

Regional Tectonics

Both of our study areas, the Reykjanes Ridge between 57° and 63°N and the Mid-Atlantic Ridge between 36° and 38°N, represent slow-spreading ridges which are oriented obliquely to the regional spreading directions of the adjacent plates, as derived from the NUVEL model (de Mets *et al.* 1990; Sloan & Patriat 1992). Both sections also lie within the influence of hot-spot activity (Fig. 1a & b). In the case of the Reykjanes Ridge, the ridge orientation is 036° whilst the Eurasian–North American plate motion is $10 \, \text{mm} \, \text{a}^{-1}$ (half-rate) in a direction of

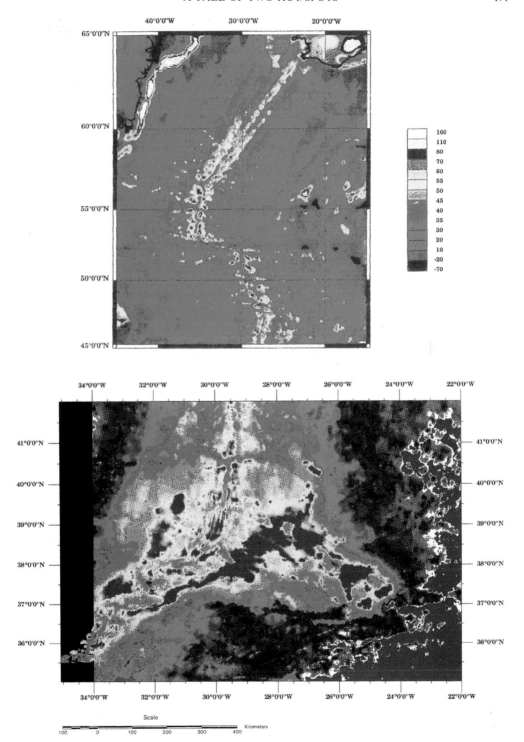

Fig. 1. (**a**) Bathymetric compilation for the Reykjanes Ridge, derived from satellite gravity data (Strange 1991). (**b**) Swath bathymetric compilation of the Mid-Atlantic Ridge close to the Azores Triple Junction (Needham *et al.* 1992).

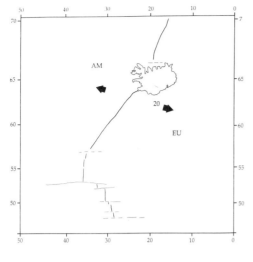

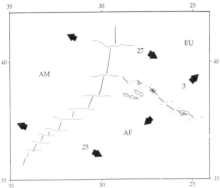

Fig. 2. Schematic representations of ridge orientations and plate tectonic motions (spreading rates and directions) at (**a**) the Reykjanes Ridge and (**b**) the Azores Triple Junction. AM, North American plate; EU, Eurasian plate; AF, African plate. Full spreading rates indicated are in mm/year.

099° (Table 1, Fig. 2a). This yields a plate-boundary–ridge obliquity angle of 27°. The Mid-Atlantic Ridge between 36° and 38°N trends 045° whilst the African–North American plate motion at this latitude is 12 mm a^{-1} (half spreading-rate) along 110°, yielding a plate-boundary–ridge obliquity angle of 25° (Table 1, Fig. 2b).

An important difference between the two study areas, readily discernible from the bathymetry, is the degree of second-order ridge segmentation which occurs (Fig. 1). Along the 800 km of the Reykjanes Ridge, second-order ridge segmentation, which normally character-izes slow-spreading ridges, is completely absent (Parson *et al.* 1993). This is seen in the broad-scale bathymetry of the ridge-crest which exhibits only a gently undulating topographic profile, becoming increasingly deeper at greater distances away from Iceland (Fig. 1a). By contrast, second-order ridge segmentation is well defined and relatively pronounced along the MAR between 36° and 38°N (Needham *et al.* 1992; Detrick *et al.* 1995). There, the ridge is characterized by a number of well developed axial rift valley segments (picked out as linear areas coloured blue, Fig. 1b), oriented roughly orthogonal to the plate spreading direction. The rift-axis is oblique to the general orientation of the plate boundary, but the obliquity is accom-modated by a series of right-stepping second-order non-transform discontinuities (NTDs) which separate segments of ridge crest 20–55 km in extent. Displacement along the NTDs ranges from 15–50 km.

Incidence of hydrothermal activity

A systematic programme studying the water column overlying the Reykjanes Ridge for hydrothermal signals included occupation of at least one conventional CTD (conductivity–temperature–depth)-nephelometer–transmisso-meter station every 15 km along the entire ridge crest between 57°45'N and 63°10'N (German 1993). At each station a 12-bottle profile of water samples was taken for dissolved Si and Mn $\pm$ CH$_4$ and H$_2$ analyses (German *et al.* 1994). Sampling was focussed in the deep portion of the water column, with bottles typically fired at 50 m intervals between 50 and 600 m off bottom, to maximize our opportunity of intercepting any neutrally-buoyant hydrothermal plume. (As described above, all previously reported stable hydrothermal plumes overlying mid-ocean ridges have occured at depths between 100 and 350 m off-bottom; Baker *et al.* 1995). Real-time nephelometer and transmissometer data were monitored continuously for evidence of anomal-ous suspended particle enrichments. Peculiar to the Reykjanes Ridge, a bottom nepheloid layer was seen at all stations, unlike studies from the Mid-Atlantic Ridge rift-valley, farther south, which have yielded uniform low background nephelometer traces indicating optically clear deep-waters (e.g. Nelsen *et al.* 1986). The transmissometer and nephelometer anomalies observed on the Reykjanes Ridge increased monotonously towards the seabed, unlike a hydrothermal plume which would exhibit anomalous optical properties in a discrete band, 50–100 m thick, 100–350 m above seabed. Further, water column samples collected from within the bottom nepheloid layer exhibited no

Table 1. *Ridge segmentation and hydrothermal activity: key characteristics*

	Reykjanes Ridge	Azores Triple Junction
Half-spreading rate	$10 \, mm \, a^{-1}$	$12 \, mm \, a^{-1}$
Spreading direction	099°	110°
Ridge orientation	036°	045°
Obliquity	27°	25°
Segmentation	Not pronounced	Pronounced second-order
Hydrothermal activity	Rare	Abundant
Hot-spot activity	Present	Present?

anomalous Mn, Si, H_2 or CH_4 anomalies, confirming that this was not a hydrothermal feature.

Within the *c.* 800 km Reykjanes Ridge study section, more detailed sampling was completed in those areas where seafloor data for swath bathymetry and sidescan sonar and/or other geophysical data had previously been collected (Kuznetsov *et al.* 1985; Parson *et al.* 1993). Sampling was targetted every 3–5 km along the crests of individual AVRs and also offset by 2–4 km, either side of these volcanic features, to east and west. It was particularly important to collect samples from across the ridge-axis as well as along axis because of concern that deep-ocean currents passing from west to east or from east to west across the ridge axis might disperse currents away from the ridge. This had not been a concern in our previous studies from the Mid-Atlantic Ridge (e.g. TAG at 26°N) because there the height of rise of the neutrally buoyant plume was constrained well within the bounding topography of the rift-valley walls (Klink-hammer *et al.* 1985, 1986). Along the Reykjanes Ridge, where a rift valley is poorly developed or absent, such is not the case.

In areas where seafloor data had not previously been obtained, a coarser sampling resolution of one station every 15 km was used. This coarser sampling resolution, nevertheless, represented at least a 33% increase on strategies employed previously along the similarly slow-spreading MAR, 11–26°N, which had yielded evidence for at least one discrete site of hydrothermal activity every 150–175 km along axis. Further, within the three 100 km sections of Reykjanes Ridge crest selected for detailed investigation in this work, the sampling frequency applied represented an increase to 300–600% of earlier work. Nevertheless, throughout the study, which comprised 175 separate CTD-nephelometer–transmissometer sampling stations (Fig. 3), evidence for high-temperature hydrothermal activity was found at only one site (German *et al.* 1994). This indicates

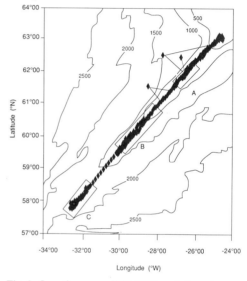

Fig. 3. Location map of 175 CTD-nephelometer-transmissometer sampling stations along the Reykjanes Ridge. Areas 'A', 'B' and 'C' represent areas of detailed earlier study using TOBI and swath bathymetry (Parson *et al.* 1993).

an anomalously low incidence of hydrothermal venting along the Reykjanes Ridge.

The one site at which evidence for hydrothermal activity was detected was the previously reported Steinahóll vent site at 63°06′N (Olafsson *et al.* 1991). An important part of our study, therefore, was to complete a systematic survey of the extent over which this plume remained detectable, as it was dispersed. Employing the same techniques applied the length of the Reykjanes Ridge between 58°N and 63°N, our survey showed that dissolved Mn, CH_4 and H_2 plume-anomalies were resolvable over distances of 7–11 km away from the source of active venting (German *et al.* 1994). The Steinahóll vent-field is located close to the summit of an AVR, in just 300 m of water, with a plume height

of 150 m. This configuration should make dispersion of the Steinahóll plume extremely susceptible to strong near-surface currents. We consider
this area to represent a good end-member
location in which to have tested the sensitivity of
the shipboard analytical techniques employed
along the entire Reykjanes Ridge, therefore,
because it is extremely unlikely that any stronger
prevailing current flow would have been encountered farther south, in increasingly deeper
waters. Given the sensitivity of the techniques
demonstrated during this work, therefore, it can
be seen that even at our coarsest spacing of one
CTD station every 15 km, no single station should
have been located more than 7.5 km from any
possible hydrothermal vent-field. Even then, if
the vent-field was at the very mid-point between
two consecutive CTD stations, any plume-
discharge from that vent-field should only have
been able to escape detection by the shipboard
techniques employed if the prevailing direction
of plume dispersal was normal to the strike of the
ridge-axis along which sampling was taken. We
believe the statistical probability of such factors
combining to prevent significant levels of 'black
smoker' hydrothermal activity from being detected to be extremely small and, indeed, negligible
within the three 100 km segments selected for
more detailed investigation.

In contrast, abundant evidence for hydrothermal activity has been detected along the
Mid-Atlantic Ridge between 36° and 38°N.
Preliminary vertical profiling for optical and
chemical plume anomalies was carried out in 1992
as part of the FARA (French American Ridge
Atlantic) programme, project FAZAR (Charlou
et al. 1993; Klinkhammer *et al.* 1993). This study
led to the discovery of the Lucky Strike
hydrothermal field at 37°17'N (Langmuir *et al.*
1993) and subsequent submersible dives, focussed initially at that location, also led to the
discovery of the Menez Gwen hydrothermal field
at 37°50'N (Fouquet *et al.* 1994). Importantly,
however, the initial FAZAR cruise also yielded
mid-water (200–500 m off-bottom) optical
(nephelometer and transmissometer) evidence
for hydrothermal plume signals in the central
AMAR segment and in the AMAR Minor and
South AMAR segments (Klinkhammer *et al.*
1993) and evidence for dissolved CH_4-rich
plumes in both the AMAR and FAMOUS
segments (Charlou *et al.* 1993).

Subsequent to those investigations, we have
made a systematic and continuous along-axis
survey through this section of the ridge-crest,
using a transmissometer mounted upon TOBI,
the SOC deep-tow sidescan sonar instrument
(German *et al.* 1995*b*). From this survey we have

identified at least seven different sources of
hydrothermal venting: in the southern Lucky
Strike segment, the North FAMOUS segment,
the southern FAMOUS segment, the central
AMAR segment, the southern AMAR segment, the AMAR Minor/South AMAR offset
and in the central South AMAR segment
(German *et al.* 1996). A brief summary map of
the local segmentation, the area covered by
TOBI insonification and the locations at which
TOBI transmissometer anomalies were detected
is shown in Fig. 4. In selected areas, the
preliminary transmissometer data have also
been ground-truthed using more conventional
CTD-nephelometer profiles to confirm the validity of the TOBI-transmissometer signals originally observed. Examples of CTD-
nephelometer profiles obtained from these three
sites, in the southern Lucky Strike segment, the
southern FAMOUS segment and the RAIN-
BOW area in the AMAR/AMAR Minor offset
are shown in Fig. 5. At both the Southern Lucky
Strike and Rainbow stations clear midwater
particle-rich anomalies are seen in the nephelometer profiles. Subsequent filtration of water
samples from these plume depths have revealed
high concentrations of material with high particulate Fe concentrations (E. Ludford, unpubl.
data) confirming their 'black smoker'-type
hydrothermal origin. In the case of the
FAMOUS profile, CTD-02 (Fig. 5b) clouding of
the optical windows of the nephelometer led to
an erroneously high background reading (see
mid-water nephelometer voltage of approx.
3.58 V versus correct background of 3.48 V
recorded at stations CTD-08 and CTD-10).
Consequently, the maximum anomaly recorded
in the FAMOUS profile at *c.*2300 m is relatively
obscured, although still exhibiting a departure
from clear-water background values. Additionally, the water column at this depth also shows
strong anomalies in dissolved CH_4 (Charlou *et
al.* 1993), total oxidizable Mn, as determined
from the OSU ZAPS probe (G. Klinkhammer,
unpubl. data), and particulate Fe concentrations
(E. Ludford, unpubl. data). All of this evidence
serves to support the validity of our TOBI-
transmissometer data-set as a reliable indicator
of the presence of black-smoker type hydrothermal plumes.

Of course, because TOBI only intercepts a
very restricted depth-range of the water column,
the level of activity represented by Fig. 1 must
nevertheless represent a minimum incidence
value. Although TOBI was typically 'flown' at
between 200 m and 500 m above the immediately
underlying seabed, (the optimum depth at which
to intercept any neutrally buoyant plume orig-

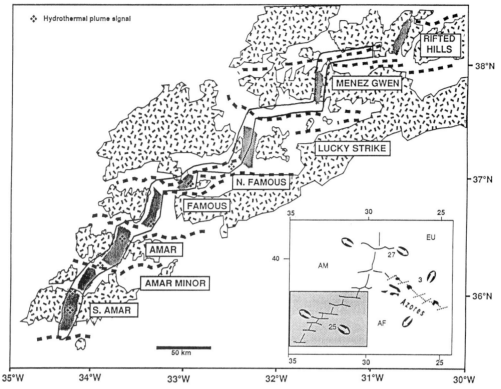

Fig. 4. Schematic illustration of the Mid-Atlantic Ridge, 36–38°N, showing the locations of the seven new sites of hydrothermal activity identified by TOBI transmissometer and their relationships to the tectonic characteristics of the underlying seabed. The area insonified by TOBI is outlined by a thick black line and the locations of the previously known hydrothermally active Lucky Strike and Menez Gwen segments are also shown at 37°17′N and 37°50′N, respectively.

inating from a source situated directly along the track of the TOBI tow-path) the roughness of the MAR seafloor topography is such that active venting may be located atop a neovolcanic ridge, off-set from the TOBI track, at a quite different elevation. A good example of this was observed during the CD89 cruise, in the South AMAR segment (German *et al.* 1995b). During a southward pass through the segment, TOBI was flown at 300 m off-bottom over the deepest portion of the rift-valley floor close to the west wall to obtain the best possible sidescan image from that area, even though we knew the deep-tow instrument to then be 150–200 m deeper than the depth at which hydrothermal transmissometer and nephelometer anomalies had previously been reported from this segment during the FAZAR expedition (Klinkhammer *et al.* 1993). During the return pass through this segment, from south to north, the TOBI instrument was towed 200–300 m shallower, along the top of a neovolcanic ridge, and the previously reported South AMAR plume was

duly intercepted. Although this represents one plume where a known signal was detected during a two-pass TOBI survey, there are many sections of ridge-crest where only a single pass was made, or where both passes may have failed to intercept the depth at which hydrothermal plume emissions were located. For this reason, we emphasize that the number of hydrothermal plume signals detected during the TOBI survey should be considered a minimum and the incidence of activity should therefore be considered to be at least one site of venting every 25–30 km along axis.

Of course, it can be argued that it is not valid to compare the continuous along-axis data set collected between 36° and 38°N on the MAR using TOBI with the punctuated point-sampling data set collected along the Reykjanes Ridge. But, even if one only considers the directly comparable survey carried out during the earlier FAZAR cruise (Klinkhammer *et al.* 1993), evidence for hydrothermal activity was still observed at least once every 150–175 km along

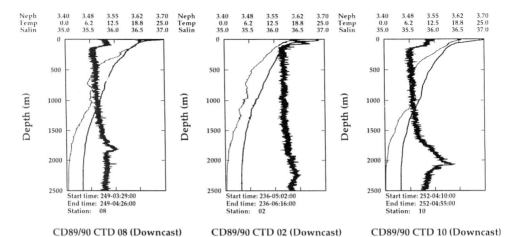

Fig. 5. Vertical profiles of salinity (thinnest line), temperature and nephelometer (heaviest line) data recorded from vertical profiles in the southern Lucky Strike area, 37°03′N 32°25′W (CTD 08), the southern FAMOUS segment, 36°38′N 33°20′W (CTD 02)and the Rainbow area in the AMAR/AMAR Minor offset, 36°16′N 33°46′W (CTD10). Note: the particle-rich nephelometer signal, 1750–1800 m, at station CTD 08 (southern Lucky Strike); the weak particle-rich nephelometer signal, 2250–2300 m, at station CTD 02 (FAMOUS); and the strong particle-rich nephelometer signal, 1700–2150 m, at station CTD 10 (Rainbow). Note also the increased background nephelometer reading (*c.* 3.58) at FAMOUS due to initial clouding of the nephelometer during this early cast (CTD 02) which may have obscured the true magnitude of the anomaly at this site when compared to the subsequent deployments at Lucky Strike and Rainbow (CTDs 08, 10; nephelometer background *c.* 3.48).

the MAR between 36° and 38°N. Even at this coarser resolution, the minimum incidence of venting detected along the MAR represents a pronounced increase over that recorded from the Reykjanes Ridge.

Hot-spot influence

The influence of hot-spot activity upon the volcanism and magmatism of both Iceland and the Azores has been well recognised for many years. Enhanced mantle potential temperatures beneath the ridge-crest have resulted in larger than normal volumes of melt generation for oceanic spreading and concomitant anomalous thickening of the overlying oceanic crust (White 1989). The combination of elevated bathymetry and coincident positive free-air gravity anomalies over these hot-spot regions of the North Atlantic has been used by Kaula (1971) to ascribe their origin to rising mantle plumes. The Reykjanes Ridge exhibits the more pronounced bathymetric hot-spot effect with a general tapering of the bathymetrically elevated crust over *c.* 800 km SW away from Iceland, to approximately 57°N (Fig. 1a). In contrast, bathymetrically elevated crust at the Mid-Atlantic Ridge southwest of the Azores tapers towards a constant width over a much shorter

distance. No anomalously shallow ridge crest is observed south of approximately 36°30′N (Fig. 1b). An alternative criterion for assessing the effect of the two hot spots can be seen by comparing the areas bound by (e.g.) the 1500 m isobath in the two regions. In the Azores region, the area bound by that isobath occupies less than one fifth of the corresponding Iceland area, even when the Iceland–Faroes and Iceland–Greenland ridges are excluded. According to both criteria, then, the Icelandic hot-spot can be said to be a much more active phenomenon than the corresponding Azores hot-spot.

The hot spot influence on the two ridge-sections can also be assessed from the geochemical effects of the enriched mantle that characterizes both the Azores and Icelandic regions (Schilling, 1973; Shirey *et al.* 1987). This is especially seen in elevated ratios of La_N/Sm_N, $^{87}Sr/^{86}Sr$ and low ϵ_{Nd}. Such signatures can be traced along the Reykjanes Ridge and the Mid-Atlantic Ridge away from their neighbouring hot-spots. Schilling (1973) reported coincident $^{87}Sr/^{86}Sr$ and La_N/Sm_N variations with distance from Iceland. Elevated ratios occurred close to Iceland but decreased progressively southwest to 59°N. Recent, more detailed studies have shown that this Icelandic geochemical signature can be traced as far south as 57°N

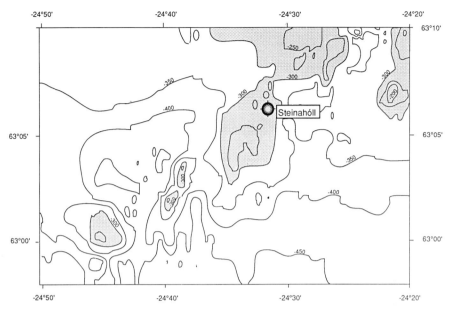

Fig. 6. Detailed bathymetric map of the Reykjanes Ridge showing the precise location of the Steinahóll hydrothermal field (63°06′N 24°32′W), located close to the axis of an approximately N–S-oriented axial volcanic ridge (AVR). Stipple indicates all areas shallower than 300 m and conveniently outlines the Steinahóll AVR, 63°02–08′N 24°30–34′W.

(Taylor *et al.* 1995). The Mid-Atlantic Ridge southwest of the Azores carries a geochemical signature characteristic of the Azores plateau (Schilling 1975). This high La_N/Sm_N ratio is traceable as far south as 34°N, although it is disrupted by the Oceanographer fracture zone at 35°30′N (Shirey *et al.* 1987).

Apart from the mantle source signatures that characterise hot-spot influences on spreading ridges, an increased volcanic flux also arises due to the elevated mantle potential temperatures beneath the hotspots which, hence, lead to higher degrees of partial melting. This also has an effect on the degree of crystal fractionation. The thicker oceanic crust in proximity to hot spots provides a longer ascent path before magmas erupt at the surface. This greater ascent path generally results in greater residence time in the crust for rising magmas, increasing their potential for crystal fractionation. In the case of Iceland, evidence of this can be seen in the strongly increasing $FeO/(FeO + MgO)$ ratios and the higher occurrence of three-phase assemblages (i.e. plagioclase, olivine and clinopyroxene) northeast along the Reykjanes Ridge towards Iceland (Schilling 1973; Taylor *et al.* 1995) reflecting an increasing dominance of eutectic fractional crytallisation. The same effect is only very poorly expressed along the Mid

Atlantic Ridge towards the Azores, where there is a slight increase in the modal proportion of clinopyroxene (Schilling 1975). In summary, the differences between the two ridges can be ascribed to a greater magmatic flux and thicker crust at the Icelandic hot-spot (Talwani *et al.* 1971) when compared to the Azores.

Fine-scale characteristics of the Reykjanes Ridge and MAR (36–38°N) ocean floor

Above, it has been demonstrated that the incidence of hydrothermal activity along the Reykjanes Ridge is apparently very rare (German *et al.* 1994). At the one site of activity that is known to occur, the Steinahóll hydrothermal field, high-frequency echo-sounding of emitted gas-rich plumes has indicated that the precise site of active venting lies close to the crest of an AVR (Fig. 6). This is very similar in geological situation to a number of other sites on the Mid-Atlantic Ridge, such as the Snake Pit at 23°N (Detrick *et al.* 1986) and Broken Spur at 29°N (Murton *et al.* 1994) hydrothermal fields which sit astride neovolcanic ridges located within the axial valley floor. Indeed, AVRs such as the one which provides the setting for the Steinahóll hydrothermal vent-field are characteristic of the entire Reykjanes Ridge. Here, the

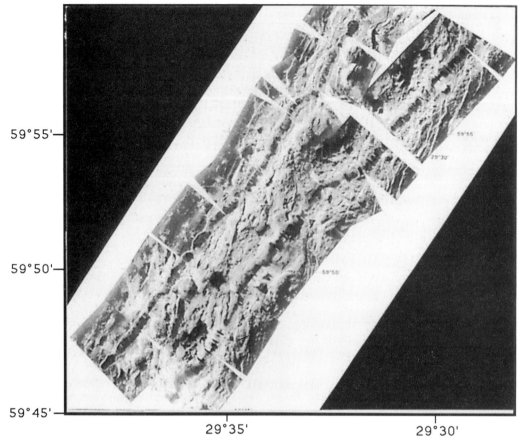

59°55'—

59°50'—

59°45'—

29°35'   29°30'

Fig. 7. (a) TOBI deep-tow sidescan sonograph and (b) line-drawn geological interpretation of a mature axial volcanic ridge situated close to 59°N on the Reykjanes Ridge. Bright tones in the sonograph correspond to strong backscattering, a function of signal incidence angle, surface roughness and seafloor physical properties. A range of volcanic constructional features are observed, including single point-source volcanoes, largely of a 'flat-topped' style, locally with collapsed calderas. Several of these exceed 1.5 km in diameter. These volcanoes are locally cut by faulting of a range of amplitudes and spacing. Also common are linear arrays of closely-spaced subordinate cones, and a widespread hummocky texture (stippled in (b)), interpreted as resulting from a compounded mounding of pillow lavas. The latter material forms the bulk of the AVRs (Axial Volcanic Ridges) prevalent along the Reykjanes Ridge, believed to relate to one of the most significant components of crustal accretion at all slow-spreading ridges.

axial volcanic ridges are abundant and arranged en echelon along the ridge axis (Parson *et al.* 1993; Murton *et al.* 1993). AVRs are believed to represent the fundamental building blocks of slow-spreading mid-ocean ridge crest and are the units which can be considered to mark a third order of ridge segmentation. On the Reykjanes Ridge especially, the AVRs appear to exhibit a cyclical life-history from initial fissure eruptions through subsequent stages of constructional growth, followed by magmatic starvation leading to tectonic dismemberment and, subsequently, renewal as magmatic supply is initiated once

more (Parson *et al.* 1993; Murton *et al.* 1993). Deep-tow sidescan data have indicated that volcanic constructional features and textures dominate the younger en echelon AVRs that characterise the Reykjanes Ridge axis, with subordinate tectonic fault-related fabrics. A good example is shown in Fig. 7 which portrays a TOBI sonograph of a mature axial volcanic ridge between 59°47'N and 59°57'N. Here, a range of volcanic constructional features are observed, including: single 'point source' volcanoes, largely of a 'flat-topped' style and locally with collapsed calderas; linear arrays of closely-

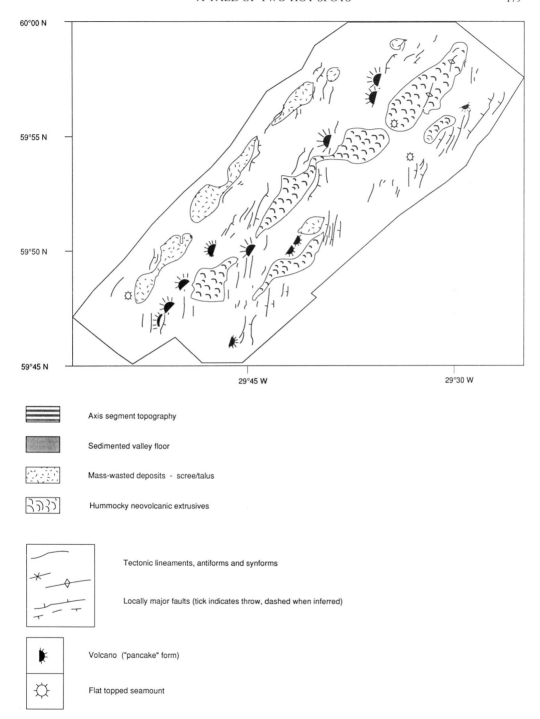

Axis segment topography

Sedimented valley floor

Mass-wasted deposits - scree/talus

Hummocky neovolcanic extrusives

Tectonic lineaments, antiforms and synforms

Locally major faults (tick indicates throw, dashed when inferred)

Volcano ("pancake" form)

Flat topped seamount

spaced subordinate cones; and a widespread hummocky texture which is interpreted to result from a compounded mounding of pillow lavas.

During our survey along the MAR between 36° and 38°N, the majority of hydrothermal fields coincided with segment terminations, rather than occurring close to segment centres (Fig. 4). TOBI insonification of these areas has

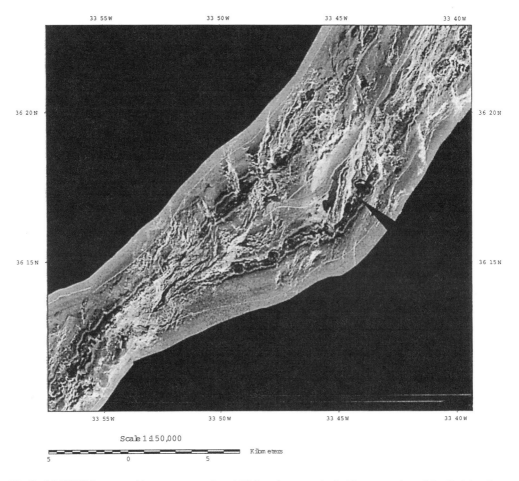

Fig. 8. (**a**) TOBI deep-tow sidescan sonograph and (**b**) line-drawn geological interpretation of the 'Rainbow' hydrothermal area, 36°10'–36°23'N, Mid-Atlantic Ridge, situated within the NTD between the AMAR and AMAR Minor segments. The acoustic pattern is dominated by linear, strongly backscattering surfaces which we interpret as complex arrays of fault scarps, often in cross-cutting and intersecting patterns. The faults cut a background, uniform, weakly-backscattering surface which we interpret as the sedimented floor of the NTD offsetting the Mid-Atlantic Ridge at this location. Broad areas of medium strength reflectivity close to the two isolated N–S ridges in the offset are likely to be mass-wasting erosional products associated with their degradation. Key as for Fig. 7.

indicated that they are characterised not by fresh volcanic/constructional features such as typify the Reykjanes Ridge, but by highly tectonised crust often overlain, at least in part, by accumulated sediments (German *et al.* 1996). A typical example is shown in Fig. 8 which portrays a TOBI sidescan sonograph and line-interpretation for the NTD situated between the southern limit of the AMAR segment and the northern end of the AMAR Minor segment, 36°10'–36°23'N, Mid-Atlantic Ridge. This tectonically dominated zone contains the Rainbow area, setting for the largest hydrothermal plume

signals yet detected in the North Atlantic and is characterised by a pervasive pattern of faulting in both north–south and northeast–southwest trends. This faulting cross-cuts both the widely sedimented axial floor and the subordinate areas of extrusive volcanics. Two relict blocks are observed within the NTD which exhibit features indicative of extensive faulting and mass-wasting. Preliminary investigations have indicated that hydrothermal venting is most probably associated with the more south-westerly of these two blocks – the 'Rainbow Ridge'.

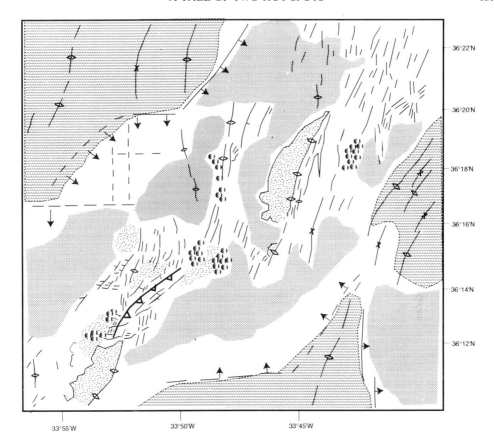

Discussion

Combining our observations of the Reykjanes Ridge and the Azores Triple Junction it seems clear that whilst there are many similarities in the regional tectonic settings of both sites, much more abundant hydrothermal activity is seen close to the Azores Triple Junction than is observed along the Reykjanes Ridge. This correlates with the much higher degree of second-order segmentation observed in the former rather than the latter. Study of our data set from the MAR between 36° and 38°N has revealed that hydrothermal venting is frequently associated with the fault structures characteristic of NTDs at the terminations of second order ridge segments (German *et al.* 1996). This is particularly borne out by sidescan sonographs in which the settings of active venting are seen to comprise highly tectonised regions of oceanic crust, populated by intersecting families of cross-cutting faults, often with an overlay of accumulated sediment, but in the complete absence of any fresh neovolcanic constructional features (Parson *et al.* 1994). This is in marked

contrast to TOBI sidescan sonographs of the Reykjanes Ridge which are dominated by the fresh eruptive features of AVRs (Parson *et al.* 1993) but from which hydrothermal activity, apparently, is largely absent (German *et al.* 1994).

It is clear, then, that hydrothermal activity in these ridge sections is not focussed primarily in those areas exhibiting the freshest and most recent neovolcanic activity. Rather, much hydrothermal circulation seems to be located within relatively old and highly fractured sections of mid-ocean ridge, indicating that an important geological control may well be the provision of pathways for fluid circulation in the underlying crust by enhanced tectonism.

Hydrothermal circulation in the upper oceanic crust (lavas and dykes) has long been established to derive from the liberation of thermal energy released from cooling magma. In its simplest conceptualisation, thermal contraction of fresh oceanic crust leads to fracturing which provides sufficient permeability for downward penetration of seawater (e.g. Lister 1974). These fluids are then heated near the interface between solidified intrusives and the cooling magma body (the 'cracking front') and advect heat out of the crust by rising buoyantly back to the ocean floor. The process of hydrothermal circulation is considered to be particularly efficient in cooling the upper oceanic crust because sheeted dyke complexes preserved in the geological record exhibit pervasive alteration and also uniformly exhibit chilled margins along their exteriors (Gass & Smewing 1973). This is taken as evidence that the upper oceanic crust at a ridge-axis is routinely cooled to temperatures less than 400°C between succesive dyke intrusions. Because this process is so efficent, and because successive dyke emplacements may be only intermittent at slow spreading ridges it is unlikely that continuous high-temperature hydrothermal fluid circulation through the upper oceanic crust should be sustained along the MAR between 36° and 38°N or along the Reykjanes Ridge.

An important tectonic setting where long-lived hydrothermal circulation might be expected on a slow-spreading ridge, however, would be in an environment where active deep-penetrating faulting allowed penetration of seawater into deeper levels of the underlying oceanic crust. We propose that the NTDs, which separate adjacent second order ridge-segments of the MAR between 36° and 38°N, may represent just such a tectonic environment. Continuous tectonic activity at these fault zones would repeatedly form new fluid pathways into the higher temperature host rocks of the upper and middle oceanic crust. Indeed, in recent work MacLeod & Manning (this volume) have reported evidence for just such deep-penetration of seawater into oceanic crustal layer 3 rocks from Hess Deep. They have used observations of low-temperature phases infilling the centres of veins initially lined with higher-temperature phases to argue that uniformly oriented segmentation-related fractures which penetrate deep, into the lower crust, have retained their permeability due to continuous tectonic extension. The uniform orientation of these structures and the longevity of fluid flow through the associated vein network (down to at least 150°C) have led MacLeod & Manning (this volume) to surmise that the structures are tectonically controlled and are intrinsically related to cross-axis displacement on the East Pacific Rise close to the end of a second-order segment.

We propose that directly comparable fractures arise in the NTDs between second-order segments of the MAR between 36° and 38°N and that it is the penetration of seawater along these fractures which gives rise to the high incidence of high temperature hydrothermal vent signals observed in the water column overlying such sites. Along the Reykjanes Ridge, no such NTDs defining second order ridge segmentation are observed north of 57°30'N. Therefore, although neovolcanic activity is abundant along the entire Reykjanes Ridge axis, constantly-active faulting which would penetrate deeply into the crust (i.e. layer 3) is not observed and the associated subsurface hydrothermal circulation cells required to support long-lived 'Black Smoker'-type activity are not established. Short-lived hydrothermal circulation might still be expected close to regions of most recent magmatic extrusion, but long-lived continuous venting would not be expected to be sustained.

Building upon this, however, the simple identification of a link in the North Atlantic between the frequency of occurrence of hydrothermal activity and degree of ridge segmentation does not explain why such pronounced differences should be observed between the Reykjanes Ridge and the MAR SW from the Azores Triple Junction. Given the close similarity in gross tectonic settings between the two areas (spreading rate, spreading direction, ridge orientation and ridge/spreading obliquity; Table 1) the most striking difference which can be identified is in the extent of hot-spot influence observed along the ridge axes. In the case of the Reykjanes Ridge the enhanced magma flux along its c. 800 km length, due to the proximity of the strongly active Icelandic hot-spot, provides sufficient volumes of melt to support abundant AVR construction, inhibiting deep brittle crustal failure and the development of second-order ridge segmentation. By contrast, increased magmatic flux along the MAR as a result of proximity to the Azores hot-spot is not as extensive. As a result, south of 38°N, pronounced second-order ridge segmentation is established and intense cross-cutting fracturing of the oceanic crust within the associated NTDs provides a tectonically active, highly permeable environment conducive to abundant high-temperature hydrothermal circulation.

As an interesting aside, independent research from the East Pacific Rise has recently led to an extremely similar set of conclusions. Haymon (this volume) has concluded that two types of hydrothermal setting can be identified along the EPR: those hosted by either (i) magma-rich or (ii) magma-starved segments. In our study, this translates directly into vent-sites that are dominated by either (i) neovolcanic or (ii) tectonically dominated terrains.

Summary and conclusions

A systematic study of the seafloor of two sections of slow spreading ridge-crest has been completed, combining swath bathymetric mapping with SOC's deep-tow sidescan sonar instrument, TOBI. The two areas which have been studied in this way are the Reykjanes Ridge southwest from Iceland, 57°45′N – 63°30′N and the Mid-Atlantic Ridge southwest from the Azores Triple Junction, 36–38°N. Complementary investigations of the water column overlying these sections of ridge crest have employed a combination of real-time optical sensors together with conventional water column sampling and shipboard analysis to prospect for hydrothermal plume signals. Comparison and integration of these different geophysical and geochemical data sets has allowed the following conclusions to be drawn.

(1) Both hydrothermal activity and second-order ridge-segmentation are much more pronounced along the Mid-Atlantic Ridge, 36–38°N, southwest from the Azores Triple Junction, than they are along the Reykjanes Ridge.

(2) Predominantly, hydrothermal activity along the MAR, 36–38°N, appears to coincide with tectonic activity at non-transform discontinuities between adjacent segments. This indicates that tectonic activity may be instrumental in controlling the location and distribution of high-temperature hydrothermal activity on slow spreading mid-ocean ridges by providing and renewing subsurface pathways for fluid circulation.

(3) Both the increased occurrence of second-order ridge segmentation and hydrothermal activity along the MAR, southwest from the Azores Triple Junction, appear to be linked to a lower magma flux associated with this hot-spot when compared to the Reykjanes Ridge southwest of the Icelandic hot-spot.

Mid-Ocean Ridge research at SOC (formerly IOSDL) has been supported by various grants from the NERC's Community Research Programme, BRIDGE (Reykjanes Ridge, Azores Triple Junction) and by EC MAST 2 contract MAS2 CT93-0070, 'MARFLUX/ATJ' (Azores Triple Junction). Swath bathymetry data for the Azores Triple Junction was collected by H. D. N. with support from the French/US FARA Program (Project SIGMA).

References

APPELGATE, B. & SHOR, A. N. 1994. The northern Mid-Atlantic and Reykjanes Ridges: Spreading center morphology between 55°55′N and 63°00′N. *Journal of Geophysical Research*, **99**, 17935–17956.

BAKER, E. T., GERMAN, C. R. & ELDERFIELD, H. 1995. Hydrothermal plumes: global distributions and geological inferences. *In*: HUMPHRIS, S. E. LUPTON, J., MULLINEAUX, L. & ZIERENBERG, R. (eds) *Seafloor Hydrothermal Systems: Physical, Chemical, Biological and Geological Interactions*. AGU Geophysical Monographs, **91**, 47–71.

CHARLOU, J. L., BOUGAULT, H., DONVAL, J. P., PELLÉ, H., LANGMUIR, C. & FAZAR SCIENTIFIC PARTY 1993. Seawater CH_4 concentration over the Mid-Atlantic Ridge from the Hayes F. Z. to the Azores Triple Junction (abstract). *EOS, Transactions of the American Geophysical Union*, **74** (Suppl.), 380.

DEMETS, C., GORDON, R. G., ARGUS, D. F. & STEIN, S. 1990. Current plate motions. *Geophysical Journal International*, **101**, 425–478.

DETRICK, R. S., HONNOREZ, J. & ODP LEG 106 SCIENTIFIC PARTY 1986. Drilling the Snake Pit hydrothermal field on the Mid Atlantic Ridge, Lat. 23°22′N. *Geology*, **14**, 1004–1007.

—, NEEDHAM, H. D. & RENARD, V. 1995. Gravity anomalies and crustal thickness variations along the Mid-Atlantic Ridge between 33N and 40N. *Journal of Geophysical Research*, **100**, 3767–3787.

FOUQUET, Y., CHARLOU, J. L., DONVAL, J. P., RADFORD-KNOERY, J., PELLÉ, H., ONDRÉAS, H., SÉGONZAC, M., COSTA, I., LOURENÇO, N. & TIVEY, M. K. 1994. Geological setting and comparison of the Menez Gwen and Lucky Strike vent fields at 37°17′N and 37°50′N on the Mid Atlantic Ridge. Preliminary results of the DIVA1 diving cruise with Nautile (abstract). *EOS, Transactions of American Geophysical Union*, **75** (suppl.), 313.

GASS, I. G. & SMEWING, J. D. 1973. Intrusion, extrusion and metamorphism at constructive plate margins: evidence from the Troodos massif, Cyprus. *Nature*, **242**, 26–29.

GERMAN, C.R. 1993. *Hydrothermal activity on the Reykjanes Ridge: an ODP site survey*. IOSDL Cruise Reports, **238**.

——, BRIEM, J., CHIN, C., DANIELSEN, M., HOLLAND, S., JAMES, R., JÓNSDOTTIR, A., LUDFORD, E., MOSER, C., ÓLAFSSON, J., M. R., P. & RUDNICKI, M. D. 1994. Hydrothermal activity on the Reykjanes Ridge: the Steinahóll vent-field at 63°06′N. *Earth and Planetary Science Letters*, **121**, 647–654.

——, BAKER, E. T. & KLINKHAMMER, G. 1995a. The regional setting of hydrothermal vents. *In*: PARSON, L. M., WALKER, C. L. & DIXON, D. R. (eds) *Hydrothermal Vents and Processes*. Geological Society, London, Special Publications, **87**, 3–16.

——, PARSON, L. M. & THE HEAT SCIENTIFIC PARTY 1995b. *Hydrothermal exploration at the Azores Triple Junction: HEAT*. IOSDL Cruise Reports, **246**.

——, —— & —— 1996. Hydrothermal exploration near the Azores Triple junction: tectonic control of venting at slow-spreading ridges? *Earth and Planetary Science Letters*, in press.

HAYMON, R. M. 1996. The response of ridge crest hydrothermal systems to segmented, episodic magma supply. *This volume*.

KAULA, W. M. 1971. Global gravity and mantle convection. *Tectonophysics*, **13**, 46–62.

KLINKHAMMER, G., ELDERFIELD, H., GREAVES, M., RONA, P. & NELSEN, T. 1986. Manganese Geochemistry near High-Temperature Vents in the

Mid-Atlantic Ridge Rift Valley. *Earth and Planetary Science Letters*, **80**, 230–240.

——, RONA, P., GREAVES, N, H, & ELDERFIELD, H. 1985. Hydrothermal manganese plumes over the Mid-Atlantic Ridge rift valley. *Nature*, **314**, 727–731.

——, CHIN, C. & WILSON, C. 1993. Surveys of the FARA section of Mid-Atlantic Ridge for Hydrothermal activity during FAZAR (abstract). *EOS, Transactions of American Geophysical Union*, **74** (Suppl.), 380.

KUZNETSOV, A. P., BOGDANOV, Y. A., SAGALEVICH, A. M., SBORSHCHIKOV, I. M., AL'MUKHAMEDOV, A. M., KUZ'MIN, M. I., PODRAZHANSKIY, A. M. & GORLOV, A. A. 1985. Some results of geological investigation of the Reykjanes Ridge from submersibles. *Oceanology*, **25**, 77–82.

LANGMUIR, C. H., FORNARI, D., COLODNER, D., CHARLOU, J. L., COSTA, I., DESBRUYERES, D., DESONIE, D., EMERSON, T., FIALA-MEDIONI, A., FOUQUET, Y., HUMPHRIS, S., SALDANHA, L., SOURS-PAGE, R., THATCHER, M., TIVEY, M., VAN DOVER, C., VON DAMM, K., WIESE, K. & WILSON, C. 1993. Geological setting and characteristics of the Lucky Strike vent-field at 37°17′N on the Mid-Atlantic Ridge (abstract). *EOS, Transactions of American Geophysical Union*, **74** (suppl.), 99.

LISTER, C. R. B. 1974. On the penetration of water into hot rock. *Geophysical Journal of the Royal Astronomical Society*, **39**, 465–609.

MACDONALD, K. C., FOX, P. J., PERRAM, L. J., EISEN, M. F., HAYMON, R. M., MILLER, S. P., CARBOTTE, S. M., CORMIER, M.-H. & SHOR, A. N. 1988. A new view of the mid-ocean ridge from the behaviour of ridge-axis discontinuities. *Nature*, **335**, 217–225.

——, SCHEIRER, D. S. & CARBOTTE, S. M. 1991. Mid-ocean ridges: discontinuities, segments and giant cracks. *Science*, **253**, 986–994.

MACLEOD, C. J. & MANNING, C. E. 1996. Influence of axial segmentation on hydrothermal circulation at fast-spreading ridges: insights from Hess Deep. *This volume.*

MURTON, B. J. & PARSON, L. M. 1993. Segmentation, volcanism and deformation of oblique spreading centres: a quantitative study of the Reykjanes Ridge. *Tectonophysics*, **222**, 237–257.

——, KLINKHAMMER, G., BECKER, K., BRIAIS, A., EDGE, D., HAYWARD, N., MILLARD, N., MITCHELL, I., ROUSE, I., RUDNICKI, M., SAYANAGI, K., SLOAN, H. & PARSON, L. 1994. Direct evidence for the distribution and occurrence of hydrothermal activity between 27°N–30°N on the Mid-Atlantic Ridge. *Earth and Planetary Science Letters*, **125**, 119–128.

NEEDHAM, H. D., DAUTEUIL, O., DETRICK, R. & LANGMUIR, C. 1992. Structural and volcanic features of the Mid-Atlantic Ridge rift zone between 40°N and 33°N (abstract). *EOS, Transactions of American Geophysical Union*, **43** (Suppl.), 552.

NELSEN, T. A., KLINKHAMMER, G. P., TREFRY, J. H. & TROCINE, R. P. 1987. Real-time observation of dispersed hydrothermal plumes using neph-

elometry: examples from the Mid-Atlantic Ridge. *Earth and Planetary Science Letters*, **81**, 245–252.

OLAFSSON, J., THORS, K. & CANN, J. 1991. A sudden cruise off Iceland. *Ridge Events*, **2**, 35–38.

PARSON, L. M., MURTON, B. J., SEARLE, R. C., BOOTH, D., EVANS, J., FIELD, P., KEETON, J., LAUGHTON, A., MCALLISTER, E., MILLARD, N., REDBOURNE, L., ROUSE, I., SHOR, A., SMITH, D., SPENCER, S., SUMMERHAYES, C. & WALKER, C. 1993. En echelon axial volcanic ridges at the Reykjanes Ridge: a life cycle of volcanism and tectonics. *Earth and Planetary Science Letters*, **117**, 73–87.

——, GERMAN, C. R., COLLER, D. W. & HEAT SCIENTIFIC PARTY. 1994. Interrelationship between volcanism and tectonism at the MAR, 38–36°N. *EOS, Transactions of American Geophysical Union*, **75** (suppl.), 658.

SCHILLING, J.-G. 1973. The Icelandic mantle plume, geochemical evidence along the Reykjanes Ridge. *Nature*, **242**, 565–578.

—— 1975. Azores mantle blob: Rare Earth Evidence. *Earth and Planetary Science Letters*, **25**, 103–115.

SEARLE, R. C. & LAUGHTON, A. S. 1981. Fine scale sonar study of tectonics and volcanism on the Reykjanes Ridge. *Oceanologica Acta*, **4** (suppl.), 5–13.

SEMPERÉ, J.-C., PURDY, G. M. & SCHOUTEN, H. 1990. Segmentation of the Mid-Atlantic Ridge between 24°N and 30°40′N. *Nature*, **344**, 427–431.

——, LIN, J., BROWN, H. S., SCHOUTEN, H. & PURDY, G. M. 1993. Segmentation and morphotectonic variations along a slow-spreading center: the Mid-Atlantic Ridge (24°00′N–30°40′N). *Marine Geophysical Researches*, **15**, 153–200.

SHIREY, J. B., BENDER, J. F. & LANGMUIR, C. H. 1987. Three component isotopic heterogeneity near the Oceanographer transform, Mid Atlantic Ridge. *Nature*, **325**, 217–223.

SLOAN, H. & PATRIAT, P. 1992. Kinematics of the North American-African plate boundary between 28 and 29N during 10 Ma: evolution of the axial geometry and spreading rate and direction. *Earth and Planetary Science Letters*, **113**, 323–341.

SPIESS, F. N. & RISE PROJECT GROUP. 1980. East Pacific Rise: Hot springs and geophysical experiments. *Science*, **207**, 1421–1433.

STRANGE, T. M. 1991. *The determination and geophysical applications of free air gravity anomalies using satellite altimetry and ship gravity*. Ph.D thesis, University of Oxford.

TALWANI, M., WINDISH, C. C., & LANGSETH, M. G. JR. 1971. The Reykjanes Ridge crest: a detailed geophysical study. *Journal of Geophysical Research*, **76**, 473–517.

TAYLOR, R., MURTON, B. J. & THIRLWALL, M. F. 1995. Petrographic and geochemical variation along the Reykjanes Ridge, 57°N–59°N. *Journal of Geological Society*, **155**, 1031–1037.

WHITE, R. S. 1989. Asthenospheric controls on magmatism in the ocean basins. *In*: SAUNDERS, A. D. & NORRY, M. J. (eds) *Magmatism in the Ocean Basins*. Geological Society, London, Special Publications, **42**, 17–29.

Influence of axial segmentation on hydrothermal circulation at fast-spreading ridges: insights from Hess Deep

C. J. MacLEOD[1] & C. E. MANNING[2]

[1]Department of Earth Sciences, University of Wales, PO Box 914, Cardiff CF1 3YE, Wales UK

[2]Department of Earth and Space Sciences, University of California, Los Angeles, CA 90024-1567, USA

Abstract: The evolution of oceanic hydrothermal systems depends heavily upon the origin, timing and spatial distribution of permeability near the magmatic heat source. At fast-spreading ridges, where permanent magma chambers are inferred, permeability in the plutonic portion of the crust is generated during cooling by time-, strain-rate- and temperature-dependent fracturing, which propagates downwards through the crust as the rocks cool and move off-axis. At major axial discontinuities, however, where magmatism may be intermittent and tectonism plays an important role, the controls on hydrothermal alteration are likely to be more complex.

Gabbros exposed at Hess Deep and sampled by the Ocean Drilling Program during Leg 147 provide the first insights from direct observation into the structural and metamorphic evolution of the lower crust near axial discontinuities. Early amphibolite-facies deformation and alteration are related to cooling, fracturing, and hydrothermal fluid flow within several kilometres of the East Pacific Rise axis. Later, dense fracturing and limited alteration occurred in the greenschist facies as the gabbros were transported into the deformation field at the 'tectonic tip' of the westward-propagating Cocos–Nazca ridge.

Although this is only one example of the possible influences of axial segmentation on hydrothermal circulation, and may represent a comparatively simplified view of other natural systems, given the potential variations in strain and strain rate coupled with the possibility of reheating events associated with renewed magmatism, we suggest that the multi-stage structural and metamorphic evolution of the Hess Deep plutonics may be a good analogue for the cooling history of the lower crust at other fast-spreading ridge axial discontinuities.

Oceanic hydrothermal systems are driven by thermal energy liberated from cooling magma. In conceptual models of these systems, seawater is drawn downward toward a cooling axial magma body and heated near the interface between solidified intrusive rocks and magma. The fluids then advect thermal energy out of the crust during rapid, buoyant rise and discharge at the ocean floor (e.g. Lowell 1975; Strens & Cann 1982, 1986; Lister 1983; Cann et al. 1985; Lowell & Burnell 1991; Lowell & Germanovich 1994). The hydrochemical evolution of these systems depends strongly on the origin, timing, and spatial distribution of permeability in and near the magmatic heat source.

Petrological constraints on pore-network evolution and fluid-rock interaction in the root zones of oceanic hydrothermal systems come from metamorphosed oceanic gabbros. Studies of these rocks have focused principally on slow spreading environments, where exposures of lower crustal lithologies are common (e.g.

Bonatti et al. 1975; Helmstaedt & Allen 1977; Mével et al. 1978; Prichard & Cann 1982; Ito & Anderson 1983; Honnorez et al. 1984; Mével 1987, 1988). It is now well established, however, that significant differences exist in the structure of oceanic crust formed at slow- and fast-spreading ridges. In slow-spreading environments, where tectonic extension may be necessary to accommodate plate separation because of insufficient magma supply, high-temperature ductile shear zones occur in the lower crust close to the ridge axis (e.g. Harper 1985; Karson et al. 1987; Karson 1990; Cannat & Mével 1991; Dick et al. 1991). It is these that provide the initial permeability, allowing seawater penetration into the gabbros at near-solidus to amphibolite-facies conditions (c. 750–550°C; Cannat et al. 1991; Stakes et al. 1991; Gillis et al. 1993b; Kelley et al. 1993). In contrast, magma chambers at fast-spreading ridges are thought to be more or less permanent (e.g. Detrick et al. 1987; Kent et al. 1990, 1993a, b; Toomey et al. 1990; Sinton

From: MacLeod, C. J., Tyler, P. A. & Walker, C. L. (eds) 1996, *Tectonic, Magmatic, Hydrothermal and Biological Segmentation of Mid-Ocean Ridges*, Geological Society Special Publication No. 118, pp. 185–198.

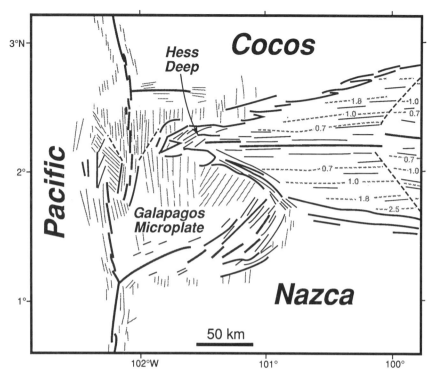

Fig. 1. Morphotectonic map of the Pacific–Cocos–Nazca triple junction and the Galapagos microplate (after Lonsdale 1988), showing the location of Hess Deep.

& Detrick 1992), and high-temperature plastic deformation in gabbros to be absent; permeability in the plutonic portion of the crust is in this case believed to be generated primarily by a temperature-dependent cracking front that propagates downwards through the crust in the wake of the magma-gabbro interface (Lister 1974). This led Mével & Cannat (1991) to hypothesize that the inception of fluid-rock interaction should take place at much lower temperatures (<500°C) at fast-spreading than slow-spreading environments.

Spreading rate may not be the only control on the physical and chemical behaviour of oceanic crust (e.g. Rona 1988; Agar 1994). On a regional scale we suggest that the cracking-front model may be of limited applicability if the segmentation of fast-spreading ridges is taken into account. The morphology of fast-spreading ridges, such as the East Pacific Rise (EPR), is far from uniform, with a hierarchy of distinct accretionary units bounded by various types of axial discontinuity (e.g. Macdonald et al. 1988, 1991). At the larger (first- and second-order) axial discontinuities (such as propagating rifts and major overlapping spreading centres

(OSCs)) the axial magma chamber is usually interrupted (e.g. Detrick et al. 1987; Kent et al. 1993a, b), and more amagmatic extension occurs (Goff et al. 1993; Carbotte & Macdonald 1994); consequently, greater tectonically induced fracture permeability and hence access of seawater to the lower crust is likely at segment ends. Macdonald et al. (1988) estimate that up to 20% of EPR crust off-axis has been influenced by the effects of second-order segmentation. Enhanced, tectonically induced seawater access to the lower crust may therefore occur across a significant proportion of fast-spread ocean floor, and segmentation may play an important role in controlling the geometry of hydrothermal circulation systems. We note, for example, recent observations from the super-fast-spreading portions of the EPR at approximately 18°30'S (Urabe et al. 1994), of hydrothermal plume signatures in the water column above OSCs rather than above segment centres, apparently occurring by means of diffuse outflow from tectonic fissures in the OSC basins (T. Urabe pers. comm. 1995).

Little direct evidence exists, however, as to the nature of the root zones of hydrothermal

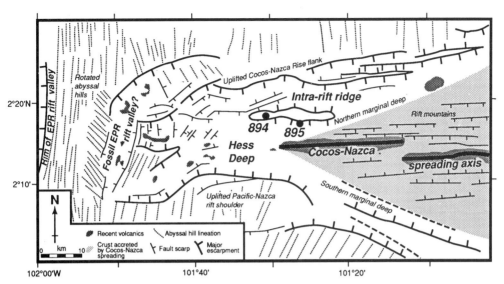

Fig. 2. Morphotectonic map of Hess Deep, modified after Lonsdale (1988) with reference to GLORIA sidescan sonar imagery originally acquired by Searle & Francheteau (1986). ODP drill sites are marked. Site 895 recovered sections of EPR shallow mantle (Gillis *et al.* 1993*a*).

circulation systems at fast-spreading ridges, largely because of the rarity of exposure of lower crustal sections on the seafloor. The few meta-gabbros that have been sampled from the fast-spreading EPR are all from areas complicated by changes in spreading rate (Mathematician Ridge: Batiza & Vanko 1985; Stakes & Vanko 1986), or from transform fault zones where mid-ocean ridge processes may be obscured by transform-related deformation from the earliest stages (Bideau *et al.* 1991). As a result, comparisons of hydrothermal processes in the plutonic portions of fast and slow spreading centres have relied primarily on the metamorphic record in the Oman ophiolite (Gregory & Taylor 1981; Stakes *et al.* 1983; Nehlig & Juteau 1988; Stakes & Taylor 1992; Nehlig 1993, 1994; Nehlig *et al.* 1994), which is generally thought to be a good analogue for fast-spread oceanic crust (Nicolas 1989; MacLeod & Rothery 1992). However, the ophiolite's complex igneous history (e.g. Lippard *et al.* 1986; Boudier & Nicolas 1988), with emplacement-related deformation apparently commencing before the complete crystallization of some ophiolite gabbros (e.g. Juteau *et al.* 1988; Gnos & Peters 1993), has had an unknown but undoubtedly complicating effect on metamorphism and hydrothermal fluid flow. Clearly, the need for *in situ* sampling of plutonic rocks from modern fast spreading ridges is paramount.

Hess Deep

Hess Deep (latitude 2°15'N, longitude 101°30'W) is a 5400 m deep rift valley lying close to the Pacific–Cocos–Nazca ridge–ridge–ridge triple junction in the equatorial eastern Pacific Ocean (Fig. 1). It has been formed by the progressive dismemberment of young lithosphere generated at the north–south-trending EPR by the westward propagation of the Cocos–Nazca spreading axis towards the triple junction (Hey *et al.* 1972, 1977; Johnson *et al.* 1976; Lonsdale 1988; MacLeod *et al.* 1996*b*). It is one of the few areas where extensive outcrops of the lower oceanic crust from a fast-spreading ridge are known to be exposed on the seafloor (Kashintsev *et al.* 1982; Francheteau *et al.* 1990; Girardeau & Francheteau 1993; Hekinian *et al.* 1993), and thus offers a rare opportunity to study the nature of fluid–rock interaction in the root zones of a fast-spreading ridge oceanic hydrothermal system.

It is important, however, to recognize the complexities and possible limitations of studying the EPR section exposed in Hess Deep; by the very fact of its exposure on the seafloor at the present day, it is likely to have had an atypical cooling and uplift history. Nevertheless, the fact that it was apparently formed at a relatively 'normal' section of the EPR (albeit perhaps close to an OSC; Lonsdale 1988) and then later dismembered by a propagating rift, suggests that it should still tell us something about processes at the larger (first- or second-order) axial discontinuities.

Although not typical of other axial discontinuities, the 90° difference in orientations between the EPR spreading axis, at which the Hess Deep lithosphere was formed, and the Cocos–Nazca axis, at which it was later deformed, is actually helpful: by determining the orientations of different vein types in the plutonic portion of the crust relative to the ridge reference frame we are able to distinguish between the different possible influences on the establishment of fracture permeability at the base of the hydrothermal circulation system.

Ocean drilling in Hess Deep

In 1992–93 Ocean Drilling Program (ODP) Leg 147 sampled the EPR lower crust and shallow mantle in the Hess Deep rift valley, as part of its 'offset drilling strategy': to obtain a composite section of the oceanic crust by drilling short, technically feasible holes in 'tectonic windows' rather than drilling a single, technically challenging hole many kilometres deep.

The prime focus of drilling was a fault-bounded horst within the Hess Deep basin, from which gabbroic rocks had been recovered by dredging and from submersibles (Francheteau *et al.* 1990; Fig. 2). The principal hole at the site, Hole 894G, was located close to the summit of this intra-rift ridge, at a depth of 3023.4 m below sea level. The hole penetrated 154.5 m below sea floor, with moderate recovery (35.8%), into a succession of gabbronorites and gabbros (with or without olivine) cut by two basaltic dykes (Gillis *et al.* 1993a). These rocks are not modally layered, and have grain sizes and textures that vary rapidly on short length scales; many have magmatic flow fabrics (MacLeod *et al.* 1996a). Comparison of the petrology and geochemistry of the drilled section with samples recovered during submersible traverses in Hess Deep (Francheteau *et al.* 1990; Hekinian *et al.* 1993) suggest that the hole was most probably sited at intermediate to high levels within the plutonic layer (Gillis *et al.* 1993a). Samples from immediately below the sheeted dyke-gabbro transition on the scarps of the northern wall of Hess Deep include oxide-bearing ferrodiorites and tonalites, and hence are significantly more evolved than those recovered at Site 894 (Natland & Dick 1996). Natland & Dick suggest that these highly fractionated lithologies mark the site of the axial melt lens, and they place the Site 894 section no more than a few hundred metres

below this level, probably corresponding to a crystallization depth of 2–3 km below the sea-floor.

Cooling and metamorphic history of high-level gabbros

All gabbroic rocks from Site 894 are metamorphosed. The extent of replacement of magmatic minerals by secondary minerals is at least 10% in Hole 894G, and ranges to >50% (Gillis *et al.* 1993a). No systematic relationship between metamorphism and lithology is observed. Metamorphism is related to the penetration of the gabbros by aqueous solutions through fractures at a range of scales. This brittle fracturing and fracture filling is extremely common in these rocks. Manning & MacLeod (1996) identify 1040 macroscopic veins in Hole 894G cores, at a density of 0.21 veins per metre. Virtually no non-mineralized fractures were recorded. With the possible exception of a few rare, cataclastically deformed zones described in Gillis *et al.* (1993a), Früh-Green *et al.* (1996) and MacLeod *et al.* (1996b), no shear displacements were observed in any of the vein types examined. This suggests that the veins resulted from tensile (mode I) brittle failure.

Crosscutting vein relationships and metamorphic grades of mineral-fluid interaction in Hole 894G gabbros present a time-integrated record of the physico-chemical evolution of hydrothermal circulation in the upper part of the EPR (layer 3) plutonic section, documenting the cooling of the section from the moment of initial sub-solidus penetration of seawater, through to its exhumation and exposure on the seafloor.

Methodology and vein classification

We investigated the nature of fracture-controlled metamorphism at Site 894 by combining petrographic and structural studies of mineral-filled fractures (veins) with borehole wall imaging data and electron microprobe analyses. Optical petrography revealed the existence of an additional microscopic vein type in addition to the macroscopic veins mentioned above. Similar data were collected for these microscopic veins to the maximum extent possible, based on observation of representative thin sections. We grouped veins on the basis of their two scales of occurrence, microscopic and macroscopic, because we found that veins at these two scales were temporally, mineralogically, and texturally distinct. A second-order subdivision of macroscopic veins based on mineral fill yielded three types: macroscopic amphibole, chlorite and zeolite–calcite veins. Chlorite veins were further subdivided into chlorite–calc-silicate veins and chlorite–smectite veins. For each vein, its location, width, length, crosscutting relations, orientation, colour, mineral fill and associated wall-rock alteration were recorded (Manning & MacLeod 1996; MacLeod *et al.* 1996*c*; Manning *et al.* 1996). Representative mineral compositions are given in Manning and MacLeod (1996). Below, we describe these veins types in order of oldest to youngest as established by crosscutting relations.

Of the 1040 macroscopic veins present in the recovered cores from Hole 894G, 601 were in core pieces that could be oriented relative to the fiducial axis of the borehole and 303 were in pieces for which the stable magnetic remanence direction has also been determined (MacLeod *et al.* 1996*c*). Orienting core pieces relative to this stable remanence direction can provide a common reference frame for comparing veins and other features in the core, assuming that all samples acquired their magnetizations under the same field conditions. This assumption is reasonable for the gabbroic rocks drilled at Site 894, which are likely to have cooled sufficiently slowly for secular variations of the Earth's magnetic field to be averaged out. Knowledge of the core magnetization direction is not, however, sufficient in itself to determine the true orientations of structures: the horizontal component of the stable magnetic remanence direction can only be equated with present-day north or south if it can be demonstrated that there has been no tectonic rotation of the sample after the magnetization was acquired. At Site 894, magnetic inclinations consistently dip at approximately $+40°$ (down toward the declination direction), contrasting markedly with the expected inclination of the axial dipole field of $+4.6°$ for this latitude, and leading us to suspect that some tectonic rotation of the Hess Deep intra-rift ridge has occurred (MacLeod *et al.* 1995, 1996*b*), and, therefore, that the stable magnetic remanence direction need not necessarily point toward north or south.

The only reliable indication of fracture orientation in geographical coordinates comes from down-hole measurements of the borehole wall. On Leg 147, Formation MicroScanner (FMS) microresistivity images were obtained for a 35 m interval (45–80 mbsf) of the wall of Hole 894G (MacLeod *et al.* 1995; Célérier *et al.* 1996). These images are oriented reliably in the geographical reference frame (Ekstrom *et al.* 1987). By comparing the distributions of veins in the core, partially restored relative to the magnetisation direction, with the distributions

(a)

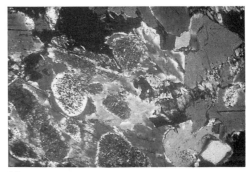

(b)

Fig. 3. (a) Microscopic amphibole veins at plagioclase-plagioclase grain boundaries (Sample 147-894G-4R-2, 33–37 cm; partly crossed polars; field of view 1 mm). (b) Secondary clinopyroxene rims (clear) on altered exsolved magmatic clinopyroxene (Sample 147-894G-18R-2, 15–17 cm; plane polarized light; field of view 2 mm).

of fractures on the logs, the FMS images allow the vein measurements to be reoriented fully to geographical coordinates. The principles and methodology of this technique are discussed in detail by MacLeod *et al.* (1992, 1994, 1995).

Vein types

Microscopic amphibole veins are the oldest veins recognized, and are responsible for grain-scale alteration of the gabbros, accounting for 10% to >50% of the modal mineralogy. They take the form of a random, interconnected network of tiny (average 17 μm wide) cracks filled by green, blue-green, and olive-green calcic amphibole, and occur pervasively throughout the Hole 894G section (Fig. 3). Manning & MacLeod (1996) show that more than 80% of grain boundaries were affected by the fluid flow at the time that this microfracture

Table 1. *Summary of vein types, metamorphic grades and postulated temperatures in Hole 894G gabbros*

Vein type	Metamorphic facies	Temperature
Microscopic amphibole	Cpx-amphibolite to amphibolite	750–600°C
	(Amphibolite to amphibolite/greenschist transition in coarse gabbros)	(750–450°C)
Macroscopic amphibole	Amphibolite to amphibolite/greenschist transition	600–450°C
Chlorite–calc-silicate	Greenschist	450–250°C
Chlorite–smectite	Sub-greenschist	<250°C
Zeolite/calcite	Zeolite	<150°C

Data and details of geothermometry in Manning & MacLeod (1996) and Früh-Green *et al.* (1996).

network was formed. They find that amphibole composition varies with the primary mineralogy of the vein wall, suggesting that material transport distances were short, and consistent with rapid reaction rates and low fluid fluxes. The presence of secondary clinopyroxene in association with the microscopic amphibole veins implies high temperatures of metamorphism. Manning & MacLeod (1996) calculate temperatures as high as 725 ± 40°C, with most mineral compositions implying amphibolite facies temperatures of 600–750°C (Table 1).

Macroscopic amphibole-filled veins cross-cut their microscopic precursors in all cases. They are light green to blue in colour, usually <1 mm wide, and range from wispy, discontinuous features to straight-walled veins continuous for several centimetres. In hand specimen they are rare (only 44 were identified in total), and frequently difficult to identify, especially in the most strongly-altered hosts. Mineral assemblages in the macroscopic veins are calcic, but differ from the microscopic vein amphiboles by having a compositional gap (between Al-rich and Al-poor types) that implies temperatures of formation in the range 450–600°C (i.e. lower amphibolite facies to amphibolite–greenschist facies transition; Manning and MacLeod 1996; Table 1).

Macroscopic chlorite–calc-silicate and chlorite–smectite veins. Chlorite-bearing veins are by far the most common in Hole 894G, with 987 veins recorded (Manning & MacLeod 1996). These veins are characterized by chlorite-lined vein walls and usually by vein centres filled either with calc-silicates (epidote and prehnite) or smectite. A distinction between the mineralogies of vein centre-fill was made during description of the cores, but this is largely arbitrary: a continuum of compositions is recognised, with some veins containing calc-silicates and also clay minerals as a late-infilling phase. Chlorite–calc-

silicate veins are, however, predominantly (though not exclusively) older than the chlorite–smectite veins. Chlorite veins of both types cut the amphibole veins where cross-cutting relationships are preserved. Brecciated wall-rock clasts and evidence for multiple generations of infill attest to repeated opening and filling, locally under high pore-fluid pressures.

Chlorite veins are characteristically continuous, straight-sided, planar features up to 3 mm in width, and are often associated with alteration haloes of white albitized plagioclase (Fig. 4). They often form parallel sets of closely-spaced veins cutting across the cores. Chlorite–calc-silicate veins ($N = 404$) are dark green to black in hand specimen, often with white prehnite centres. Chlorite–smectite veins ($N = 583$) are characterized by soft green, yellow or brown smectites in vein centres. The core is often broken along these planes of weakness. Late zeolites and calcite commonly fill residual pore space.

Temperature estimates for the formation of the chlorite–calc-silicate vein assemblage (Früh-Green *et al.* 1996) range as high as approximately 400–450°C. Estimates of 300–400°C are consistent with the assemblage chlorite + albite + actinolite in the alteration haloes. Veining and alteration must have continued down to 150–250°C, as determined from the compositions of clay mineralogies and oxygen isotope data from secondary plagioclase and calcite vein samples (Table 1).

Macroscopic zeolite and/or calcite veins. White veins filled by zeolite and calcite are the youngest hydrothermal veins observed in Hole 894G. They tend to have either zeolite or calcite fill, but rarely both. Neither are common ($N = 5$). The veins are similar in width to the other macroscopic types. Little wall-rock alteration is associated with these veins, implying that water-rock reaction was limited. Temperature estimates for this latest phase of veining are as low as 150°C (Früh-Green *et al.* 1996; Table 1).

Fig. 4. Typical appearance of planar, macroscopic chlorite veins in core section. These are composite veins, up to 2 mm wide, with chlorite rims and prehnite and clay minerals in their centre. Note the white alteration haloes of secondary plagioclase, and dark spots of clays/amphibole/talc *etc.* after olivine (Sample 147-894G-BR-1, piece 4, 14–44 cm).

Basaltic dykes

The two dykes that cut the Hole 894G section are porphyritic (plagioclase ± olivine ± spinel) and of primitive composition, similar to those dredged from the nearby Cocos–Nazca Ridge

(Gillis *et al.* 1993*a*). The dykes are fine-grained throughout, and have pronounced chilled margins. No evidence is seen within the dyke rock for the pervasive amphibolite-facies metamorphism seen in the gabbros; most of the alteration minerals are clays, prehnite, chlorite and zeolite (Gillis *et al.* 1993*a*, Lécuyer *et al.* 1993). Although the orientation of the dyke margins is unknown (no stable remanence direction was obtained upon demagnetisation; Gillis *et al.* 1993*a*), the above evidence argues strongly that the dykes were intruded late in the history of the Hole 894G gabbro section, at a time at which it had moved off axis and cooled substantially; hence, the dykes most probably represent an early sign of renewed magmatism associated with Cocos–Nazca Ridge propagation. Gillis *et al.* (1993*a*) report atypical sulphide minerals and textures in the gabbroic country rocks immediately adjacent to the dykes, and suggest that they were remobilised by the thermal effects of the intrusion; otherwise, however, no thermal effect of the intrusion could be detected in the gabbros.

Hole 894G vein orientations

Vein orientations, sub-divided by mineralogy and oriented with respect to the stable magnetic remanence direction, are presented in Fig. 5. The relatively small number of amphibole veins restored in this way ($N = 7$) means that their orientations are poorly constrained from a statistical standpoint, but the data we do have show no preferred orientation. The chlorite-bearing veins, in contrast, show a very marked preferred orientation ($N = 291$). Their strikes are predominantly oriented in a direction approximately 110° clockwise from the magnetic declination direction, and they have a moderate south-southwestward dip. The chlorite–calc-silicate vein orientations are slightly more clustered than the chlorite–smectite veins. The few reoriented zeolite–calcite veins ($N = 5$) have similar strikes to the chlorite veins, as do the few cataclastic shear zones for which orientation data are available (grouped here, because of their mineralogy, with the chlorite–calc-silicate or chlorite–smectite vein types).

Details of processing and interpretation of FMS data from Hole 894G are discussed by Célérier *et al.* (1996) and will not be reiterated here. The logs were of low quality because of the poor hole conditions, yet MacLeod *et al.* (1995) and Célérier *et al.* (1996) were successful in detecting a large number of low-resistivity, sub-planar to planar features on the FMS images. There is little doubt that these represent fractures, probably open and water-filled. In

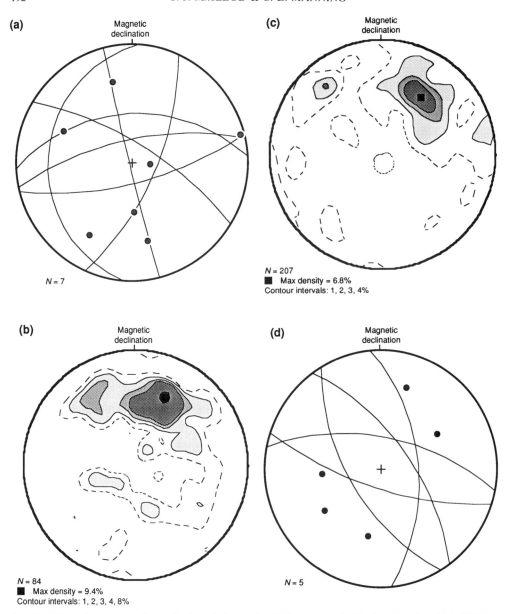

Fig. 5. Orientations of hydrothermal veins relative to the stable magnetic declination direction, Hole 894G, sub-divided by mineralogy. (**a**) Macroscopic amphibole veins; (**b**) chlorite–calc-silicate veins; (**c**) chlorite–smectite veins; and (**d**) zeolite/calcite veins. Lower hemisphere, equal area projections of poles to planes (contoured in (**b**) and (**c**); shown with great circles of planes in (**a**) and (**d**)).

Fig. 6, fractures are plotted as poles to planes in lower hemisphere, equal-area stereographic projections, oriented relative to true north. These data yield a preferred pole to plane of 354.4°; 39.7° ($N = 177$), corresponding to a plane dipping at 50.3° southwards, with strike of 084.4°. The marked preferred east-west strike is notable, being parallel to the trend of the

Cocos–Nazca rift and of the intra-rift ridge at Site 894.

The veins measured in the core are predominantly planar (lined) fractures from the centre of the borehole. The low-resistivity features identified on the FMS images are believed to be planar fractures on the borehole wall. Because of the poor quality of the FMS images, it is not possible

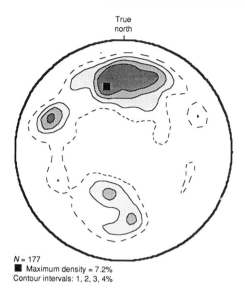

True
north

N = 177
■ Maximum density = 7.2%
Contour intervals: 1, 2, 3, 4%

Fig. 6. Orientations of fractures on the wall of Hole 894G, with respect to true north (FMS images, 45–80 m below seafloor).

to match the features on a 1:1 basis; however, the FMS fracture data considered together may be compared with the core fracture (vein) data from the same (45–80 mbsf) depth interval (MacLeod *et al.* 1995). The distributions of fractures on both cores and logs are very similar: both show marked preferred orientations, with similar dips, and a maximum strike direction *c.* 15% of the total. There must be little doubt that the same features are being compared. The difference between the two is that whereas the FMS data are recorded relative to true north, the core data are measured relative to the stable magnetisation direction, which was acquired when the gabbro section cooled, and which need not be north. Matching the two data sets (Fig. 7) implies that the stable magnetic remanence of the Hole 894G gabbros points towards the NNW. This is most readily explicable as a vertical-axis component of rotation of *c.* 30° anticlockwise (assuming normal polarity: MacLeod *et al.* 1996*b*), in addition to the *c.* 35° northward tilt inferred from the steeply dipping inclinations (see above). The implications for the tectonic evolution of the Hess Deep are discussed by MacLeod *et al.* (1996*b*). In Fig. 5, therefore, the magnetic declination directions shown actually point NNW, and the planar chlorite vein sets (Fig. 5b and c) are (relative to true north) east–west-striking, dipping moderately to steeply south. They parallel the orientation of the Hess Deep rift valley and Cocos–

Nazca spreading centre, and are therefore most plausibly related to the tectonics of Cocos–Nazca rifting (see below).

Discussion

Cooling history of Hess Deep gabbros

Leg 147 drilling verified the first-order hypothesis that the crustal section and underlying shallow mantle lithologies now exposed in Hess Deep were formed by seafloor spreading at the north-south trending EPR (e.g. Gillis *et al.* 1993*a*; MacLeod *et al.* 1996*a*). This lithosphere was transported eastward away from the EPR ridge axis and became influenced by extensional tectonism caused by amagmatic rifting in advance of the westward-propagating Cocos–Nazca spreading centre.

Seawater influx into the plutonic portion of the oceanic crust, commenced soon after crystallisation, and while the host was at a very high temperature (estimated at up to 750°C; Manning & MacLeod 1996). Pervasive influx of water along apparently randomly oriented microfracture networks and grain boundaries gave rise to significant mineralogical and isotopic alteration of the section, mostly at temperatures of 600–750°C (Manning & MacLeod 1996; Lécuyer *et al.* 1993). This permeability was probably created by tensile brittle failure in response to local stresses induced upon subsolidus cooling and thermal contraction of the gabbro as it was transported away from the ridge axis (e.g. Lister 1974). Comparison with numerical simulations of heat flow and material transport (e.g. Phipps Morgan & Chen 1993; Henstock *et al.* 1993) suggests that this early fracture permeability must have been established within several tens of thousands of years after axial magma emplacement and within a few kilometres of the ridge axis (here spreading at 65 km a^{-1} half rate) (Manning & MacLeod 1996).

At the point at which the upper plutonic section had cooled to the upper greenschist facies (a temperature of approximately 450°C) it became influenced for the first time by the effects of Cocos–Nazca rifting. Development of a dense array of east–west, steeply dipping, tensile (mode I) fractures occurred almost certainly in response to the north–south, far-field extensional stresses expected for Cocos–Nazca spreading. The moment at which this changeover occurred is less well constrained by theoretical studies than for the higher-temperature veining; however, it may be estimated at several to tens of kilometres away from the axis, and tens to a couple of hundred thousand years

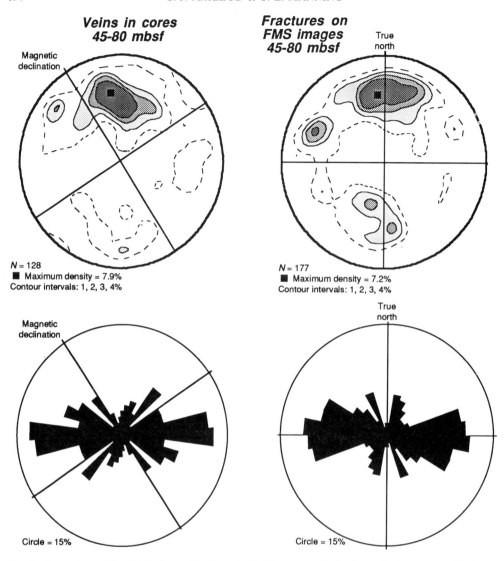

Fig. 7. Comparison of the distributions of fractures in FMS images and hydrothermal veins measured in the recovered core (undifferentiated) from the interval 45–80 m below seafloor, Hole 894G. Matching the core and logging data shows that the magnetic declination direction of the core samples is not directed towards north but north-northwest; thus, that the section has suffered a vertical axis component of rotation of approximately 30° anticlockwise, and that the chlorite veins really have an east-west strike in geographical co-ordinates.

after axial magmatism. This is roughly comparable to the distance east of the present-day EPR axis that the first east–west lineaments are observed today (approximately 25 km, from GLORIA sidescan sonar records: Searle & Francheteau 1986).

The observation of lower-temperature phases in the centres of veins whose walls are lined by higher-temperature phases suggests that the east–west fractures were not sealed, but maintained their permeability, presumably in response to continued extension. All subsequent fluid circulation and cooling of the section (documented down to approximately 150°C) appears to have been controlled strongly by this permeable fracture network.

Implications for hydrothermal circulation at fast-spreading ridge axial discontinuities

The early randomly-oriented amphibole vein systems are compatible with the Lister (1974) cracking front model for the generation of initial permeability at the root zones of hydrothermal systems at fast-spreading ridges (Mével & Cannat 1991), although we show that this can take place at higher temperatures (up to 750°C) than envisaged by Mével & Cannat. This cracking front model may indeed be the predominant mechanism for the initial failure of the gabbros away from ridge segment terminations.

At Hess Deep, however, we see very clearly the influence that far-field stresses associated with ridge axial segmentation can have upon the geometry of hydrothermal circulation and permeability structure of the lower crust. We suggest that this kind of two- (or multi-) stage cooling history may be of general applicability to fast-spread crust that has been influenced by segmentation. This being so, it should be recognised that the two-stage structural and metamorphic evolution documented at Hess Deep is nevertheless likely to be a comparatively simplified view for many axial discontinuities. The timing of tensile brittle failure and initial temperature of seawater access through the tectonically induced fracture system will depend critically upon the offset of the ridge axes. Variations in strain and strain rate may also be important and, most particularly, the extent of intrusion of new magmatic material into the older crust will have a significant effect on the cooling history.

References

AGAR, S. M. 1994. Rheological evolution of the ocean crust: A microstructural view. *Journal of Geophysical Research*, **99**, 3175–3200.

BATIZA, R. & VANKO, D.A. 1985. Petrologic evolution of large failed rifts in the eastern Pacific: Petrology of volcanic and plutonic rocks from the Mathematician Ridge area and the Guadalupe Trough. *Journal of Petrology*, **26**, 564–602.

BIDEAU, D., HEBERT, R., HEKINIAN, R. & CANNAT, M. 1991. Metamorphism of deep-seated rocks from the Garrett ultrafast transform (East Pacific Rise near 13°25′S). *Journal of Geophysical Research*, **96**, 10079–10099.

BONATTI, E., HONNOREZ, J., KIRST, P. & RADICATI, F. 1975. Metagabbros from the Mid-Atlantic Ridge at 6°N: contact-hydrothermal-dynamic metamorphism beneath the axial valley. *Journal of Geology*, **83**, 61–78.

BOUDIER, F. & NICOLAS, A. 1988. The ophiolites of Oman: introduction. *Tectonophysics*, **151**, vii–viii.

CANN, J. R., STRENS, M. R. & RICE, A. 1985. A simple magma-driven thermal balance model for the formation of volcanogenic massive sulphides. *Earth and Planetary Science Letters*, **76**, 123–134.

CANNAT, M. & MÉVEL, C. 1991. Stretching of the deep crust at the slow-spreading Southwest Indian Ridge. *Tectonophysics*, **190**, 73–94.

——— & STAKES, D. S. 1991. Normal ductile shear zones at an oceanic spreading ridge: tectonic evolution of Site 735 gabbros. *Proceedings of the ODP, Scientific Results*, **118**. College Station, TX (Ocean Drilling Program), 415–430.

CARBOTTE, S. M. & MACDONALD, K. C. 1994. Comparison of seafloor tectonic fabric at intermediate, fast and super fast spreading ridges: influence of spreading rate, plate motions, and ridge segmentation on fault patterns. *Journal of Geophysical Research*, **99**, 13609–13631.

CÉLÉRIER, B., MACLEOD, C. J. & HARVEY, P. K. 1996. Constraints on the geometry and fracturing of Hole 894G, Hess Deep, from Formation Micro-Scanner logging data. *Proceedings of the ODP, Scientific Results*, **147**. College Station, TX (Ocean Drilling Program), 329–345.

DETRICK, R. S., BUHL, P., VERA, E., MUTTER, J., ORCUTT, J., MADSEN, J. & BROCHER, T. 1987. Multi-channel seismic imaging of a crustal magma chamber along the East Pacific Rise. *Nature*, **326**, 35–41.

DICK, H. J. B., MEYER, P. S., BLOOMER, S., KIRBY, S., STAKES, D. & MAWER, C. 1991. Lithostratigraphic evolution of an in-situ section of oceanic layer 3. *Proceedings of the ODP, Scientific Results*, **118**. College Station, TX (Ocean Drilling Program), 439–538.

EKSTROM, M. P., DAHAN, C., CHEN, M.-Y., LLOYD, P. & ROSSI, D. J. 1987. Formation imaging with microelectrical scanning arrays. *Log Analysts*, **28**, 294–306.

FRANCHETEAU, J., ARMIJO, R., CHEMINÉE, J. L., HEKINIAN, R., LONSDALE, P. & BLUM, N. 1990. 1 Ma East Pacific Rise oceanic crust and uppermost mantle exposed by rifting in Hess Deep (equatorial Pacific Ocean). *Earth and Planetary Science Letters*, **101**, 281–295.

FRÜH-GREEN, G. L., PLAS, A. & DELL'ANGELO, L. N. 1996. Mineralogic and stable isotope record of polyphase alteration of East Pacific Rise upper crustal gabbros (Hess Deep, ODP Site 894). *Proceedings of the ODP, Scientific Results*, **147**. College Station, TX (Ocean Drilling Program), 235–254.

GILLIS, K. M., MÉVEL, C., ALLAN, J. F., ET AL. 1993a. *Proceedings of the ODP, Initial Reports*, **147**. College Station, TX, (Ocean Drilling Program).

———, THOMPSON, G. & KELLEY, D. S. 1993b. A view of the lower crustal component of hydrothermal systems at the Mid-Atlantic Ridge. *Journal of Geophysical Research*, **98**, 19597–19619.

GIRARDEAU, J. & FRANCHETEAU, J. 1993. Plagioclase-wehrlites and peridotites on the East Pacific Rise (Hess Deep) and the Mid-Atlantic Ridge (DSDP Site 334): Evidence for magma percolation in the oceanic upper mantle. *Earth and Planetary Science Letters*, **115**, 137–149.

GNOS, E. & PETERS, TJ. 1993. K-Ar ages of the metamorphic sole of the Semail ophiolite: implications for ophiolite cooling history. *Contributions to Mineralogy and Petrology*, **113**, 325–332.

GOFF, J. A., MALINVERNO, A., FORNARI, D. J. & COCHRAN, J. R. 1993. Abyssal hill segmentation:

quantitative analysis of the East Pacific Rise flanks 7°S–9°S. *Journal of Geophysical Research*, **98**, 13 851–13 862.

GREGORY, R. T. & TAYLOR, H. P., JR. 1981. An oxygen isotope profile in a section of Cretaceous oceanic crust, Samail ophiolite, Oman: evidence for $\delta^{18}O$ buffering of the oceans by deep (>5 km) seawater-hydrothermal circulation at mid-ocean ridges. *Journal of Geophysical Research*, **86**, 2737–2755.

HARPER, G. D. 1985. Tectonics of slow spreading mid-ocean ridges and consequences of a variable depth to the brittle/ductile transition. *Tectonics*, **4**, 395–409.

HEKINIAN, R., BIDEAU, D., FRANCHETEAU, J. CHEMI-NEE, J. L., ARMIJO, R., LONSDALE, P. & BLUM, N. 1993. Petrology of East Pacific Rise crust and upper mantle exposed in Hess Deep (eastern equatorial Pacific). *Journal of Geophysical Research*, **98**, 8069–8094.

HELMSTAEDT, H. & ALLEN, J. M. 1977. Metagabbronorite from DSDP Hole 334: an example of high temperature deformation and recrystallization near the Mid-Atlantic Ridge. *Canadian Journal of Earth Sciences*, **14**, 886–898.

HENSTOCK, T. J., WOODS, A. W. & WHITE, R. S. 1993. The accretion of oceanic crust by episodic sill injection. *Journal of Geophysical Research*, **98**, 4143–4161.

HEY, R., DEFFEYES, K. S., JOHNSON, G. L. & LOWRIE, A. 1972. The Galapagos triple junction and plate motions in the East Pacific. *Nature*, **237**, 20–22.

——, JOHNSON, G. L. & LOWRIE, A. 1977. Recent plate motions in the Galapagos area. *Geological Society of America Bulletin*, **88**, 1385–1403.

HONNOREZ, J., MÉVEL, C. & MONTIGNY, R. 1984. Geotectonic significance of gneissic amphibolites from the Vema fracture zone, equatorial Mid-Atlantic Ridge. *Journal of Geophysical Research*, **89**, 11 379–11 400.

ITO, E. & ANDERSON, A. T., JR. 1983. Submarine metamorphism of gabbros from the Mid-Cayman Rise: petrographic and mineral constraints on hydrothermal processes at slow-spreading ridges. *Contributions to Mineralogy and Petrology*, **82**, 371–388.

JOHNSON, G. L., VOGT, P. R., HEY, R., CAMPSIE, J. & LOWRIE, A. 1976. Morphology and structure of the Galapagos Rise. *Marine Geology*, **21**, 81–120.

JUTEAU, T., ERNEWEIN, M., REUBER, I., WHITE-CHURCH, H. & DAHL, R. 1988. Duality of magmatism in the plutonic sequence of the Sumail Nappe, Oman. *Tectonophysics*, **151**, 107–135.

KARSON, J. A. 1990. Seafloor spreading on the Mid-Atlantic Ridge: Implications for the structure of ophiolites and oceanic lithosphere produced in slow-spreading environments. *In*: MALPAS, J., MOORES, E. M., PANAYIOTOU, A. & XENOPHONTOS, C. (eds) *Ophiolites: Oceanic Crustal Analogues*. Geological Survey Department, Cyprus, 547–555.

——, THOMPSON, G., HUMPHRIS, S. E., EDMOND, J. M., BRYAN, W. B., BROWN, R. R., WINTERS, A. T., POCKALNY, R. A., CASEY, J. F., CAMPBELL, A. C., KLINKHAMMER, G., PALMER, MR., KINZLER, R. J. & SULANOWSKA, M. M. 1987. Along-axis

variations in seafloor spreading in the MARK area. *Nature*, **329**, 681–685.

KASHINTSEV, G. L., KUZMIN, M. I. & POPLITOV, E. N. 1982. Composition and structure of the oceanic crust in the vicinity of the Hess Basin (Pacific Ocean). *Geotectonics*, **16**, 512–520.

KELLEY, D. S., GILLIS, K. M. & THOMPSON, G. 1993. Fluid evolution in submarine magma-hydrothermal systems at the Mid-Atlantic Ridge. *Journal of Geophysical Research*, **98**, 19579–19596.

KENT, G. M., HARDING, A. J. & ORCUTT, J. A. 1990. Evidence for a smaller magma chamber beneath the East Pacific Rise at 9°30′N. *Nature*, **344**, 650–653.

——, —— & —— 1993a. Distribution of magma beneath the East Pacific Rise between the Clipperton Transform and the 9°17′N Deval from forward modeling of common depth point data. *Journal of Geophysical Research*, **98**, 13945–13969.

——, —— & —— 1993b. Distribution of magma beneath the East Pacific Rise near the 9°03′N overlapping spreading center from forward modeling of common depth point data. *Journal of Geophysical Research*, **98**, 13971–13995.

LÉCUYER, C., GRUAU, G., REYNARD, B., MÉVEL, C. & THE LEG 147 SCIENTIFIC PARTY 1993. High-temperature diffusion of seawater through gabbroic rocks at Hess Deep Rift Valley. *EOS*, **74 (43)**, 653.

LIPPARD, S. J., SHELTON, A. W. & GASS, I. G. 1986. *The Ophiolite of Northern Oman*. Geological Society, London, Memoirs, **11**.

LISTER, C. R. B. 1974. On the penetration of water into hot rock. *Geophysical Journal of the Astronomical Society*, **39**, 465–609.

—— 1983. The basic physics of water penetration into hot rocks. *In*: RONA, P. A., BOSTROM, K., LAUBIER, L. & SMITH, K. L. JR. (eds) *Hydrothermal Processes at Seafloor Spreading Centers*. Plenum, New York, 141–168.

LONSDALE, P. 1988. Structural pattern of the Galapagos Microplate and evolution of the Galapagos triple junction. *Journal of Geophysical Research*, **93**, 13 551–13 574.

LOWELL, R. P. 1975. Circulation in fractures, hot springs, and convective heat transport on mid-ocean ridge crests. *Geophysical Journal of the Royal Astronomical Society*, **40**, 351–365.

—— & BURNELL, D. K. 1991. A numerical model for magma-hydrothermal boundary layer heat transfer in the oceanic crust. *Earth and Planetary Science Letters*, **104**, 56–69.

—— & GERMANOVICH, L. N. 1994. On the temporal evolution of high-temperature hydrothermal systems at ocean ridge crests. *Journal of Geophysical Research*, **99**, 565–575.

MACDONALD, K. C., FOX, P. J., PERRAM, L. J., EISEN, M. F., HAYMON, R. M., MILLER, S. P., CARBOTTE, S. M., CORMIER, M.-H & SHOR, A. N. 1988. A new view of the mid-ocean ridge from the behaviour of ridge-axis discontinuities. *Nature*, **335**, 217–225.

——, SCHEIRER, D. S. & CARBOTTE, S. M. 1991. Mid-ocean ridges: discontinuities, segments and giant cracks. *Science*, **253**, 986–984.

MACLEOD, C. J. & ROTHERY, D. A. 1992. Ridge axial

segmentation in the Oman ophiolite: evidence from along-strike variations in the sheeted dyke complex. *In*: PARSON, L. M., MURTON, B. J. & BROWNING, P. (eds) *Ophiolites and their Modern Oceanic Analogues*. Geological Society, London, Special Publications, **60**, 39–63.

——, BOUDIER, F., YAOUANCQ, G. & RICHTER, C. 1996a. Gabbro fabrics from ODP Site 894, Hess Deep: implications for magma chamber processes at the East Pacific Rise. *Proceedings of the ODP, Scientific Results*, **147**. College Station, TX (Ocean Drilling Program), 317–328.

——, CÉLÉRIER, B., FRÜH-GREEN, G. L. & MANNING, C. E. 1996b. Tectonics of Hess Deep: a synthesis of drilling results from Leg 147. *Proceedings of the ODP, Scientific Results*, **147**. College Station, TX (Ocean Drilling Program), 461–475.

——, ——, HARVEY, P. K. & THE ODP LEG 147 SCIENTIFIC PARTY 1995. Further techniques for core reorientation by core-log integration: application to structural studies of the lower oceanic crust in Hess Deep, eastern Pacific. *Scientific Drilling*, **5**, 77–86.

——, MANNING, C. E., BOUDIER, F., CÉLÉRIER, B., KENNEDY, L. A., KELSO, P., KIKAWA, E., PARISO, J. & RICHTER, C. 1996c. Data report: structural measurements from Sites 894 and 895, Hess Deep. *Proceedings of the ODP, Scientific Results*, **147**. College Station, TX (Ocean Drilling Program), 515–529.

——, PARSON, L. M., SAGER, W. W. & THE ODP LEG 135 SCIENTIFIC PARTY 1992. Identification of tectonic rotations in boreholes by the integration of core information with Formation MicroScanner and Borehole Televiewer images. *In*: HURST, A., GRIFFITHS, C. M. & WORTHINGTON, P. F. (eds) *Geological Applications of Wireline Logs II*. Geological Society, London, Special Publications, **65**, 235–246.

——, —— & —— 1994. Reorientation of core using the Formation MicroScanner and Borehole Televiewer: application to structural and palaeomagnetic studies with the Ocean Drilling Program. *Proceedings of the ODP, Scientific Results*, **135**. College Station, TX, (Ocean Drilling Program), 301–312.

MANNING, C. E. & MACLEOD, C. J. 1996. Fracture-controlled metamorphism of Hess Deep gabbros, Site 894: constraints on the root zones of mid-ocean ridge hydrothermal systems at fast-spreading centers. *Proceedings of the ODP, Scientific Results*, **147**. College Station, TX, (Ocean Drilling Program), 189–212.

——, ——, FRÜH-GREEN, G., KELLEY, D. S. & LECUYER, C. 1996. Metamorphic veins from Site 894: data report. *Proceedings of the ODP, Scientific Results*, **147**. College Station, TX, (Ocean Drilling Program), 497–513.

MÉVEL, C. 1987. Evolution of oceanic gabbros from DSDP Leg 82: influence of the fluid phase on metamorphic crystallizations. *Earth and Planetary Science Letters*, **83**, 67–79.

—— 1988. Metamorphism of oceanic layer 3, Gorringe Bank, eastern Atlantic. *Contributions to Mineralogy and Petrology*, **100**, 496–509.

—— & CANNAT, M. 1991. Lithospheric stretching and hydrothermal processes in oceanic gabbros from slow-spreading ridges. *In*: PETERS, T., NICOLAS, A. & COLEMAN, R. G. (eds) *Ophiolite Genesis and Evolution of the Oceanic Lithosphere*. Kluwer, Dordrecht, 293–312.

——, CABY, R. & KIENAST, J. R. 1978. Amphibolite facies conditions in the oceanic crust: example of amphibolitized flaser-gabbro and amphibolites from the Chenaillet ophiolite massif (Haute Alpes, France). *Earth and Planetary Science Letters*, **39**, 98–108.

NATLAND, J. H. & DICK, H. J. B. 1996. Melt migration through high-level gabbroic cumulates of the East Pacific Rise at Hess Deep: inferences from rock textures and mineral compositions. *Proceedings of the ODP, Scientific Results*, **147**. College Station, TX, (Ocean Drilling Program), in press.

NEHLIG, P. 1993. Interactions between magma chambers and hydrothermal systems: Oceanic and ophiolitic constraints. *Journal of Geophysical Research*, **98**, 19621–19633.

—— 1994. Fracture and permeability analysis in magma-hydrothermal transition zones in the Samail ophiolite (Oman). *Journal of Geophysical Research*, **99**, 589–601.

—— & JUTEAU, T. 1988. Flow porosities, permeabilities and preliminary data on fluid inclusions and fossil thermal gradients in the crustal sequence of the Sumail ophiolite (Oman). *Tectonophysics*, **151**, 199–221.

——., JUTEAU, T., BENDEL, V. & COTTEN, J. 1994. The root zones of oceanic hydrothermal systems: constraints from the Samail ophiolite (Oman). *Journal of Geophysical Research*, **99**, 4703–4713.

NICOLAS, A. 1989. *Structures of Ophiolites and Dynamics of Oceanic Lithosphere*. Kluwer, Boston.

PHIPPS MORGAN, J. & CHEN, Y. J. 1993. The genesis of oceanic crust: Magma injection, hydrothermal circulation, and crustal flow. *Journal of Geophysical Research*, **98**, 6283–6297.

PRICHARD, H. M. & CANN, J. R. 1982. Petrology and mineralogy of dredged gabbro from Gettysburg Bank, eastern Atlantic. *Contributions to Mineralogy and Petrology*, **79**, 46–55.

RONA, P. A. 1988. Hydrothermal mineralization at oceanic ridges. *Canadian Mineralogist*, **26**, 431–465.

SEARLE, R. C. & FRANCHETEAU, J. 1986. Morphology and tectonics of the Galapagos triple junction. *Marine Geophysical Research*, **8**, 95–129.

SINTON, J. M. & DETRICK, R. S. 1992. Mid-ocean ridge magma chambers. *Journal of Geophysical Research*, **97**, 197–216.

STAKES, D. S. & TAYLOR, H. P. JR. 1992. The northern Semail ophiolite: an oxygen isotope, microprobe, and field study. *Journal of Geophysical Research*, **97**, 7043–7080.

—— & VANKO, D. A. 1986. Multistage hydrothermal alteration of gabbroic rocks from the failed Mathematician Ridge. *Earth and Planetary Science Letters*, **79**, 75–92.

——, MÉVEL, C., CANNAT, M. & CHAPUT, T. 1991. Metamorphic stratigraphy of Hole 735B. *Pro-*

ceedings of the ODP, Scientific Results, **118**. College Station, TX, (Ocean Drilling Program).

——, TAYLOR, H. P. JR. & FISHER, R. L. 1983. Oxygen-isotope and geochemical characterization of hydrothermal alteration in ophiolite complexes and modern oceanic crust. *In*: GASS, I. G., LIPPARD, S. J. & SHELTON, A. W. (eds) *Ophiolites and Oceanic Lithosphere*. Geological Society, London, Special Publications, **13**, 199–204.

STRENS, M. R. & CANN, J. R. 1982. A model of hydrothermal circulation in fault zones at mid-ocean ridge crests. *Geophysical Journal of the Royal Astronomical Society,* **71**, 225–240.

—— & —— 1986. A fracture-loop thermal balance model of black smoker circulation. *Tectonophysics,* **122**, 307–324.

TOOMEY, D. R., PURDY, G. M., SOLOMON, S. C. & WILCOCK, W. S. D. 1990. The three-dimensional seismic velocity structure of the East Pacific Rise near latitude 9°30′N. *Nature,* **347**, 639–645.

URABE, T., BAKER, E. T. & THE RIDGE FLUX GROUP, 1994. An overview of multi-disciplinary cruise of *R/V Melville* to superfast-spreading East Pacific Rise, 13.5°S–18.5°S. *EOS,* **75 (44)**, 320.

Hydrothermal activity and segmentation in the Magnitogorsk–West Mugodjarian zone on the margins of the Urals palaeo-ocean

V. V. ZAYKOV[1], V. V. MASLENNIKOV[1], E. V. ZAYKOVA[1] & R. J. HERRINGTON[2]

[1]Institute of Mineralogy, Urals Branch Russian Academy of Sciences, Miass, Chelyabinsk district, 456301 Russia

[2]Department of Mineralogy, The Natural History Museum, Cromwell Road, London, SW7 5BD, UK

Abstract: The Magnitogorsk–West Mugodjarian zone is interpreted to have developed as a marginal rift to the main Urals palaeo-ocean. In this zone, seven distinct tectonic segments each with a strike extent between 140 and 220 km have been identified. These segments can be distinguished on the basis of volcanic facies and features of hydrothermal activity. In the three southernmost segments the volcanics comprise basaltic and rhyolite–basalt complexes. The former contain massive sulphides of Cyprus-type and cherty iron-formation whilst the latter contain massive sulphides of 'Uralian' type, cherty iron-formation and manganese oxide deposits. In the two central segments rhyolite–basalt complexes are developed associated with giant massive sulphide deposits (Sibay >100 million tonnes massive sulphides). In the two northernmost segments rhyolite–basalt and andesite–basalt complexes are developed, both associated with cherty iron-formation and manganese oxide deposits. It is proposed that the variations in frequency, size and composition of the metalliferous deposits between the segments are a result of differences in the sequence of Palaeozoic oceanic development. Rifting started during the Eifelian in the south (current position) and in Late Givetian times in the north. This diachronous rifting and tectonic development has influenced the major geological features of the segments.

Within the internal zones of the Uralian fold belt, remnants of Palaeozoic oceanic and island arc complexes are preserved (Zonenshain *et al.* 1984). Many of these fragments of former marine sequences contain submarine hydrothermal mineral deposits (e.g. massive sulphides, hydrothermal cherts), which show a heterogeneous distribution through the fold belt. This article discusses the manifestation of hydrothermal activity found in one of the most extensive of the internal tectonic zones, namely the Magnitogorsk–West Mugodjarian (MWM) 'synclinorium' (Fig. 1). This synclinorium formed a broad rift zone which developed during the Devonian on the margin of the Urals palaeo-ocean. The oceanic volcanic sequences are interpreted as having formed in an inter-arc setting (Zonenshain *et al.* 1984; Zaykov 1991).

This paper proposes that the setting, composition, and volume of hydrothermal activity within the MWM zone should be re-examined in the light of the possible original segmentation of the rift zone. This factor has been ignored in the literature to date, despite the description of numerous cross-faults active during Devonian volcanism and sedimentation (Frolova & Bu-

rikova 1977; Seravkin 1986). Evidence for this early segmentation is important in making any thorough comparison between ancient and modern oceanic ridge–arc hydrothermal systems.

Geological position of the hydrothermal deposits

The broader aspects of regional geology and tectonic history in the southern Urals have been discussed in works by Zonenshain *et al.* (1984) and Koroteev (1986). The main features and origin of the massive sulphide deposits formed in the marginal sea of the Urals palaeo-ocean have been discussed by numerous authors elsewhere, e.g. Prokin (1977), Smirnov (1988), Baranov *et al.* (1988), Zaykov (1991), Maslennikov (1991), Kontar (1992), Ivanov & Prokin (1992), Zlotnik-Hotkevich *et al.* (1993).

The Magnitogorsk–West Mugodjarian zone is the best preserved palaeo-oceanic structural unit of the south Urals (Zonenshain *et al.* 1984; Zaykov 1991). Currently, the interpreted palaeo-rift takes the form of a large synclinal structure (synclinorium). In the centre, the

From: MacLeod, C. J., Tyler, P. A. & Walker, C. L. (eds) 1996, *Tectonic, Magmatic, Hydrothermal and Biological Segmentation of Mid-Ocean Ridges*, Geological Society Special Publication No. 118, pp. 199–210.

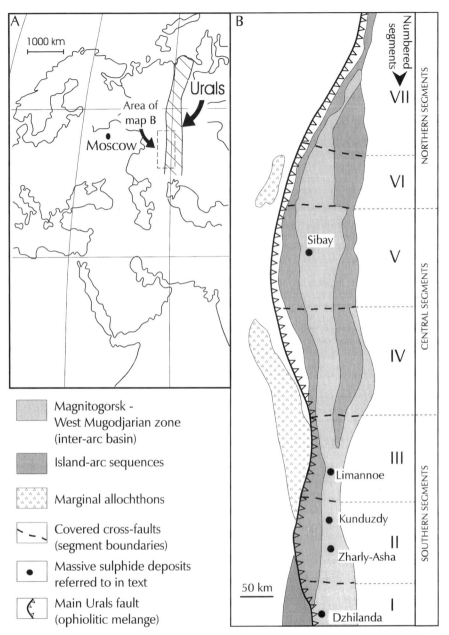

Fig. 1. (a) Locality map of southern Urals. (b) Key structural elements of the southwestern Urals Palaeozoic island arc – inter-arc sequences (after Zaykov 1991).

stratigraphy is largely sub-horizontal, whilst at the margins bedding dips between 30 and 60° towards the centre. The strata broadly span the Lower Devonian to Lower Carboniferous, younging from west to east. The southern part of the MWM zone is marked by a truly oceanic complex of Middle Devonian (Eifelian – Give-tian) age. Here tholeiitic (MORB) basalts show evidence for deep water formation (>3000 m), interpreted by Zonenshain *et al.* (1984) as having moderate ocean spreading rates comparable to those at the Gorda Ridge. Further north, the two flanks of the MWM zone comprise ridges of back inter-arc complex volcanic rocks which are

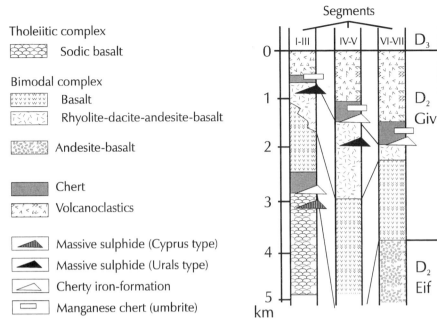

Tholeiitic complex

- Sodic basalt

Bimodal complex

- Basalt
- Rhyolite-dacite-andesite-basalt

- Andesite-basalt

- Chert
- Volcanoclastics

- Massive sulphide (Cyprus type)
- Massive sulphide (Urals type)
- Cherty iron-formation
- Manganese chert (umbrite)

Fig. 2. Stratigraphic columns for segments I–VII.

Lower to Middle Devonian age in the west and are bounded by the main Urals fault zone. In the east, the sequence is Middle to Upper Devonian in age with some Lower Carboniferous volcanics. The volcanic sequences in both areas are largely composed of two distinctive complexes: bimodal rhyolite–basalt and calc alkaline andesite–basalt. A further tholeiitic basaltic sequence is developed only in the southern part of the MWM zone, which is believed to represent oceanic crust (Zonenshain *et al.* 1984). The volcanic complexes are generally flanked by locally derived breccias and conglomerates. In the northern and southern parts of the zone, the boundaries between these volcanic ridges and the margins of the zone are complicated by faulting and are often marked by the presence of serpentinite bodies.

The host volcanics for the hydrothermal deposits are both tholeiitic sodic basalts and bimodal rhyolite–basalts as shown graphically in Figs 2 and 3. The lower parts of the bimodal series are represented by basaltic volcanics which pass upward into a more variable series of basalts, some andesites, dacites, and rhyolites. Actual host strata for mineralization include bands of jaspers, siliceous rocks, siltstones and hyaloclastic sandstones. Sea-water depth for the sulphide deposits can be interpreted from the lack of carbonates in the sequence as around 3000 m, below or near to the carbonate compen-

sation depth for the Palaeozoic (Zonenshain *et al.* 1984). The hydrothermal activity can be broadly divided into a high-temperature type, expressed largely as massive sulphide deposits and a low-temperature type characterized by cherty iron-formation and siliceous manganese deposits, a situation somewhat similar to the broad division of modern hydrothermal fields (Ivanov & Prokin 1992; Rona & Scott 1993).

Although sulphide deposits in the zone occur in both basalt and rhyolite-basalt complexes, the latter are the host to the largest proportion of the volcanogenic massive sulphide deposits. In the case of the basaltic complex hosted ores, the mineralogy is dominated by chalcopyrite–pyrite resembling typical Cyprus-type sulphide bodies. In the rhyolite–basalt complexes the ores are more polymetallic and consist of chalcopyrite, sphalerite, pyrite with minor bornite, galena, sulphosalts and tellurides. Russian geologists (e.g. Kontar 1992) classify such deposits as a distinct 'Urals' type, showing features of both Cyprus and Kuroko type sulphide bodies (Ivanov & Prokin 1992). Practically all the sulphide occurrences are accompanied by strata of submarine gossanites formed from the seafloor oxidation products of sulphide ores.

The submarine palaeohydrothermal fields in the Urals are rather unusual for the fossil record as they show some well preserved features of their seafloor vent origin. Black smoker chimney

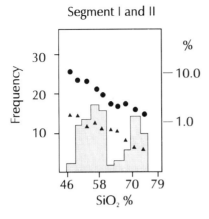

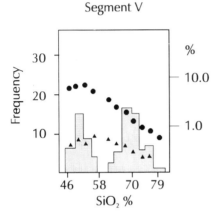

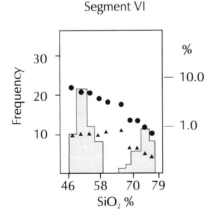

● MgO % (log scale)   ▲ TiO₂ % (log scale)

Fig. 3. Histograms of silica, magnesia and titania analyses for the bimodal complex (rhyolite–basalt) rocks.

fragments and well preserved, completely sulphidized vent-mouth fauna are recorded at Sibay (Kuznetsov *et al.* 1990). Also common are the reworked products of the sea floor destruction and oxidation of former sulphide masses (Zaykov *et al.* 1993).

In the footwall, evidence for palaeohydrothermal activity takes the form of conformable zones of metasomatic alteration, indicative of the rocks having formed hydrothermal fluid reservoirs and conduits. Hanging-wall strata show intense alteration in places caused by continued hydrothermal activity. In deposits where carbonates are present in the hanging wall, hydrothermal karst has developed. In the hanging-wall strata, buried mineralization is also indicated by the presence of mineralized xenoliths in later lavas and intrusive bodies, and reworked mineralized fragments in volcanoclastic units.

Segmentation

In the exposed sections of the MWM zone, seven distinct segments have been recognized with strike lengths of between 140 and 220 km (see Fig. 1). The segments can be distinguished by differences in volcanism and hydrothermal activity. Cross faults, highlighted by lithological offsets, parallel dyke swarms, topographic depressions and sharp offsets in published gravity data, separate the segments.

The existence of these segments during volcanism, sedimentation and hydrothermal activity is demonstrated by: (1) a distinct facies contrast between volcanic complexes of different segments; (2) compositional differences in the Middle Devonian bimodal rhyolite–basalt complex which hosts the bulk of the hydrothermal mineralization; (3) differences in the thicknesses of Late Givetian volcanoclastic deposits. These features are shown graphically in Figs 2 to 4 and in Table 1. Preliminary lithogeochemical data on the basalts of the southern segments show them to have higher average TiO₂ and MgO contents. The average content of SiO₂ in the rocks of the bimodal suites also show some differences.

The cross structures which separate the segments have widths of between 10 and 30 km. Key evidence for segmentation occurring during the Mid-Devonian is the temporal correlation of cross-cutting dyke swarms which parallel the segment faults with the MWM zone bimodal volcanism. Later Givetian age sediments are largely unaffected by the segment structures (Zaykov 1991).

Continued segmentation during the Late

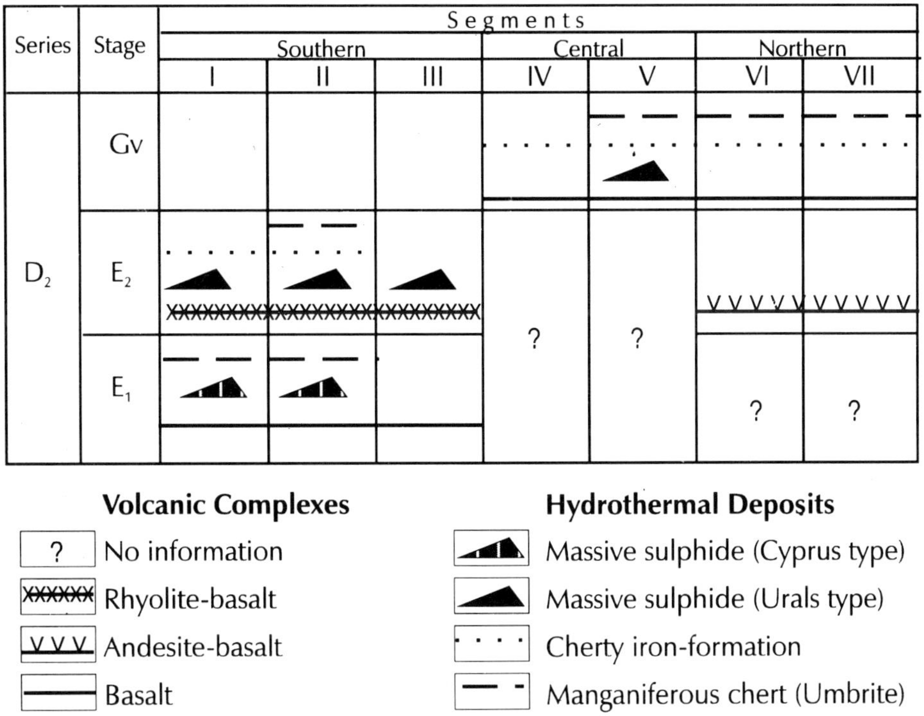

Series	Stage	Segments						
		Southern			Central		Northern	
		I	II	III	IV	V	VI	VII
D₂	Gv				· · · · ·	· · · ◣	· · · ·	· · · · ·
	E₂	· · · · · ◣	· · · ◣	◣	?	?	∨∨∨∨∨∨∨∨∨	
		✕✕✕✕✕✕✕✕✕✕✕✕✕✕✕✕✕✕✕✕✕✕✕						
	E₁	◢◣	◢◣				?	?

Volcanic Complexes

?	No information
✕✕✕✕	Rhyolite-basalt
∨∨∨	Andesite-basalt
———	Basalt

Hydrothermal Deposits

◢◣	Massive sulphide (Cyprus type)
◣	Massive sulphide (Urals type)
· · · ·	Cherty iron-formation
— -	Manganiferous chert (Umbrite)

Fig. 4. Schematic representation of comparative volcanic and hydrothermal activity in the various segments.

Devonian is demonstrated by contrasts in sedimentary facies. In the northern and central segments, grey siliciclastic sediments, indicative of deep-sea conditions, are characteristic. In the south, rudaceous sediments of reworked volcanic detritus, interpreted as shallow-water sediments, dominate. Such facies contrasts persisted into the Carboniferous, reflected by the general dominance of coal measures in the southern segments, whilst carbonate facies are prevalent in the other segments.

The segment structures are not directly indicated by current geomorphology, but are expressed as differences in the degree of erosion of Palaeozoic sequences. The southern segments have large outcrops of Middle Devonian rocks whereas Upper Devonian and Carboniferous rocks are preserved in only synclines. The central and northern segments are less eroded and in most cases Carboniferous sequences crop out.

Hydrothermal deposits

The southern segments (I, II, III)

Basaltic complex. The basaltic complex rocks are host to both sulphide and cherty iron-formation deposits (Fig. 5). The former are well developed in two areas, located close to the boundary of segments I and II. The sulphides occur some 200–300 m below the top of the basaltic unit, at the change of aphyric basalts to fine-porphyritic plagioclase andesitic basalts with abundant amygdales (Zaykov *et al.* 1993).

Ore bodies in the Zharly–Asha deposit (Fig. 1) take the form of mounds, measuring some 200–300 m in diameter with a maximum thickness of 10–15 m. The ore mineralogy is pyrite, sphalerite–chalcopyrite–pyrite and chalcopyrite–pyrite with both massive and brecciated structures. Clastic sulphide ores form thin interlayers on the flanks of the more massive lenses. In the footwall basalts, a funnel-shaped alteration pipe with quartz–chlorite and quartz–chlorite–carbonate alteration zones is developed (Zaykov 1991).

Cherty iron-formation (jasperites) form lenses and small mounds and are found distributed along the full length of segments I and II. Jasperites are concentrated in the lower part of a generally red siliciclastic unit which overlies the basalts. This unit is around 50 m thick but can be subdivided into basal, middle and top sections. The basal part can contain 2–7 m thick tabular bodies of scarlet and dark-lilac hematite–quartz

Table 1. *Chemical compositions of volcanic rocks from the MWM zone*

Complex	No. of segment	Range SiO_2	No of analyses	SiO_2	TiO_2	Al_2O_3	Fe_2O_3	FeO	MnO	MgO	CaO	Na_2O	K_2O	P_2O_5	Total
Basaltic, dykes	I–III	46–52	93	49.05	1.65	14.10	4.59	7.78	0.21	8.18	8.78	3.50	0.16	0.21	98.21
		52–58	9	54.96	1.40	14.20	3.96	7.05	0.16	6.45	6.01	4.33	0.38	0.22	99.12
Basaltic, lavas	I–II	46–52	92	49.02	1.76	14.62	6.19	6.28	0.23	7.11	7.61	3.37	0.25	0.20	96.74
		52–58	39	54.22	1.70	14.12	6.98	4.95	0.16	6.00	4.85	4.71	0.35	0.27	98.35
Rhyolite–basaltic, lavas	I–III	46–52	9	50.27	0.99	15.73	3.58	6.57	0.21	8.05	6.22	4.06	0.31	0.32	96.31
		52–58	33	55.23	0.83	14.71	4.48	5.81	0.15	4.89	5.62	4.08	0.48	0.15	96.43
		58–64	18	60.73	0.77	14.03	3.20	5.39	0.14	3.36	3.72	4.57	0.61	0.23	96.75
		64–70	7	67.19	0.56	13.39	4.35	3.87	0.85	1.11	3.17	4.16	0.51	0.14	99.30
		70–79	19	72.86	0.28	11.82	1.80	3.09	0.07	0.34	1.65	4.62	0.46	0.10	97.09
Rhyolite–basaltic, lavas	V	46–52	58	49.14	0.55	15.22	3.64	8.20	0.18	7.87	7.78	3.10	0.40	0.07	96.15
		52–58	39	55.03	0.63	14.69	3.79	7.14	0.18	5.92	5.80	3.71	0.42	0.08	101.12
		58–64	8	59.48	0.75	14.33	3.47	6.12	0.14	4.79	3.61	5.10	0.67	0.15	98.61
		64–70	38	65.75	0.51	13.21	2.16	4.67	0.10	1.85	2.23	5.06	0.34	0.14	96.02
		70–79	66	74.26	0.28	11.42	1.42	3.71	0.08	0.94	1.37	4.60	0.51	0.10	100.11

Compiled partial analytical data (excludes water, CO_2) from Zaykov, V. V.; Korinevsky, V. A. & Medeov, E. M. (previously unpublished data).

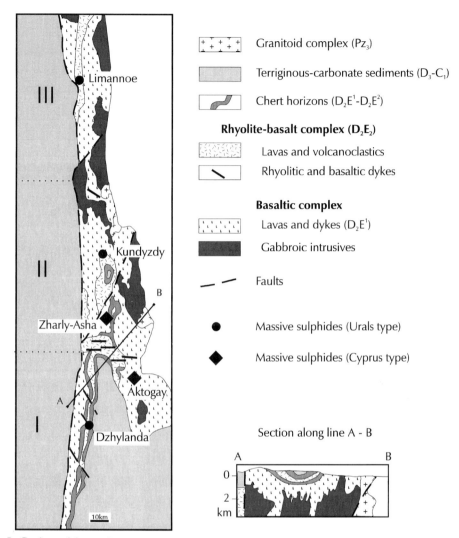

Fig. 5. Geology of the southern segments showing key massive sulphide deposits.

rocks which may be fine grained and homogeneous, spotted or show breccia textures. Thin red-brown siliciclastic layers occur. In the middle and top sections of the unit, mounds and lenses of banded and nodular hematite–quartz are found up to 1–3 m thick and some tens of metres in strike length. At the top and on flanks of these mounds medium coarse sediments cemented by hematite and chlorite formed from the reworked products of these mounds. The jasperites form linear groups of deposits up to 2–3 km wide and 5–8 km long developed along both the margins and axes of the MWM zone. A hydrothermal origin for the jasperites is pro-

posed based on chemistry, petrographic features, general morphology, structure and their stratigraphic association (Zaykov 1991).

Rhyolite–basalt complex. In the bimodal rhyolite–basalt complex hydrothermal mineralization is more varied. Massive sulphide mineralization occurs in three main areas (Dzhilanda, Limmannoe and Zharly–Asha–Kunduzdy, see Figs 1 & 5). The mineralization occurs in the middle part of the bimodal complex (Fig. 2). The volcanics hosting the sulphide deposits are related to tectonic depressions of the order of 2–3 km in width and 5–10 km in length which

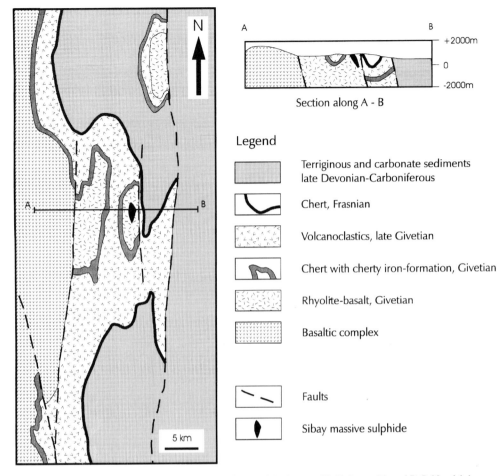

Fig. 6. Geology of the western part of segment V (unpublished map of I. V. Lennykh and P. I. Nozdrin).

controlled the volcanism. Sulphide bodies take two main forms, tabular-bedded and simple mounds.

The Dzhilanda deposit contains multiple sulphide bodies which have a thickness of up to 55 m and 200–400 m in diameter. They are composed of fine-grained sphalerite–pyrite–chalcopyrite ores with massive, spotted, brecciated and banded structures. Other minor phases include galena, marcasite, arsenopyrite, pyrrhotite, tennantite, tellurides and native gold. The Limmannoe deposit shows some mineralogical zoning, expressed in a lower chalcopyrite–pyritic and chalcopyrite–sphalerite–pyrite facies passing upward into a sphalerite–pyrite facies. Footwall alteration forms bedded zones of 200–300 m thickness below the sulphide and comprises quartz–sericite, quartz–chlorite–sericite and quartz–

chlorite zones. The sulphide bodies of the Kunduzdy deposit are lens shaped, showing relics of the original sea-floor sulphide mounds. The axial parts of the lenses are formed of massive and brecciated chalcopyrite–sphalerite–pyrite ores. On the flanks they pass into more layered ores of similar compositions and these pass into sediments containing sulphide clasts and gossanites.

Chert horizons occur in the same ore fields, but are situated higher up stratigraphically (Fig. 2). These cherts form a package of red siliciclastic units, similar to the jasperites in the basaltic complex rocks. Manganese oxide deposits are also found in small depressions, confined to the cross structure between segments I and II. These oxide bodies have a thickness of 5–10 m and are composed of alternating pyrolusite–psilomelane and siliciclastic layers.

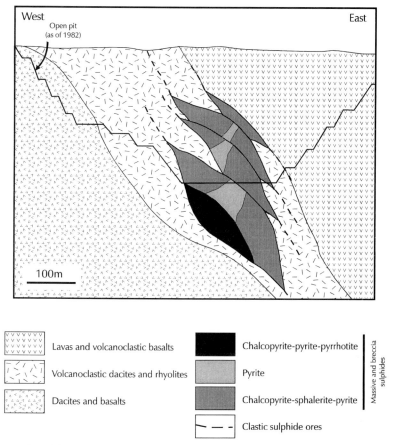

Lavas and volcanoclastic basalts	Chalcopyrite-pyrite-pyrrhotite
Volcanoclastic dacites and rhyolites	Pyrite
Dacites and basalts	Chalcopyrite-sphalerite-pyrite
	Clastic sulphide ores

Massive and breccia sulphides

Fig. 7. Cross-section of the Sibay, multi-tiered massive sulphide deposit (after Zhabin 1977 and Maslennikov 1991).

The central segments (IV, V)

Hydrothermal sulphide deposits are hosted in the middle portions of the bimodal complex rocks in the centre segment V (see Figs 2 & 6). The hydrothermal field formed in a small rift (width 1–3 km), cutting a large volcanic dome complex (Maslennikov 1991). This rift hosts the giant (>100 million tonnes sulphide) Sibay massive deposit.

The main Sibay massive sulphide body takes the form of three stacked lenses (Fig. 7) with thicknesses of between 20 and 40 m, which are connected at the apices (Zhabin 1977; Smirnov 1988; Maslennikov 1991). The total thickness of this complex body ranges up to 110 m, and it has a diameter of up to 700 m down dip. The mounds are mineralogically zoned with pyrite ores in the axial zone passing into assemblages of chalcopyrite–pyrite and finally sphalerite–chalcopyrite–pyrite on the flank. A cyclic structure of the

individual lenses is indicated by a regular alternation of textures and mineralogical ore types. The lower mound is somewhat different, being made up of two chalcopyrite–siderite–pyrite lenses, which contain early pyrrhotite and magnetite. At the top of the lens siderite disappears, and the ores become vuggy with chalcopyrite or sometimes sphalerite infilling the vugs. These textures are similar to vuggy and drusy textured ores that are forming at the tops of many modern 'black smoker' mounds. Brecciated sulphides flank and overlie the massive sulphide lenses and here chalcopyrite–pyrite fragments are cemented by quartz, pyrite and aggregates of sphalerite.

In the roof of the mounds, faunal communities and interbedded units containing fragments of sulphidised fauna are found (Kuznetsov et al. 1990). One community developed over an area of 10 m^2 is formed of pyrite ore, entirely replacing tube worms and bivalve shells. The central part comprises accumulations of tubes which measure up to 30–40 cm long and between 0.2 and 1.0 cm across. Tubes cross-cut bedding in the surrounding sulphides, supporting the view that they have been fossilized in a life position. The tube-worm fossils are similar to the often thick and bushy aggregations seen in the modern hydrothermal systems and resemble the modern vestimentiferans, particularly *Riftia*. Mineralogical analysis showed the fossil tubes to consist mainly of pyrite and sphalerite. The bivalves apparently filled a similar ecological niche to modern *Calyptogena*, which is recorded at both hot vents and cold seeps (Turner & Lutz 1984; Juniper & Sibuet 1987).

Cherty iron formation and manganese oxide deposits are more widespread. They formed at the end of rhyolite–basalt volcanism, when deposition of a generally red siliciclastic chert unit in both the central and boundary regions of the segments occurred (Fig. 2). These deposits form groups in several areas, and are up to 30 m thick and 100–150 m in diameter. Actual fossil vent sites are marked by travertine-like jasperites containing tabular clasts of manganese oxide. Macroscopic textures are mainly massive, brecciated, and aggregated. The vent proximal bodies are flanked by jasperitic units formed as a result of later degradation of the central mounds. More distally the clasts in these flanking units are manganese-rich tabular fragments.

The northern segments (VI, VII)

Hydrothermal deposits in these segments are very much less common in these segments. Minor jasperite bodies are developed throughout the stratigraphic section of the bimodal rhyolite–basalt complex. The main controls to distribution seem to be located on palaeo-scarps along flanks of volcanic ridges fringing the MWM zone.

The largest deposits, 4–6 m thick and developed over several tens of metres of strike, consist of hematite and quartz with both massive and brecciated texture. The latter include fragments of hyaloclastic volcanics replaced by iron oxides. Malachite is sometimes present in oxidized material as kidney-shaped segregations, crusts, films and dendrites.

Manganese oxide deposits (umbrites) are developed at the base of the grey and red siliciclastic units, which together with interbedded carbon-bearing siltstones form beds of up to 2–3 m thick. The mineralogy of the manganese ore is largely psilomelane and braunite, sometimes rhodonite and rhodochrosite. Braunite is

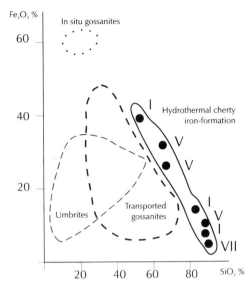

Fig. 8. Bulk composition of iron-bearing sediments from the MWM zone (from Zaykov 1991). Roman numerals refer to analyses in Table 2 indicating segment location.

considered to be a primary mineral reworked into the umbrite. Manganese carbonates are also primary minerals, whilst the silicates (rhodonite, bustamite) are formed during diagenesis and metamorphism.

The composition of hydrothermal deposits

The sulphide deposits show only a minor variation in composition within the segments. The ores mainly comprise chalcopyrite–sphalerite–pyrite and pyrite ores although pyrrhotite ores occur in the Sibay deposit of the central segment. Abundant oxidation products are characteristic of the roofs and flanks of all the sulphide ore bodies. A characteristic feature of the south segment sulphide bodies (Dzhilanda, Kunduzdy and Limmannoe) is a marked cobalt enrichment.

Jasperites of the tholeiite basaltic complexes appear to be more iron rich compared to jasperites of the bimodal rhyolite–basalt complexes. However, the *in situ* sea-floor gossan zones (gossanites) formed after sulphides are rather variable in composition. They contain between 40 and 80% Fe_2O_3 and 0.1 to 1.5% P_2O_5, but more significantly Cu and Zn from 0.1 to 1% and elevated values of Au, Ag, and rare earths (up to 0.03%), particularly Eu and Ce (Zaykov 1991). The major element chemistries of the various metalliferous sediments are shown on Fig. 8 and Table 2.

Table 2. Chemical compositions of chert formations from MWM zone (Zaykov 1991)

No	No. of samples	SiO_2	TiO_2	Al_2O_3	Fe_2O_3	FeO	MnO	MgO	CaO	Na_2O	K_2O	P_2O_5	S	CO_2	Total
S1	6	82.38	<0.01	0.20	13.12	2.62	0.05	0.08	0.61	0.17	0.07	0.13	–	0.64	100.07
S2	7	46.59	0.31	3.99	36.37	2.01	0.25	1.26	3.57	0.14	0.47	0.58	–	4.24	98.88
S3	13	87.41	0.04	0.80	5.50	2.57	0.08	0.55	1.06	0.16	0.08	0.16	–	1.45	99.7
S4	2	80.41	0.03	1.30	14.36	1.64	0.02	0.12	1.16	–	–	0.06	0.23	1.17	100.27
S5	10	64.50	0.04	1.46	24.44	2.60	0.09	0.62	1.99	0.18	0.23	0.11	–	0.53	97.02
S6	8	87.19	0.05	0.70	10.29	1.28	0.18	0.13	0.09	0.15	0.16	0.09	–	0.37	101.96
S7	2	64.04	0.05	0.55	30.06	2.20	0.58	0.11	0.10	0.20	0.25	0.16	–	0.28	98.58
S8	5	91.90	0.06	1.67	2.66	1.45	0.16	0.66	0.19	0.10	0.36	0.05	–	0.42	99.68

Sample localities and age: Basaltic complex D_2E age: S1–3 (segment I) S4 (segment II) Rhyolite–basalt D_2Giv age: S5–7, 9–12 (segment V), S8 (segment VII)

Conclusions

(1) Differences between the segments of the Magnitogorsk–West Mugodjarian Zone of the Urals palaeo-ocean probably reflect differences in the sequence of the initial rift opening. Rifting started during the Eifelian in the south (current position), with formation of oceanic crust, which started in the late Givetian times in the north. This has influenced the important geological features of the segments, the width of the zone basin, and basement architecture.

(2) The progressive spreading was accompanied by a change of volcanic facies. In the southern segments these are oceanic tholeiitic basalts followed by bimodal rhyolites–basalts; in the central segments dominantly bimodal rhyolite–basalts and in the northern segments the volcanics are calc alkaline andesite-basalts and bimodal rhyolites–basalts.

(3) The stratigraphic setting of hydrothermal activity shows a lateral age zonation in line with the development of the volcanics. The sulphide deposit ages change from Eifelian in the south to Late Givetian in the north.

(4) A broad field of massive sulphide deposits in the central part of the segments appears to be divided by the segment cross-structures.

(5) The density of distribution of the sulphide bodies significantly varies along the course of the Urals rift. The largest sulphide orebodies (in some cases estimated at many hundreds of millions of tonnes) formed in the central segments. In the southern segments the ore bodies are an order of magnitude smaller, whilst the northern segments are generally unproductive.

The Russian authors thank the Russian Foundation of Fundamental Researches (project 93-05-9181), International Scientific Foundation (project NOH000). We acknowledge the help of O. Telenkov, I. Sinjakovskaja, S. Maslennikova, S. Tesalina and H. Belogub. R. J. H. thanks the Natural History Museum Interdisciplinary Research Fund for fieldwork support. Finally, the authors are very grateful for the support of V. V. Koroteev and V. N. Anfilogov. C. Stanley, G. Cressey and R. Symes are thanked for their comments on the manuscript.

References

BARANOV, E. N., SCHTEINBERG, A. D. & KARPUKHINA, V. S. 1988. A genetic model and exploration criteria for buried massive sulphide deposits of the Verkhneuralsky area, southern Urals, USSR. In: Proceedings of the 7th Quadrennial IAGOD Symposium. E. Schweizerbart'sche Verlagsbuchhandlung (Nägele u. Obermiller), D-7000 Stuttgart1, 449–460.

FROLOVA, T. I. & BURICOVA, I. A. 1977. *The geosynclinal volcanism*. Moscow University.

IVANOV, S. N. & PROKIN, V. A. (eds) 1992. *Copper-pyritic deposits of Urals. Formation conditions*. UrO RAN, Ekaterinburg.

JUNIPER, S. K. & SIBUET, M. 1987. Cold seep benthic communities in Japan subduction zones: spatial organization, trophic strategies and evidence for temporal evolution. *Marine Ecology, Progress Series*, **40**, 115–126.

KONTAR, E. S. 1992. Cu, Zn and Pb sulphide deposits through geological history and in the structure of the Urals. *CSM Quarterly Review*, **92(1)**, 43–54.

KOROTEEV, V. A. 1986. The ophiolites of the Eastern slope of the South Urals. *Tectonophysics*, **127**, 361–369.

KUZNETSOV, A. P., MASLENNIKOV, V.V., ZAYKOV, V.V. & ZONENSHAIN, 1990. Fossil hydrothermal vent fauna in Devonian sulfide deposits of the Uralian ophiolites. *Deep-Sea Newsletter (Denmark)*, **17**, 9–10.

MASLENNIKOV, V.V. 1991. *The lithological control of copper-pyritic ores (by an example of Sibay and Oktyabrsk deposits of Ural)*. UrO AN SSSR, Sverdlovsk.

PROKIN, V. A. 1977. *Location relationships of massive sulphide deposits in the South Urals*. Nedra, Moscow.

RONA, P. A. & SCOTT, S. D. 1993. A special issue on sea-floor hydrothermal mineralization: New perspectives. *Economic Geology*, **88**, 1933–1976.

SERAVKIN, I B. 1986. *The volcanism and massive sulphide deposits of the South Urals*. Nauka Publishing House, Moscow.

SMIRNOV, V. I. (ed.) 1988. *Copper-pyritic deposits of Urals, Geological structure*. UrO AN SSSR, Sverdlovsk.

TURNER, R. D. & LUTZ, R. A. 1984. Growth and distribution of mollusca at deep-sea vents and seeps. *Oceanus*, **27**, 55–62.

ZAYKOV, V. V. 1991. *Volcanicity and massive sulphide hills of palaeooceanic margins (Urals and Siberia massive sulphide-bearing zones for example)*. Nauka Publishing House, Moscow.

——, MASLENNIKOV, V. V. & ZAYKOVA, E. V. 1993. *Volcanism and metalliferous deposits of the Devonian island arc system of the Southern Urals*. UrO RAN, Ekaterinburg.

ZHABIN, A. G. 1977. Peculiarities of ore body of ore deposition hydrothermal-sedimentary facies. *Geology of Ore Deposits*, **1**, 51–69.

ZLOTNIK-HOTKEVITCH, A. G., MIRLIN, E. G., ELYANOVA, E. A., SHIRAI, E. P., AGEEVA, S. T., MIRONOV, YU. V., ZORINA, YU. G. & KUZNETSOV, A. G. 1993. Comparative Classification of Formation Settings and Sulphide Ores. *Resource Geology Special Issue*, **17**, 160–168.

ZONENSHAIN, L. P., KORINEVSKI, V. G., KAZMIN, V. G., PECHERSKI, D. M., KHAIN, V. V. & MATVEENKOV, V. V. 1984. Plate tectonic model of the South Urals development. *Tectonophysics*, **109**, 95–135.

Ocean-ridge segmentation and vent tubeworms (Vestimentifera) in the NE Pacific

EVE C. SOUTHWARD[1], VERENA TUNNICLIFFE[2], MICHAEL B. BLACK[2,3],
DAVID R. DIXON[4] & LINDA R.J. DIXON[1]

[1]*Marine Biological Association, Citadel Hill, Plymouth PL1 2PB, UK*

[2]*School of Earth and Ocean Sciences/Biology, University of Victoria, Victoria, BC, Canada V8W 2Y2*

[3]*Present address: Center for Theoretical and Applied Genetics, Cook College, Rutgers University, New Brunswick, NJ 08903-0231, USA*

[4]*Plymouth Marine Laboratory, Citadel Hill, Plymouth PL1 2PB, UK*

Abstract: Vestimentiferan tube worms are important components of the hydrothermal vent and cold seep communities of the Pacific Ocean. The distribution and geographic intraspecies variation of *Ridgeia piscesae* and *Lamellibrachia barhami* were examined in the region of the Explorer (1 site), Juan de Fuca (33 sites) and Gorda (2 sites) Ridges and the nearby Cascadia Subduction zone. Isozyme electrophoresis, DNA restriction fragment length polymorphism (RFLP) and DNA sequencing techniques have been used. *Ridgeia piscesae* is widespread at hot vent sites along the ridges. The transform offset between Explorer and Juan de Fuca does not appear to impede gene flow (isozyme and RFLP data). Hydrographic conditions close to the Juan de Fuca Ridge favour bidirectional along-axis transport of planktonic larvae and some cross-axis transport. Rapid colonization by *R. piscesae* has been observed at a new vent site on CoAxial Segment of the Juan de Fuca Ridge, supporting the idea of a pool of larvae in the water overlying the ridge. The 360 km offset between southern Juan de Fuca and Gorda Ridges is associated with a significant level of genetic differentiation (shown by RFLP) which indicates some interruption to larval dispersal at this scale. The occurrence of the cold-seep species *Lamellibrachia barhami* is confirmed at one hydrothermal site, the sedimented Middle Valley on Juan de Fuca Ridge. Vestimentiferans appear to have a tremendous dispersal, location and adaptation capability. It is only at the thousand kilometre scale, that complete barriers to dispersal occur.

Geological features of the underlying terrain can influence range subdivision and speciation of organisms, but assessment of the role of crustal events in terrestrial and shallow water ecosystems is complicated by climate and latitudinal effects. Identification of the major factors controlling evolution and ecology of deep-sea organisms is somewhat easier. Species restricted to ridge crests, such as those at hydrothermal vents, are likely to respond strongly to geological features. The phenomenon of ridge segmentation is interesting as vent species' ranges tend to transgress segment boundaries.

Owing to ridge linearity the hot vent habitat has little 'lateral' dimension. Physical discontinuity arising from segmental offsets means that populations are disjunct. Additionally, the geological behaviour of each segment is relatively independent, because of varying activity of separate magma chambers (MacDonald *et al.* 1991). Longevity of hydrothermal activity at any one field and the frequency of habitat disruption by volcanic and tectonic events are also important influences on biology and biogeography. Is it possible to detect population responses to such segmental characteristics in a species that inhabits several segments of the same ridge? One can predict two likely 'responses'. The first is a classic phenomenon in linearly distributed species: species clines or gradual changes in characteristics from one end of the range to the other. A cline indicates genetic communication between adjacent populations but little connection between the ends of the range. A second response in ecological or behavioural characters may occur because of different conditions of venting among segments (Van Dover & Hessler 1990; Baker & Hammond 1992). The size, age, vigour and depth of a ventfield all influence habitat and community structure which, in turn, define the niche boundaries and competitor–predator interactions of a species.

From: MacLeod, C. J., Tyler, P. A. & Walker, C. L. (eds) 1996, *Tectonic, Magmatic, Hydrothermal and Biological Segmentation of Mid-Ocean Ridges*, Geological Society Special Publication No. 118, pp. 211–224.

211

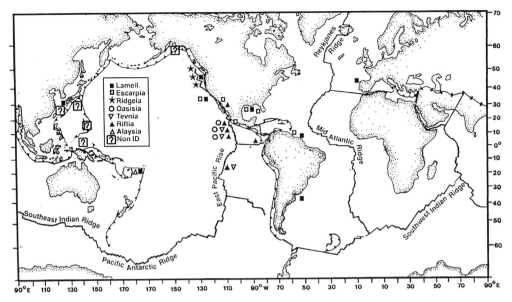

Fig. 1. World distribution of the genera of vestimentiferan tube-worms. Lamell, *Lamellibrachia*; Non ID, not identified.

In the absence of some unifying force, populations in isolated habitats, such as hydrothermal vent fields, tend to diverge genetically because of random drift and local bottlenecks. Such events can lead to significant genetic differentiation if effective population sizes are small, relative rates of increase are low and migrant exchange is limited (Wright 1951). Recent genetic studies of vent organisms indicate considerable gene flow for East Pacific Rise (EPR) populations of the brooding amphipod *Ventiella sulfuris* (France *et al.* 1992), western Pacific populations of *Bathymodiolus* spp. (Moraga *et al.* 1994), EPR populations of *B. thermophilus* (Craddock *et al.* 1995), EPR populations of the tube worm *Riftia pachyptila* and vesicomyid clams (Black *et al.* 1994; Trivedi *et al.* 1994; Vrijenhoek *et al.* 1994) and EPR populations of alvinellid polychaetes (Jollivet *et al.* 1995). However, populations of a vent snail, *Alviniconcha* sp., show great genetic differentiation between adjacent back-arc basins in the western Pacific (Denis *et al.* 1993).

This paper examines published and new information on tubeworms of the NE Pacific for evidence of variation that may reflect ridge segmentation.

Vestimentiferans

In the eastern Pacific Ocean vestimentiferan tubeworms often dominate hydrothermal com-

munities, making a major contribution to biomass and providing a framework for colonization by other organisms. Information is available on their distribution, physiology, ecology and taxonomy (Jones 1985; Childress & Fisher 1992; Southward *et al.* 1995). Recently there has been some intraspecific and intergeneric analysis at the molecular level (Williams *et al.* 1993; Trivedi *et al.* 1994; Southward *et al.* 1995). Vestimentiferans contain chemosynthetic symbiotic bacteria which enable them to make efficient use of bacterial carbon fixation, with the oxidation of sulphide as an energy source (Childress & Fisher 1992). This restricts them to situations where both oxygen and dissolved sulphide are available, in both hot and cold effluents. Four genera (*Riftia*, *Tevnia*, *Oasisia* and *Ridgeia*) are known at hydrothermal sites on the spreading ridges of the eastern Pacific (Jones 1985; Van Dover *et al.* 1990; Tunnicliffe 1991; Black *et al.* 1994) (Fig. 1). Cold seeps on the continental margin are inhabited by species of *Lamellibrachia* and *Escarpia* (Webb 1969; Jones 1985).

Lamellibrachia is the most widespread vestimentiferan genus: it has been reported from 'cold' sites on the east and west margins of both the Pacific and Atlantic Oceans, and also from two hydrothermal sites, the Lau Back-arc Basin in the western Pacific, and Middle Valley on the Juan de Fuca Ridge (Van der Land & Nørrevang 1975; Suess *et al.* 1985; Mañé-Garzón & Mon-

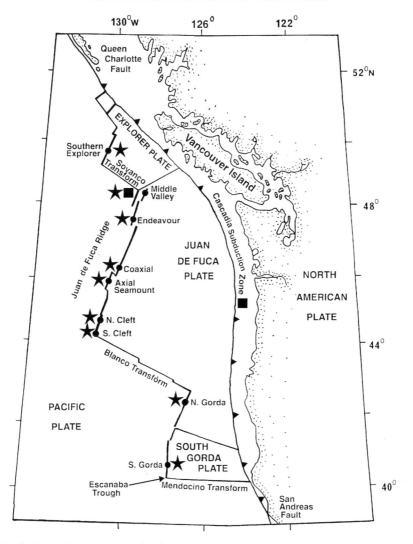

Fig. 2. Distribution of *Ridgeia piscesae* (star) and *Lamellibrachia barhami* (square) in the Northeast Pacific.

téro 1986; MacDonald *et al.* 1989; Hashimoto *et al.* 1989; Jollivet *et al.* 1990; E. C. Southward 1991; Dando *et al.* 1992; Williams *et al.* 1993). Several genera inhabit the back-arc basin hydrothermal fields of the western Pacific: *Lamellibrachia* and *Alaysia* in the Lau Basin (E. C. Southward 1991), a slender spiral tube-worm, similar to *Alaysia*, photographed in the Okinawa Basin (Kim & Ohta 1991), and a new genus in the Manus Basin (Southward & Galkin in preparation). There are no vestimentiferans at the Mid-Atlantic Ridge hydrothermal sites (Segonzac 1992; Van Dover 1995), although vestimentiferans inhabit cold seeps on the Atlantic margins (Jollivet *et al.* 1990; Dando *et al.* 1992).

The regional nature of the distribution of the genera is apparent from Fig. 1. The discontinuity imposed by the North American Plate between the Northeast Pacific ridge system and the East Pacific Rise, separates the eastern Pacific tubeworms, with *Riftia*, *Tevnia* and *Oasisia* on the EPR and *Ridgeia* on the Juan de Fuca Ridge (Jones 1985).

Geographical distribution of *Ridgeia piscesae* and *Lamellibrachia barhami*

The two species of *Ridgeia*, *R. piscesae*, and *R. phaeophiale* (Jones 1985) were amalgamated by Southward *et al.* (1995) under the name *R.*

piscesae, following examination of allozymes and morphology of worms from more than 50 vent sites on the Juan de Fuca Ridge. The original distinction of two species on the basis of obturacular saucer number and tube colour (Jones 1985) was based on specimens that represent the extremes of a range of characteristics related to size, predation and vent-fluid conditions. It was concluded that though *Ridgeia piscesae* is highly variable in appearance this is a result of the effects of differences in living conditions. Stable carbon isotope ratios differ between sites, apparently related to environmental effects on bacterial autotrophy (Southward *et al.* 1994). *Ridgeia piscesae* settles gregariously and forms small clusters or large bushes of tubes, on rock surfaces or the sides of chimneys, exposed to vent fluids. Recorded ranges for temperature and sulphide are 5–60°C and up to 920 μmol kg^{-1} sulphide (unpublished data).

A population of *Lamellibrachia* at Middle Valley on the Juan de Fuca Ridge, was identified from photographs as another variant of *Ridgeia piscesae* (Juniper *et al.* 1992, fig. 7A). Specimens collected on later cruises are clearly similar to *Lamellibrachia barhami* from the continental slope off Oregon and California, as described by Webb (1969) and Jones (1985). Comparative DNA studies show their distinction from *Ridgeia* and attribution to *Lamellibrachia* (Williams *et al.* 1993). The Middle Valley specimens correspond closely to the slope specimens of *L. barhami* in size, shape, number of branchial sheath lamellae, and tube characters. This unusual occurrence of *Lamellibrachia* on a mid-ocean ridge must be related to the sedimented nature of the Middle Valley site and may reflect the proximity of the slope subduction zone. Middle Valley is covered by hemipelagic and turbidite sediments but active hydrothermal discharge has produced mounds and sulphide outcrops. *Ridgeia piscesae* grows in bushy clusters up to 1 m high around minor sources of warm venting (Juniper *et al.* 1992). *Lamellibrachia* occurs in flatter, muddy areas between mounds, where rock outcrops are small and scattered and there is no obvious venting, though patches of black sediment indicate seepage of sulphidic fluids. The sinuous tubes emerge from the mud surface singly or in straggly groups, close to low outcrops or boulders that perhaps provide an attachment point for the unseen lower end of the tube. This habitat is different from that of *R. piscesae*, but photographs show that *Lamellibrachia* tubes sometimes occur within a metre or so of *Ridgeia* clumps. *Ridgeia piscesae* has been found at all known hydrothermal vent-fields from the South Explorer Ridge to the Southern Gorda Ridge (Fig. 2), including more than 50 vents on six segments of the Juan de Fuca Ridge. Its depth range is 1470–3320 m (Jones 1985; Tunnicliffe & Fontaine 1987; Tunnicliffe & Juniper 1990; Milligan & Tunnicliffe 1994; Juniper *et al.* 1992; Southward *et al.* 1995). *Lamellibrachia barhami* has been reported from the Californian slope at about 32°N in 1125–1847 m depth (Webb 1969) and from the Oregon slope at about 44°N in 1400–2050 m (Suess *et al.* 1985). This paper adds Middle Valley (Juan de Fuca Ridge) to the known distribution of *L. barhami*. The *Lamellibrachia* sp. report from cold seeps in the Guaymas Basin (Simoneit *et al.* 1990) may be incorrect; recent Guaymas collections appear to be *Escarpia spicata* (M. Black, unpublished observation).

Ridgeia morphology

Southward *et al.* (1995) report the range of variation in tube form, size and colour in *Ridgeia piscesae* on the Juan de Fuca Ridge. This variation appears to be related to habitat, particularly to vent fluid conditions. There is no gradual change in morphology from one end of the ridge to the other.

New material from the Explorer Ridge and Gorda Ridge, collected in 1994 (details in Table 1) extends comparison. *Ridgeia piscesae* tubes from the Southern Explorer Ridge match forms found on the Juan de Fuca Ridge. However, specimens attributed to *Ridgeia* from the Escanaba Trough on the Southern Gorda Ridge include a tube form with very large funnel shaped flanges (Fig. 3), beyond the known range of variation for the species. Another form, with small tube flanges, has bright orange branchial filaments in life, instead of the usual transparent filaments coloured red by contained blood. An orange particulate deposit is present in the epidermal cells of the filaments, presumably a mineral precipitate.

Molecular evidence

Techniques appropriate to detection of geographic intraspecies variation, isozyme electrophoresis, DNA restriction fragment analysis and DNA sequencing (Hillis & Moritz 1990), have now been applied to vestimentiferan species problems. The first has been used with large numbers of specimens from many sites, while the second and third have been used recently for analysis of DNA from smaller numbers of specimens. The DNA analyses are still in

Table 1. *New vestimentiferan collections*

Dive No.	Latitude	Longitude	Depth	*Ridgeia*	*Lamellibrachia*
1992 ROPOS Cruise July '92					
Juan de Fuca Ridge – Middle Valley					
199	48°44.72′N	128°42.5′W	2420m	+	+
1994 Atlantis II cruise 94-065					
Oregon Subduction Zone					
2797	44°40.53′N	125°17.41′W	2089m	–	+
North Gorda Ridge					
2799	42°45.32′N	126°42.15′W	2720m	+	–
South Gorda Ridge–Escanaba Trough					
2800	41°00.03′N	127°27.41′W	3317m	+	–
Juan de Fuca Ridge – Middle Valley					
2803	48°27.40′N	128°42.52′W	2416m	+	+
2805	48°27.25′N	128°42.66′W	2417m	+	–
2006	48°27.39′N	128°42.52′W	2419m	+	+
South Explorer Ridge					
2087	49°45.61′N	130°15.51′W	1814m	+	–

This material supplements collections reported in Southward *et al.* (1995).

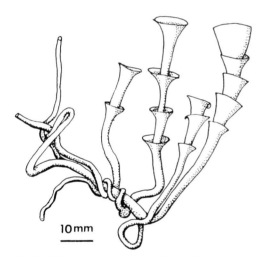

Fig. 3. *Ridgeia piscesae*: unusual form of tube from Escanaba Trough (South Gorda Ridge).

progress, but we include some preliminary results here.

Isozyme electrophoresis

The *Ridgeia* electrophoresis survey (Black 1991) consisted of 33 collections, with 9–40 specimens each, from four venting regions along the Juan de Fuca Ridge: Cleft Segment (Cleft), Axial Seamount (Axial), the Endeavour Segment (EndV) and Middle Valley (MidVal). These collections included populations matching both described species, *Ridgeia piscesae* Jones and R.

phaeophiale Jones, and three putative species of differing morphology. A total of 1025 specimens were scored for full genotypic complement at 15 presumptive gene loci in eleven enzymes, using standard methods of horizontal starch gel electrophoresis (Murphy *et al.* 1990). Details of methods are given in Black (1991) and Southward *et al.* (1995). The latter paper examines the results for evidence of multiple species. Five loci (*Fba-1*, *Fuh-1*, *Idh-2*, *Ldh-1* and *Tpi-1*) were fixed for the same electromorph in all collections. Conventional Chi-square analysis showed that most loci met expectations of Hardy–Weinberg equilibrium. Allele frequencies were remarkably similar across all collections. Nei's unbiased genetic distances (Nei 1972) for all pairwise estimates indicated no genetic support for differentiation of multiple species within the collections (Black 1991; Southward *et al.* 1995).

We now examine these data for population variation within *Ridgeia piscesae*. Analysis of the collections as separate populations justifies pooling of data for ten polymorphic loci for analysis of regional differentiation and gene flow (Table 2). Allele frequencies are very similar in all populations, with the exception of a frequency reversal for the two *Mpi-1* alleles in Middle Valley specimens. Significance of genetic variance among populations F_{ST} was tested using the relationship $\chi^2 = 2 N_T F_{ST}$ with $(s - 1)$ degrees of freedom where N_T is the total number of individuals sampled across s populations (Workman & Niswander 1970). Five of the 10 polymorphic loci show significant heterogeneity in gene frequencies, and the overall F_{ST} is also

Table 2. Ridgeia piscesae: *Allele frequencies at 10 polymorphic loci*

Locus	Population				F_{ST}
	MidVal	EndV	Axial	Cleft	
N	(70)	(141)	(395)	(419)	(1025)
demes	2	4	12	15	33
Aat-1					
110	0.829	0.918	0.870	0.897	
100	0.171	0.082	0.130	0.103	0.011*
Est-1					
100	0.757	0.833	0.743	0.764	
087	0.164	0.113	0.168	0.143	
076	0.050	0.028	0.066	0.058	
065	0.029	0.025	0.023	0.035	0.005
Gpi-1					
108	0.714	0.709	0.734	0.736	
100	0.286	0.284	0.261	0.254	
090	0.000	0.007	0.005	0.010	0.001
Idh-1					
100	0.979	0.979	0.968	0.983	
083	0.021	0.021	0.032	0.017	0.001
Lap-1					
100	0.793	0.812	0.837	0.808	
069	0.207	0.188	0.163	0.192	0.002
Lap-2					
100	0.979	0.954	0.935	0.895	
092	0.021	0.046	0.065	0.105	0.017*
Mdh-2					
122	0.943	0.986	0.970	0.952	
100	0.057	0.014	0.030	0.048	0.008†
Mpi-1					
100	0.150	0.745	0.795	0.749	
080	0.850	0.255	0.205	0.251	0.298*
Mpi-2					
100	0.621	0.773	0.819	0.780	
093	0.379	0.227	0.181	0.220	0.030*
Tpi-2					
200	0.971	0.989	0.987	0.986	
100	0.029	0.011	0.013	0.014	0.003
Mean F_{ST}					0.070*

* Significant at $\alpha < 0.001$.
† Significant at $0.001 < \alpha < 0.005$.
Pooled samples from 4 vent regions on the Juan de Fuca Ridge.
N = number of individuals surveyed; demes = the number of discrete collections from each vent region pooled for these calculations; F_{ST} = genetic variance among populations.

significantly different from zero. However, Bonferroni adjustment of the significance level ($\alpha' = 0.001$) for the multiple Hardy-Weinberg tests resulted in only one significant difference, that for *Mpi-1*. This emphasizes the marked difference in allele frequencies in the Middle Valley sample at this locus.

Levels of genetic variability are relatively high (Table 3) and consistent with the high levels of genetic variation recorded for many other marine invertebrates (Nevo *et al.* 1983). The significance of the mean fixation indices (F_{IS}) was tested using the relationship $\chi^2 = F^2N$ with one degree of freedom, where N is the mean sample size of the four populations (Li & Horvitz 1953). Significant values in Table 2 result from the heterozygote deficiencies evident in the unpooled collections, which most likely reflect a combined effect of slight inbreeding and Wahlund effect. Observed heterozygosity levels in both pooled and unpooled analyses are always slightly lower than expected under random mating.

Table 3. Ridgeia piscesae: *measures of genetic variability*

Population	(N)	$P_{0.95}$	H_{obs}	H_{exp}	F_{IS}
MidVal	70	46.7	0.149 [0.046]	0.160 [0.046]	0.069
EndV	141	40.0	0.108 [0.032]	0.139 [0.042]	0.223*
Axial	395	46.7	0.114 [0.031]	0.147 [0.041]	0.224*
Cleft	419	46.7	0.115 [0.030]	0.156 [0.042]	0.263*
Means		46.7	0.116 [0.031]	0.154 [0.043]	0.191*

* Significant at $\alpha < 0.005$.
$P_{0.95}$, proportion of polymorphic loci, defined as the most common allele not exceeding 0.95 in frequency; H_{obs}, mean heterozygosity per locus from direct count of individuals; H_{exp}, unbiased estimate of heterozygosity per locus expected under random mating; F_{IS}; mean fixation index.

Gene flow in *Ridgeia piscesae*

Two traditional gene flow models can be applied to vent populations. The 'stepping-stone' model is appropriate where gene flow between neighbouring populations greatly exceeds long range dispersal. Correlation in gene frequencies decreases exponentially as the number of steps between populations increases resulting in 'isolation by distance' (Kimura & Weiss 1964). Rates of gene flow (effective migration rate, M) are estimated from gene frequency data using among population genetic variance (F_{ST}) according to the relationship $M = (1/F_{ST} - 1)/4$ (Slatkin 1993). This model predicts that a species with limited larval dispersal should have a negative relationship between estimates of gene flow and geographical distance.

Wright's 'island model' may be a better approximation where adult populations contribute equally to a mixed pool of larvae that disperse randomly throughout the taxon's range. Gene frequencies may differ substantially among island populations as a consequence of random genetic drift, however no decrease in gene flow with geographical distance is expected. Thus the relationship between gene flow and geographical distance can help determine modes and rates of dispersal (Sokal & Wartenberg 1983; Slatkin 1993). Table 4 presents estimated effective migration rates, M, for *Ridgeia piscesae* along the Juan de Fuca Ridge. The value of M is equivalent to the pseudoparameter N_m, where N is the average effective population size and m is the average proportion of migrants in each population. Values of N_m greater than 1 are sufficient to counter differentiation of local populations at equilibrium (Wright 1931, 1943). The values in Table 4 must

be viewed with caution, as estimates of gene flow greater than 10 are highly suspect in any strict sense, other than indicating very high levels of migrant exchange and hence essential panmixia. Nevertheless, the high estimates for the three ridge segments (Cleft, Axial and Endeavour), relative to the Middle Valley site, a failed rift offset to the northeast of the main Juan de Fuca Ridge axis, are interesting. Factors such as larval transport to this northernmost site, Middle Valley, and larval recruitment in sedimented conditions may be at play.

These gene-flow data do not reflect any 'isolation by distance' signal, as was recorded for EPR populations of *Riftia pachyptila* (Black *et al.* 1994). They are comparable to the finding by Jollivet *et al.* (1995) that the vent polychaete *Paralvinella grasslei* shows the same magnitude of genetic varation among populations separated by 1000 km along the EPR as among populations within one ridge segment. The Juan de Fuca Ridge is a smaller system and it seems that distance alone does not pose any barrier. Processes of recruitment at vent sites are probably more complex than could be explained by either of the models above. Metapopulation processes will greatly complicate these scenarios and confound interpretations of population structure and dynamics (Slatkin 1977; Wade & McCauley 1988). High rates of local extinction and recolonization from a limited gene pool will rapidly lead to genetic homogeneity within the metapopulation (Maruyama & Kimura 1980; Gilpin 1991). Of interest in this regard is the absence of rare alleles among any of the original 33 populations analysed, a phenomenon also reported in *Riftia pachyptila* (Black *et al.* 1994). Frequent population extinctions and founder effects (i.e. genetic bottlenecks) should rapidly

Table 4. Ridgeia piscesae: *estimates of gene flow*

Population	MidVal	EndV	Axial	Cleft
MidVal	–	0.0809	0.0916	0.0776
EndV	2.84	–	0.0035	0.0021
Axial	2.48	70.46	–	0.0018
Cleft	2.97	118.21	141.18	–

Standardized variances, G_{ST} (Nei 1973, above diagonal) and estimated effective migration rates, M (below diagonal). Values computed using the DIST program (Slatkin 1993) from multilocus averages of pair-wise F_{STS}, as estimated by G_{ST} method (Nei & Chesser 1983).

purge rare alleles, but if local populations have high intrinsic rates of increase then hetero-zygosity resulting from remaining common alleles will not diminish substantially (Nei *et al.* 1975).

Restriction enzyme fragment length polymorphism

Different parts of the eukaryotic genome evolve at different rates. Coding sequences (genes) evolve very slowly (over millions of years), whereas non-coding regions (gene introns, spacers) evolve much more quickly (Avise 1994). These non-coding sequences accumulate DNA sequence changes without incurring a selection penalty. They represent potentially neutral markers for use in population genetics studies, assuming that an interruption to gene flow between populations eventually results in genetic differentiation owing to a combination of genetic drift and the accumulation of random mutations. This study concentrated on the intergenic spacer region (IGS, formerly called the 'large non-transcribed spacer') separating the 28S and 18S genes (Suzuki *et al.* 1987) in the ribosomal DNA (rDNA) repeat unit of *Ridgeia piscesae*. In keeping with the majority of other species, the genes coding for ribosomal RNA in *R. piscesae* occur as tandem arrays separated by non-coding spacers (D.R. Dixon *et al.* in preparation).

High molecular-weight genomic DNA was purified by proteinase K digestion followed by repeated phenol/chloroform extractions (Sambrook *et al.* 1989). All DNA samples were treated with RNAse at the time of extraction. After ethanol precipitation, pelleted DNA was vacuum dried and then dissolved in TE (10 mM Tris-HCl, 1 mM EDTA, pH 8) prior to storage at either 4°C or −20°C. Homologous ribosomal DNA (rDNA) probes were manufactured by polymerase-chain-reaction amplification of selected parts of the rDNA repeat unit of *Ridgeia piscesae* 18S and 28S genic regions:

(Holland *et al.* 1991; Williams *et al.* 1993) and cloned by conventional bacterial methods. Probe labelling and detection were accomplished using a non-radioactive method (digoxygenin, Boehringer-Mannheim; Kessler *et al.* 1990). DNA fragments were sorted according to size by gel electrophoresis on 1% agarose gel in TBE buffer (90 mM Tris-HCl, 0.89 mM boric acid, 2.5 mM EDTA, pH 8.3). Running conditions were 30–33 volts for 16–18 hours at room temperature. Gels were stained with ethidium bromide and viewed under UV-light to check the digest quality prior to vacuum blotting the DNA onto a nylon membrane (Gene-screen, Dupont) for probing.

For *Ridgeia piscesae*, 3 μg aliquots of DNA from different individuals were cut with restriction enzymes, either singly or in combination, according to the manufacturers' instructions, and blotted and probed as described above. The enzyme cutting sites within the rDNA repeat unit were mapped (D. R. Dixon *et al.* in preparation), and the restriction fragment patterns from replicated individuals from different sites examined for evidence of rDNA polymorphism (RFLPs).

Out of the 14 restriction enzymes analysed, three in particular yielded significantly high levels of polymorphism in their cutting patterns: *Eco*RI, *Sal*I, and *Cla*I. Restriction mapping revealed that virtually all the polymorphic cleavage sites were in the IGS, a putatively neutral region of the genome. The three enzymes had a basic genotype comprising between 1 and 8 bands that was present at all localities, plus an additional 8 (*Eco*RI and *Cla*I) or 17 (*Sal*I) genotypes which in most cases had a more restricted distribution. The higher number of polymorphic genotypes for *Sal*I (Table 5) reflects the greater number of cutting sites, which spanned the entire IGS, whereas *Eco*RI and *Cla*I had fewer cleavage sites, that were restricted to the proximal end of this hypervariable region.

Two levels of genetic variation were detected

Table 5. *rDNA genotype frequencies obtained using three restriction endonucleases in seven populations of* Ridgeia piscesae

Endonuclease		Population							
		Expl	MidVal	EndV1	EndV2	Axial	Cleft	NGor	SGor
*Eco*RI	N	23	26	23	33	16	44	27	22
	A	0.91	0.96	0.65	0.58	0.94	0.91	0.96	0.95
	B				0.06		0.09	0.04	
	C			0.22	0.18				
	D				0.03				
	E	0.09	0.04	0.09					
	F			0.04	0.12				
	G				0.03				
	H					0.06			
	I								0.05
*Sal*I	N	23	18	22	21	12	23	26	22
	A	0.30	1.00	1.00	0.76	1.00	0.96	0.88	0.68
	B	0.09							
	C	0.17							
	D	0.30						0.04	
	E	0.04							
	F	0.04							
	G				0.05				
	H				0.05				
	I				0.05				
	J				0.05				
	K				0.05		0.04		
	L							0.04	
	M	0.06						0.04	
	N								0.04
	O								0.04
	P								0.10
	Q								0.10
	R								0.04
*Cla*I	N	23	20	18	23	12	20	27	22
	A	0.61	0.55	0.56	0.35	0.33	0.35	0.52	0.23
	B	0.26	0.45	0.33	0.45	0.66	0.60	0.11	0.18
	C			0.11	0.13				
	D				0.04				
	E	0.13			0.08				
	F				0.04				
	G						0.05		
	H							0.07	0.05
	I							0.30	0.54

during this study: intra vent and inter ridge scale. Rather surprisingly, given the liklihood of a relatively long-lived planktonic larval stage (C. M. Young *et al.* 1996), the two samples of worms collected at different heights on the same chimney (referred to as Endv1 and Endv2 in Table 5) at Endeavour segment, Juan de Fuca Ridge, showed significant differences in the frequency of rDNA polymorphism for *Sal*I (Chi-square, $df = 2$, $P < 0.05$), but not for the other two enzymes, which suggests either settlements of genetically distinct larval cohorts at different times during the evolution of the chimney, or post-settlement selection, acting through the process of genetic linkage, favouring an otherwise rare cluster of alleles under different microecological conditions. The rDNA polymorphism frequency data for seven distinct vent fields (three ridges, seven segments; Table 5) was subjected to hierarchical analysis using Weir & Cockerham's coancestry θ (Weir & Cockerham 1984) for analysing population structure. There were significant differences in the frequency of rDNA polymorphism across the three ridges: Explorer, Juan de Fuca and Gorda, which indicated restricted gene flow between them. The average multi-endonuclease θ estimate (0.147) is high and significantly different from zero (t-test $= 4.36$, df $= 2$, $0.05 > P > 0.01$). While Gorda Ridge made the largest contribution to this variation, the population on Explorer Ridge made an appreciable, though not significant, addition ($\theta = 0.121$), which reduced to only 0.032 when the Gorda Ridge was considered separately, indicating that the 360 km long Blanco Fracture Zone separating the southern Juan de Fuca from the Gorda Ridge represents a major interruption to larval dispersal in *Ridgeia piscesae*.

When the endonucleases were analysed separately, excluding Gorda Ridge from the analysis had no effect on the θ value for *Eco*RI (0.19, $0.05 > P > 0.01$), whereas this caused a significant reduction in the θ value for *Cla*I (0.026, P.0.05; previously 5.1, $P < 0.01$). This indicated that nearly all the genetic differentiation between Gorda and the Juan de Fuca/Explorer Ridges was a result of polymorphism involving this enzyme. For *Sal*I, the θ value showed a small but not significant reduction when Explorer Ridge was excluded from the analysis (0.51, $0.05 > P > 0.01$), which suggests a slight divergence between the Explorer and Juan de Fuca Ridges also.

DNA samples from several *Lamellibrachia* specimens from two collections at Middle Valley were compared with a single specimen from the Oregon Subduction Zone (Table 1). No differences were found between the RFLPs for the Oregon Subduction zone specimen and the replicated Middle Valley specimens, thus indicating a single species of *Lamellibrachia*. In contrast, marked differences in band size and fragment number were recorded for each of the three different endonucleases when *Lamellibrachia barhami* was compared with *Ridgeia piscesae*, confirming the discrimination of the method.

DNA sequencing

A phylogenetic analysis of vestimentiferan DNA sequence data will be presented elsewhere (M. B. Black, pers. comm.). DNA isolated from two individuals from each *Ridgeia* collection (Table 1) from South and North Gorda Ridge, Cleft Segment, Endeavour Segment and Middle Valley (Juan de Fuca Ridge) and South Explorer Ridge has been sequenced for a 710 b.p. fragment of the mitochondrial cytochrome c oxidase subunit I (*COI*) gene (Folmer *et al.* 1994). While the *COI* gene is probably too conservative to differentiate closely related species, all of the *Ridgeia* sequences form a monophyletic clade with less than 0.2% sequence divergence among individuals for this gene region. Two individuals

of *Lamellibrachia* from the Oregon subduction Zone and two from Middle Valley (Table 1) show only five nucleotide substitutions for this same gene region (three of which are third position synonymous site changes) between sites, counted across all four individuals sequenced. Future sequencing of less conserved gene regions will provide greater resolution of the relationship among the *Ridgeia* morphs from the three ridge axes.

Dispersal and colonization

Vestimentiferans disperse as small planktonic larvae, but the free-living larvae have yet to be captured. Very recently the rearing of vestimentiferan larvae *in vitro* has demonstrated comparatively long lived non-feeding ciliated larvae in two cold seep species with small eggs (Young *et al.* 1996). The eggs of *Ridgeia piscesae* are similar in size, about 100 μm in diameter, and are, apparently, spawned freely into the water after internal fertilization. There is no evidence of incubation. Newly settled postlarvae have been found attached to the tubes of older individuals (Southward 1988; Jones & Gardiner 1989). The small size of *R. piscesae* postlarvae ($c. 150 \times 40$ μm) indicates little or no increase in biomass between egg and settlement. Their form suggests a free swimming stage similar to an annelid trochophore, with a large anterior ciliated ring like a prototroch, that is continuous with the ciliated mouth and gullet (in the postlarva). The larval food is probably small bacteria, if the larva feeds at all. After settlement the ciliated mouth region enlarges, the body lengthens and the gut has a digestive function. Small bacteria occur in the lumen (Southward 1988). Symbiotic bacteria become apparent in the gut epithelial cells at about 300 μm juvenile length. The midgut then develops into the trophosome, holding the multiplying symbionts, and becomes closed off from the exterior. Symbiotic bacteria have not been found in the eggs of vestimentiferans examined (Cary *et al.* 1993) and it may be that the larva has to acquire new symbionts from the environment at the time of settlement. However, there is evidence for host/symbiont specificity in several invertebrates with sulphur-oxidizing bacteria (Distel *et al.* 1988; A. J. Southward 1991) and none of the symbionts has yet been cultivated *in vitro*.

After their release the *Ridgeia piscesae* eggs may be carried to 250–350 m above the bottom in the hydrothermal plume. Oscillating tidal currents could hold larvae close to their point of release. Vertical recirculation and return to the

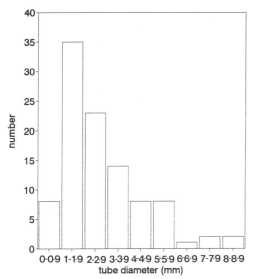

Fig. 4. Size distribution (tube diameter) of sample of *R. piscesae* from CoAxial Segment. *n* = 101.

same vent field is one possible route for the larvae (German & Sparks 1993). Along-axis currents measured at Cleft, Axial and Endeavour segments have residual rates of *c.* $2 \, cm \, s^{-1}$, northward on the west side of the ridge and southward on the east; there is some evidence of cross axis flow (Cannon & Pashinski 1990; Cannon *et al.* 1991, 1995; Thomson *et al.* 1990, 1992). A study of the Endeavour Ridge plume (Thomson *et al.* 1992) indicates that inert particles are carried laterally as much as 2.5 km from the vent site, while heat and salt anomalies show influence of the plume as much as 5 km from the vent. Thus, there are possibilities for larvae to be dispersed some kilometres from their origin, in either direction along the Juan de Fuca Ridge as well as to a lesser extent off-axis. Evidence for rapid colonization by *Ridgeia piscesae* at a new vent site has come from CoAxial segment, following a seismic event in June 1993. A visit to the site within two weeks of the eruption found new lavas at 'Flow Site' and vigorous fluid emissions at 'Floc Site' (Embley *et al.* 1995). The fluids contained large quantities of flocculated material of organic origin along with high temperature precipitates (Juniper *et al.* 1995) and thermophilic bacteria (Holden, pers. comm.). As the surface emissions were cool (max 55°C), the flocs were apparently flushed from a subsurface source at the abrupt onset of venting. No macrofauna were visible at either site after the eruption or during a second survey three months later. A possible source of colonis-

ing tubeworms was located at a small vent 20 km south of Floc Site.

In July 1994, *Ridgeia piscesae* and other organisms were present at Floc Site. Two small collections of *Ridgeia* were retrieved near markers deployed on vacant basalts in October 1993. The largest worm collected was 40 cm long. Figure 4 illustrates the size distribution of one collection. The worms greater than 4 mm tube diameter had reached sexual maturity and therefore the population may have begun to reproduce. However, nearly half the population was very small, which indicates continued recruitment from outside, but the low numbers in the smallest size category (post-settlement) may have resulted from loss during collection, since earliest recruits do not adhere well to the substratum of adult tubes.

This rapid colonization, within a few months of the eruption, supports the idea of a pool of larvae in the water overlying the ridge, ready to settle when they reach a suitable site. In fact, since *R. piscesae* was one of only two species to colonise in abundance, it must be regarded as one of the better opportunists of the northeast Pacific vents, perhaps providing needed habitat for later colonizing species.

Conclusions

Distribution and genetic data, taken together, indicate the tremendous dispersal capability of both *Lamellibrachia barhami* and *Ridgeia piscesae* within the NE Pacific region. *L. barhami* has presumably reached the Juan de Fuca Ridge from the continental slope, the nearest known site being off northern Oregon (400 km from Middle Valley), but similar seep sites probably exist elsewhere in the subduction zone, which extends northward to the Queen Charlotte Island region (Fig. 2). The population of *L. barhami* at Middle Valley is probably large enough to be self-sustaining, but the genetic data suggest that there is still gene flow between the Ridge and Slope populations. The absence of *L. barhami* from other ridge segments indicates that it is unable to colonize bare rock sites, and needs sediment-covered habitats. *Ridgeia piscesae* shows great capacity to colonise the vent sites from Explorer Ridge south to Gorda Ridge. The segmentation along the Juan de Fuca Ridge does not pose any barriers to interchange of larvae between vent fields and segments, although there is evidence of possible selection for particular genotypes on an extremely local scale in some cases. The segments have different magmatic characteristics and different surface topographies. The extensive isozyme analysis of

33 populations along the Juan de Fuca Ridge showed no major differentiation from segment to segment. There is evidence of reduced gene flow to the northernmost site at Middle Valley. Compared to the isozyme findings, the results of the rDNA analysis for *Ridgeia piscesae* revealed a higher level of genetic differentiation in different parts of the species' range. Perhaps this is not really surprising, given the neutral nature of these markers and the greater sensitivity of molecular markers in general compared to protein variation, which in a number of cases has been shown to be under the constraining influence of balancing selection (e.g. Karl & Avise 1992).

For *Ridgeia piscesae* the transform fault barriers in the NE Pacific are small. No significant genetic differentiation occurs across the offset of 160 km between the Juan de Fuca and Explorer Ridges. At the southern end of the Juan de Fuca Ridge the Blanco Fracture Zone interposes an offset of 360 km to the Gorda Ridge; some genetic differentiation between the ridges is shown by the RFLP analysis, representing a major interruption to larval dispersal. The Blanco Fracture Zone includes palaeohydrothermal systems (Hart *et al.* 1990) that may have provided intermittent opportunities for vestimentiferan colonisation in the past, which may not be happening today.

Only at a larger scale (about 1000 km), such as the Mendocino Fracture Zone plus the subduction of the Pacific plate under California, does a major barrier to vestimentiferan species occur. Here *Ridgeia piscesae* reaches its southern limit.

We are grateful for help in the preparation of this paper from A. J. Southward and L. Franklin. D. Jollivet and M. Carr helped with the statistical analysis of the RFLP data. R. A. Lutz, the chief scientist on the 1994 Atlantis II cruise, generously provided much of the new vestimentiferan material employed in this study. Funding has been provided by the Natural Science and Engineering Research Council of Canada, the Natural Environment Research Council of the UK (Grant GR3/8470), and the Leverhulme Trust.

References

AVISE, J. C. 1994. *Molecular Markers, Natural History and Evolution*. Chapman & Hall, New York.
BAKER, E. T. & HAMMOND, S. R. 1992. Hydrothermal venting and the apparent magmatic budget of the Juan de Fuca Ridge. *Journal of Geophysical Research*, **97**, B3, 3443–3456.
BLACK, M. B. 1991. *Genetic (allozyme) variation in Vestimentifera (*Ridgeia spp.*) from hydrothermal vents of the Juan de Fuca Ridge (Northeast Pacific Ocean)*. MSc. thesis, University of Victoria, BC, Canada.
——, LUTZ, R. A. & VRIJENHOEK, R. C. 1994. Gene

flow among vestimentiferan tube worm (*Riftia pachyptila*) populations from hydrothermal vents in the eastern Pacific. *Marine Biology*, **120**, 33–39.
CANNON, G. A. & PASHINSKI, D. J. 1990. Circulation near Axial Seamount, Juan de Fuca Ridge. *Journal of Geophysical Research*, **95**, 12823–12828.
——, —— & LEMON, M. R. 1991. Mid-depth flow near hydrothermal venting sites on the southern Juan de Fuca Ridge. *Journal of Geophysical Research*, **96**, 12815–12831.
——, —— & STANLEY, T. J. 1995. Fate of event hydrothermal plumes on the Juan de Fuca Ridge. *Geophysical Research Letters*, **22**, 163–166.
CARY, S. C., WARREN, W., ANDERSON, E. & GIOVANNONI, S. J. 1993. Identification and localization of bacterial endosymbionts in hydrothermal vent taxa with symbiont-specific polymerase chain reaction amplification and in situ hybridization techniques. *Molecular Marine Biology and Biotechnology*, **2**, 51–62.
CHILDRESS, J. J. & FISHER, C. R. 1992. The biology of hydrothermal vent animals: physiology, biochemistry and autotrophic symbioses. *Oceanography and Marine Biology, Annual Review*, **30**, 337–441.
CRADDOCK, C., HOEH, R., GUSTAFSON, R., LUTZ, R., HASHIMOTO, J. & VRIJENHOEK, R. C. 1995. Evolutionary relationships among deep sea mussels (Bivalvia: Mytilidae) from hydrothermal vents and cold-water methane/sulfide seeps. *Marine Biology*, **121**, 477–485.
DANDO, P. R., SOUTHWARD, A. J., SOUTHWARD, E. C., DIXON, D. R., CRAWFORD, A. & CRAWFORD, M. 1992. Shipwrecked tube worms. *Nature*, **356**, 667.
DENIS, F., JOLLIVET, D. & MORAGA, D. 1993 Genetic deparation of two allopatric populations of hydrothermal snails, *Alvini-concha* spp., (Gastropoda) from two south western Pacific back-arc basins. *Biochemical Systematics and Ecology*, **21**, 431–440.
DISTEL, D. L., LANE, D. J., OLSEN, G. J., GIOVANNONI, S. J., PACE, B., PACE, N. R., STAHL, D. A. & FELBECK, H. 1988. Sulfur-oxidizing bacterial endosymbionts: analysis of phylogeny and specificity by 16S rRNA sequences. *Journal of Bacteriology*, **170**, 2506–2510.
EMBLEY, R. W., CHADWICK, W. W. JR., JONASSON, I. R., BUTTERFIELD, D. A. & BAKER, E. T. 1995. Initial results of the rapid response to the 1993 CoAxial event: relationships between hydrothermal and volcanic processes. *Geophysical Research Letters*, **22**, 143–146.
FOLMER, O., BLACK, M., HOEH, W., LUTZ, R. & VRIJENHOEK, R. 1994. DNA primers for amplification of mitochondrial Cytochrome c Oxidase subunit I from diverse metazoan invertebrates. *Molecular Marine Biology and Biotechnology*, **3**, 294–299.
FRANCE, S. C., HESSLER, R. R. & VRIJENHOEK, R. C. 1992. GENETIC DIFFERENTIATION BETWEEN SPATIALLY DISJUNCT POPULATIONS OF THE DEEP-SEA, HYDROTHERMAL VENT ENDEMIC AMPHIPOD *Ventiella sulfuris*. *Marine Biology*, **114**, 551–556.
GERMAN, C. R. & SPARKS, R. J. 1993. Particle recycling

in the TAG hydrothermal plume. *Earth and Planetary Science Letters*, **116**, 129–1344.

GILPIN, M. E. 1991. The genetic effective size of a metapopulation. *Biological Journal of the Linnean Society*, **42**, 165–175.

HART, R., HOEFS, J. & PYLE, D. 1990. Multistage hydrothermal systems in the Blanco Fracture Zone. *In*: MCMURRAY, G. R. (ed.) *Gorda Ridge*. Springer-Verlag, 51–75.

HASHIMOTO, J., OHTA, S., TANAKA, T., HOTTA, H., MATSUZAWA, S. & SAKAI, H. 1989. Deep-sea communities dominated by the giant clam. *Calyptogena soyoae*, along the slope foot of Hatsushima Island, Sagami Bay, Central Japan. *Palaeogeography, Palaeoclimatology, Palaeoecology*, **71**, 179–192.

HILLIS, D. M. & MORITZ, C. (eds) 1990. *Molecular Systematics*. Sinauer Associates, Sunderland, Mass.

HOLLAND, P. W. H., HACKER, A. M. & WILLIAMS, N. A. 1991. A molecular analysis of the phylogenetic affinities of *Saccoglossus cambrensis* Brambell & Cole (Hemichordata). *Philosophical Transactions of the Royal Society of London*, **B332**, 185–189.

JOLLIVET, D., FAUGÈRES, J. C., GRIBOULARD, R., DESBRUYÈRES, D. & BLANC, G. 1990. Composition and spatial organization of a coldseep community on the Barbados accretionary prism: tectonic, geochemical and sedimentary context. *Progress in Oceanography*, **244**, 425–445.

——, DESBRUYÈRES, D., BONHOMME, F. & MORAGA, D. 1995. Genetic differentiation of deep-sea hydrothermal vent alvinellid populations (Annelida: Polychaeta) along the East Pacific Rise. *Heredity*, **74**, 376–391.

JONES, M. L. 1985. On the Vestimentifera, new phylum: six new species, and other taxa, from hydrothermal vents and else where. *Biological Society of Washington Bulletin*, **6**, 117–158.

—— & GARDINER, S. L. 1989. On the early development of the vestimentiferan tube worm *Ridgeia* sp. and observations on the nervous system and trophosome of *Ridgeia* sp. and *Riftia pachyptila*. *Biological Bulletiin, Woods Hole*, **177**, 254–276.

JUNIPER, S. K., TUNNICLIFFE, V. & SOUTHWARD, E. C. 1992. Hydrothermal vents in turbidite sediments on a northeast Pacific spreading center: organisms and substratum at an ocean drilling site. *Canadian Journal of Zoology*, **70**, 1792–1809.

——, MARTINEU, P., SARRAZIN, J. & GELINAS, Y. 1995. Microbial-mineral floc associated with nascent hydrothermal activity on CoAxial Segment, Juan de Fuca Ridge. *Geophysical Research Letters*, **22**, 179–182.

KARL, S. A. & AVISE, J. C. 1992. Balancing selection at allozyme loci in oysters: implications from nuclear RFLPs. *Science*, **256**, 100–102.

KESSLER, C., HOLTKE, H. J., SEIBL, R., BURG, J. & MUHLEGGER, K. 1990. Non-radioacive labelling a detection of nucleic acids: a novel DNA labelling and detection system based on dioxygenin:antidioxygenin ELISA principle. *Biological Chemistry Hoppe Seyler*, **371**, 917–927.

KIM, D. S. & OHTA, S. 1991. Submersible observations and comparison of the biological communities of the two hydrothermal vents on the Iheya Ridge of the Mid-Okinawa Trough. *Technical Reports presented at the 7th Symposium on Deep-sea Research using the Submersible 'SHINKAI 2000' System*. JAMSTEC. 221–233 [in Japanese].

KIMURA, M. & WEISS, W. H. 1964. The stepping-stone model of genetic structure and the decrease of genetic correlation with distance. *Genetics*, **49**, 561–576.

LI, C. C. & HORVITZ, D. G. 1953. Some methods of estimating the inbreeding coefficient. *American Journal of Human Genetics*, **95**, 107–117.

MACDONALD, I. R., BOLAND, G. S., BAKER, J. S., BROOKS, J. M., KENNICUTT, M. C. & BIDIGARE, R. R. 1989. Gulf of Mexico hydrocarbon seep communities II. Spatial distribution of seep organisms and hydrocarbons at Bush Hill. *Marine Biology*, **101**, 235–247.

MACDONALD, K. C., SCHEIRER, D. S. & CARBOTTE, S. M. 1991. Mid-ocean ridges: discontinuities, segments and giant cracks. *Science*, **253**, 986–994.

MAÑÉ-GARZÓN, F. & MONTERO, R. 1986. Sobre una nueva forma de verme tubicola – *Lamellibrachia victori* n.sp. (Vestimentifera) – Proposicion de un nuevo phylum: Mesoneurophora. *Revista Biologica de Uruguay*, **8**, 1–28.

MARUYAMA, T. & KIMURA, M. 1980. Genetic variability and effective population size when local extinction and recolonization of subpopulations are frequent. *Proceedings of the National Academy of Sciences, USA*, **77**, 6710–6714.

MILLIGAN, B. N. & TUNNICLIFFE, V. 1994. Vent and non-vent faunas of Cleft Segment, Juan de Fuca Ridge, and their relations to lava age. *Journal of Geophysical Research*, **99**, B3, 4777–4786.

MORAGA, D., JOLLIVET, D. & DENIS, F. 1994. Genetic differences across the Western Pacific populations of the hydrothermal vent bivalve *Bathymodiolus* spp. and the eastern Pacific (13°N) population of *Bathymodiolus thermophilus*. *Deep Sea Research*, **41**, 1551–1567.

MURPHY, R. W., SITES, J. W., JR., BUTH, D. G. & HAUFLER, C. H. 1990. Isozyme electrophoresis. *In*: HILLIS, D. & MORITZ, C. (eds) *Molecular Systematics*. Sinauer Associates. Sunderland, Mass, USA, 45–125.

NEI, M. 1972. Genetic distance between populations. *American Naturalist*, **106**, 283–292.

—— & CHESSER, R. K. 1983. Estimation of fixation indices and gene diversities. *Annals of Human Genetics*, **47**, 253–259.

——, MARUYAMA, T. & CHAKRABORTY, R. 1975. The bottleneck effect and genetic variability in populations. *Evolution*, **29**, 1–10.

NEVO, E., BEILES, A. & BEN-SHLOMO, R. 1983. The evolutionary significance of genetic diversity: Ecological, demographic and life history correlates. *Lecture Notes in Biomathematics*, **53**, 13–213.

SAMBROOK, J., FRITSCH, E. F. & MANIATIS, T. 1989. *Molecular Cloning. A Laboratory Manual*, vol. 1, 2nd edition. Cold Spring Harbor Laboratory Press.

SEGONZAC, M. 1992. Les peuplements associes a

l'hydrothermalisme oceanique du Snake Pit (dorsale médioatlantique; 23°N, 3480 m): composition et microdistribution de la mégafaune. *Comptes Rendus de l'Academie des Sciences, Paris*, **314**, (III) 593–600.

SIMONEIT, B. R. T., LONSDALE, P. F., EDMOND, J. M.& SHANKS, W. C. 1990. Deep-water hydrocarbon seeps in Guaymas basin, Gulf of California. *Applied Geochemistry*, **5**, 41–49.

SLATKIN, M. 1977. Gene flow and genetic drift in a species subject to local extinctions. *Theoretical Population Biology*, **12**, 253–262.

—— 1993. Isolation by distance in equilibrium and non-equilibrium populations. *Evolution*, **47**, 264–279.

SOKAL, R. R. & WARTENBERG, D. E. 1983. A test of spatial autocorrelation analysis using an isolation by distance model. *Genetics*, **105**, 219–237.

SOUTHWARD, A. J. 1991. Effect of temperature on autotrophic enzyme activity of bacteria symbiotic in clams and tube worms. *Kieler Meeresforschung*, Sonderheft **8**, 245–251.

——,SOUTHWARD, E. C., SPIRO, B., RAU, G. H. & TUNNICLIFFE, V. 1994. $^{13}C/^{12}C$ of organisms from Juan de Fuca Ridge hydrothermal vents: a guide to carbon and food sources. *Journal of the Marine Biological Association of the UK*, **74**, 265–278.

SOUTHWARD, E.C. 1991. Three new species of Pogonophora, including two vestimentiferans, from hydrothermal sites in the Lau Back-arc Basin (Southwest Pacific Ocean). *Journal of Natural History*, **25**, 859–881.

——1988. Development of the gut and segmentation of newly settled stages of *Ridgeia* (Vestimentifera): implications for relationship between Vestimentifera and Pogonophora. *Journal of the Marine Biological Association of the UK*, **68**, 465–487.

——, TUNNICLIFFE, V. & BLACK, M. 1995. Revision of the species of *Ridgeia* from northeast Pacific hydrothermal vents, with a redescription of *Ridgeia piscesae* Jones (Pogonophora: Obturata = Vestimentifera). *Canadian Journal of Zoology*, **73**, 282–295.

SUESS, E., CARSON, B., RITGER, S. D., MOORE, J. C., JONES, M. L., KULM, L. D. & COCHRANE, G. R. 1985. Biological communities at vent sites along the subduction zone off Oregon. *Biological Society of Washington Bulletin*, **6**, 475–484.

SUZUKI, H., MORIWIKI, K. & NEVO, E. 1987. Ribosomal DNA (rDNA) spacer polymorphism in mole rats. *Molecular Biology and Evolution*, **4**, 602–610.

THOMSON, R. E., ROTH, S. E. & DYMOND, J. 1990. Near-inertial motions over a mid-ocean ridge: effects of topography and hydrothermal plumes. *Journal of Geophysical Research*, **95**, C5, 7261–7278.

——, DELANEY, J. R., McDUFF, R. E., JANECKY, D. R. & McCLAIN, J. S. 1992. Physical characteristics of the Endeavour Ridge hydrothermal plume during July 1988. *Earth and Planetary Science Letters*, **111**, 141–154.

TRIVEDI, A. K., BLACK, M. B. & VRIJENHOEK, R. C. 1994. Allozyme diversity within and among tube worm (Vestimentifera) species from deep-sea hydrothermal vents of the East Pacific Rise *Isozyme Bulletin*, **27**, 59.

TUNNICLIFFE, V. 1991. The biology of hydrothermal vents: ecology and evolution. *Oceanography and Marine Biology, Annual Review*, **29**, 319–407.

——& FONTAINE, A. R. 1987. Faunal composition and organic surface encrustations at hydrothermal vents on the southern Juan de Fuca Ridge. *Journal of Geophysical Research*, **92**, 11 303–11 314.

—— & JUNIPER, S. K. 1990. Dynamic nature of the hydrothermal vent habitat and the nature of sulfide chimney fauna. *Progress in Oceanography*, **24**, 1–13.

VAN DER LAND, J. & NØRREVANG, A. 1975. The systematic position of *Lamellibrachia* (Annelida, Vestimentifera). *Kongelige Danske Videnskabernes Selskab, Biologiske Skrifter*, **21**(3), 1–102.

VAN DOVER, C. L. 1995. Ecology of Mid-Atlantic Ridge hydrothermal vents. *In*: PARSON, L. M., WALKER, C. L. & DIXON, D. R. (eds) *Hydrothermal Vents and Processes* Geological Society, London, Special Publications, **87**, 257–294.

—— & HESSLER, R. R. 1990. Spatial variation in faunal composition of hydrothermal vent communities on the East Pacific Rise and Galapagos Spreading Center. *In*: McMURRAY, G. R. (ed.) *Gorda Ridge*. Springer-Verlag, 253–264.

——, GRASSLE, J. F. & BOUDRIAS, M. 1990. Hydrothermal vent fauna of Escanaba Trough (Gorda Ridge). *In*: McMURRAY, G. R. (ed.) *Gorda Ridge*. Springer-Verlag, 285–287.

VRIJENHOEK, R. C., SCHUTZ, S. J., GUSTAFSON, R. G. & LUTZ, R. A. 1994. Cryptic species of deep-sea clams (Mollusca, Bivalvia, Vesicomyidae) in hydrothermal vent and cold-seep environments. *Deep Sea Research*, **41**, 1171–1189.

WADE, M. J. & McCAULEY, D. E. 1988. Evolution and recolonization: their effects on the genetic differentiation of local populations. *Evolution*, **42**, 995–1005.

WEBB, M. 1969. *Lamellibrachia barhami* gen. nov., sp. nov., (Pogonophora), from the northeast Pacific. *Bulletin of Marine Science*, **19**, 18–47.

WEIR, B. S. & COCKERHAM, C. C. 1984. Estimating F-statistics for the analysis of population structure. *Evolution*, **38**, 1358–1370.

WILLIAMS, N. A., DIXON, D. R., SOUTHWARD, E. C. & HOLLAND, P. W. H. 1993. Molecular evolution and diversification of the vestimentiferan tube worms. *Journal of the Marine Biological Association of the UK*, **73**, 437–452.

WORKMAN, P. L. & NISWANDER, J. D. 1970. Population studies on Southwestern Indian tribes. II. Local genetic differentiation in the Papago. *American Journal of Human Genetics*, **22**, 24–49.

WRIGHT, S. 1931. Evolution in Mendelian populations. *Genetics*, **16**, 97–159.

—— 1943. Isolation by distance. *Genetics*, **28**, 114–138.

—— 1951. The genetical structure of populations. *Annals of Eugenics*, **15**, 323–354.

YOUNG, C. M., VÁQUEZ, E., METAXAS, A. & TYLER, P. A. 1996. Embryology of vestimentiferan worms from deep-sea methane/sulphide seeps. *Nature*, **381**, 514–516.

Plate tectonic history and hot vent biogeography

VERENA TUNNICLIFFE[1], C. MARY R. FOWLER[2] & ANDREW G. MCARTHUR[3]

[1]*School of Earth & Ocean Sciences, University of Victoria, Victoria, BC, Canada V8W 2Y2*

[2]*Department of Geology, Royal Holloway, University of London, Egham, Surrey TW20 0EX, UK*

[3]*Department of Biology, University of Victoria, Victoria, BC, Canada V8W 2Y2*

Abstract: Since the 1977 discovery of hydrothermal vents and their extraordinary fauna, vents have been discovered along mid-ocean ridges in the eastern Pacific, on the Mid-Atlantic Ridge and on the spreading centres behind western Pacific subduction zones. The dependence of these habitats on plate activity suggests that plate history may play a strong role in their evolution. As few of the species known from vents are found in other marine habitats, an examination of biogeographic patterns with respect to plate history is feasible. An examination of taxon overlap with other sulphide-rich habitats (seeps and whale carcasses) suggests that a minority of faunal elements have common ancestors. Proximity of vent sites to each other does reflect regional similarities but it is the distance along ridge pathways and vicariant events that appear most important. An understanding of the tectonic history of our oceans provides a framework for predicting the composition of vent communities in unexplored areas.

The present-day distribution of a species reflects two phenomena: historical events that have shaped its range and the ability of the species to get around. Dispersability is important when examining the ecology and evolution of single species. For certain marine faunas, such as those found on seamounts, the role of dispersal is very important in community evolution (Briggs 1974; Johannesson 1988; Parker & Tunnicliffe 1994). However, geological or environmental events that influence all species in a regional fauna tend to have ubiquitous effects evident in the patterns of organism distributions. Thus the concept of vicariance has become important in biogeography: interpretation of organism patterns with respect to the formation and disappearance of range barriers. For instance, the range of North American placental mammals expanded into South America as the Isthmus of Panama formed in the late Pliocene (Marshall *et al.* 1982).

The movement of tectonic plates underlies many of the phenomena governing vicariating faunas. The magmatic activity at mid-ocean ridges and their segmentation also affects the distribution and abundance of hydrothermal venting and the associated faunas. One of the startling features of vent communities is that most of the species are new to science (Newman 1985; Grassle 1986; Tunnicliffe 1992). Perhaps this novelty is not unusual when considering the habitat conditions in which the animals have evolved: low oxygen and high sulphide levels, high heavy metal concentrations and large, fluctuating temperature ranges. In addition, the habitat is linear (along ridge crests) and can be locally unpredictable. Tunnicliffe (1991) and Childress & Fisher (1992) review studies on conditions and adaptations.

Vent communities represent excellent subjects for biogeography. They are wholly dependent upon ridge spreading, thus reconstructions of past ridge positions also locates likely positions of past vent communities. They are easy to recognize and document. Their recent discovery and the coordinated efforts of systematists mean that there is much greater consistency in identification of related groups of organisms for sites throughout the globe than is usual in biology. At present, the data set is still relatively small but one purpose of this paper is to illustrate areas of potential interest.

Seeps and whale carcasses

It is tempting to deal with hot vent communities alone due to ease of site definition and information access. However, there are potential relationships with other habitats that could confuse interpretations based on ridge histories

From: MacLeod, C. J., Tyler, P. A. & Walker, C. L. (eds) 1996, *Tectonic, Magmatic, Hydrothermal and Biological Segmentation of Mid-Ocean Ridges*, Geological Society Special Publication No. 118, pp. 225–238.

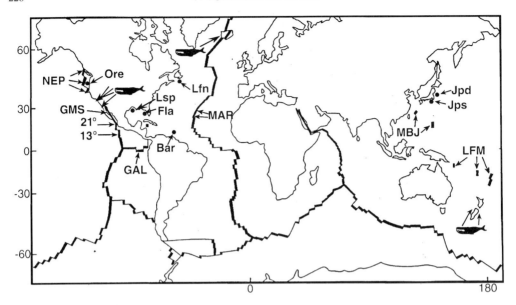

Fig. 1. Position of spreading ridges of today. Locations of vent sites included in this analysis are indicated in capital letters; site names can be found on Table 1. Seep locations are indicated in small letters and the whale locations by the icon.

alone. At continental margins, where compacting or subducting sediments generate sulphide and methane, seeping fluids attract high densities of animals, many of whom have specific adaptations to exploit potential chemical energy (Fisher 1990). The similarity of these organisms to those of hot vents has been noted, particularly in the animals that harbour symbiotic bacteria, such as the vestimentiferan tubeworms (Southward *et al.* this volume). Whale skeletons are another deep-sea sulphide source. They release reduced compounds for several decades from the lipid-rich mammal bones (Smith *et al.* 1989). These whale 'islands' have been cited as potential stepping-stones for vent organisms dispersing from one site to the next. Thus, if vent animals can occupy habitats other than ridges, we must look beyond ridge history to explain modern distribution patterns.

Plate reconstruction in the oceans

Reversals of the Earth's magnetic field and the record of that field in the permanent magnetization of the upper part of the oceanic crust as it forms along the mid-ocean ridges, provide the main method of dating the oceanic lithosphere. Marine magnetic anomalies (about 1% of the total magnetic field in magnitude) run parallel to axes of the mid-ocean ridges, are generally symmetric about the axes of the ridges, are offset by fracture zones and are a few tens of kilometres in width. These magnetic anomalies result from the permanent magnetization of the upper part of the oceanic crust as it cools. The magnitude and direction of the magnetization of the oceanic crust depend upon the latitude at which the crust was formed and are unaffected by later movements and position. Magnetic anomalies provide a powerful dating and position recorder for palaeo mid-ocean ridges. As the reversal time scale is known, individual anomalies can be attributed to a particular reversal and the measured anomaly can be used to determine the latitude and orientation of the mid-ocean ridge when the crust acquired its permanent magnetization. Thus, ideally, it is possible to gain information about the orientation, as well as the latitude and age, of the mid-ocean ridge at which any piece of oceanic lithosphere was formed.

The oldest oceanic lithosphere, in the northwestern Pacific, is of Jurassic age, so magnetic anomaly data can be used only to trace the past positions of the mid-ocean ridges back to that time. Earlier mid-ocean ridge positions must be based on geological evidence and continental magnetic data. Full reconstruction of the geological history of an ocean is possible only if all the oceanic lithosphere is preserved. When subduction has occurred, as in the Pacific, magnetic anomaly information is lost and recon-

Present

65 Ma

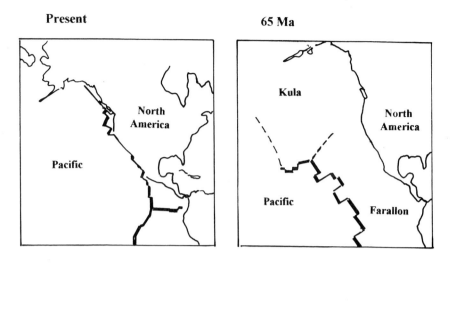

56 Ma

110 Ma

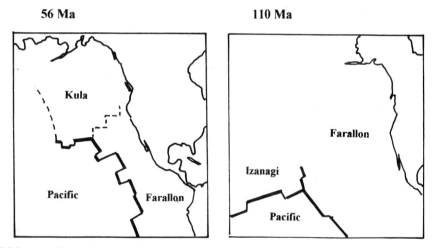

Fig. 2. Mid-ocean ridge positions for the northeast Pacific based on palaeomagnetic data. (After Atwater 1989).

struction becomes much more difficult. Detailed mapping of the oceanic magnetic anomalies and knowledge of the reversals of the Earth's field are now at such a level that the movements of the plates are fairly well known through the Cenozoic. However, knowledge of plates and their motions decreases with increasing age. Our knowledge of the locations of the Cretaceous and Jurassic Pacific and Indian mid-ocean ridge systems is far from complete.

We identify some aspects of plate history since the Jurassic that could have influenced relationships among regional fauna inhabiting the mid-ocean ridges. The present configuration of mid-ocean ridges suggests that pathways from the eastern Pacific to western Pacific to Atlantic occur through the southern Pacific and Indian Oceans, while the northeastern Pacific and the western Pacific back-arc basins are isolated (Fig. 1). If the main migration pathway for vent fauna is along the ridges, then past plate and ridge position may be reflected in the distribution patterns of taxa at present-day vents.

Eastern Pacific

The Tertiary evolution of the northeastern Pacific region is complicated but reasonably well understood. From about 80 Ma until about 55 Ma, there were four plates in the northeastern Pacific region: Pacific, Farallon, Kula and North American (e.g. Atwater 1989). The Pacific–Farallon, Farallon–Kula and Kula–Pacific ridges all met at a triple junction in the northeastern Pacific (Fig. 2). In the late Cretaceous and early Tertiary the Farallon plate was very large indeed and the Pacific–Farallon ridge was some 10 000 km in length. The Pacific plate lay to the west of the Pacific–Farallon ridge and the Farallon plate to the east. However as the Farallon plate was being subducted beneath western North America faster than new Farallon lithosphere was being formed at the ridge, it was inevitable that at some time the ridge would reach the subduction zone. This happened at 30 Ma: the Farallon plate was divided and the San Andreas Fault was born. Today the northern part of the Farallon plate is represented by the Juan de Fuca plate which is still being subducted beneath North America along the Cascadia subduction zone. The southern part of the Farallon plate also continued to be subducted beneath North America but it gradually broke up into smaller plates. At about 25 Ma it split into two (Cocos and Nazca) when an east–west mid-ocean ridge (the Cocos–Nazca ridge or Galapagos spreading centre) formed. The Cocos plate then subdivided at about 12 Ma, producing the small Rivera plate, so that now there are three plates (Rivera, Cocos and Nazca) to the east of the mid-ocean ridge known as the East Pacific Rise. Initial rifting in the Gulf of California started at about 14 Ma (Stock & Hodges 1989; Nicholson *et al.* 1994) and Baja California joined the Pacific plate.

Cross-Pacific

North Pacific. The Kula plate has now been all but subducted beneath North America (Lonsdale 1988). We know that it existed because the E–W magnetic anomalies created at the Pacific–Kula ridge are preserved in the Pacific plate. There is some uncertainty about its westward extension, although the Pacific–Kula ridge was certainly an east–west ridge in the northeastern Pacific (Fig. 2). Osozawa (1992, 1994) proposed that the Kula–North New Guinea and the Pacific–North New Guinea ridges were subducted beneath Japan in the late Cretaceous and early Tertiary. Thus we assume that a mid-ocean ridge system crossed the

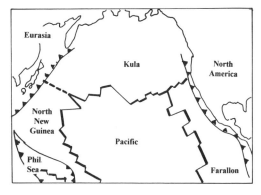

Fig. 3. Possible Eocene northern Pacific plate boundaries. (After Osozawa 1994 and Jolivet *et al.* 1989).

northern Pacific in the early Tertiary (Fig. 3). Hibbard & Karig (1990) proposed that there was a mid-ocean ridge just south of Japan in the mid-Tertiary.

Between 84 Ma and 124 Ma there was a Magnetic Quiet Zone when no reversals of the Earth's magnetic field occurred. There is therefore a gap of some 40 Ma in the mid-Cretaceous when we cannot use marine magnetic anomalies to locate the mid-ocean ridge system. The Kula plate formed during this interval, possibly by breaking off the Pacific plate or the Farallon plate. Cretaceous reconstructions of the northern Pacific require the presence of the Izanagi plate which had mid-ocean ridges as its boundaries with the Farallon and Pacific plates (Fig. 2).

South Pacific. There are currently four mid-ocean ridges in the south Pacific: the East Pacific Rise, the Chile Rise, the Pacific-Antarctic ridge and the South East Indian ridge. There is also the compressional boundary between the Australian and Pacific plates which passes through New Zealand and north to the Tonga Trench. There have been active mid-ocean ridges in the southern Pacific through the Tertiary and Cretaceous (e.g. Weissel *et al.* 1977) and from the late Cretaceous until the early Tertiary there was also an active mid-ocean ridge in the Tasman Sea west of New Zealand (Stock & Molnar 1987). By the late Jurassic there were three mid-ocean ridges and plates (Pacific, Farallon and Phoenix) in the SE Pacific (Barker *et al.* 1991) (Fig. 4).

Philippine Sea. In the early Tertiary, the North New Guinea plate lay to the southwest of the Pacific plate (Seno & Maruyama 1984) and their mutual plate boundary was the Pacific–North New Guinea Ridge. The North New Guinea

30 Ma   **60 Ma**

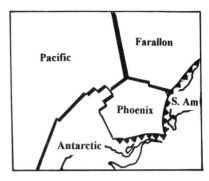

Fig. 4. Mid-ocean ridge and subduction zone positions for the southeast Pacific. The 100 Ma positions are similar to 60 Ma. (After Barker *et al.* 1991).

plate has since been completely subducted beneath northern Australia. There a set of symmetric east–west magnetic anomalies in the West Philippine Basin on the Philippine plate dating from 60–36 Ma (Hilde & Lee 1984). This east–west mid-ocean ridge was trapped behind the Philippine subduction zone (Hilde *et al.* 1977) when it was initiated in the late Eocene (*c.* 43 Ma). There is discussion about which ridge was trapped: Le Pichon *et al.* (1985) proposed that it was the Kula–North New Guinea Ridge, Gordon & Jurdy (1986) ruled out the Kula–Pacific Ridge and Jolivet *et al.* (1989) proposed that it was part of the Pacific–North New Guinea Ridge. Nevertheless, there is agreement that, in the late Eocene, part of an east–west mid-ocean ridge became trapped on the newly formed Philippine plate (Fig. 3).

Pacific–Atlantic connection

The only direct mid-ocean ridge pathway to the Mid-Atlantic Ridge from the Pacific today is south via the Pacific–Antarctic, South East Indian and South West Indian Ridges. However there were four possible previous connections between the Atlantic and the Pacific.

The first was a seaway through central America at Panama until closure about 5 Ma ago, but no mid-ocean ridge connection. Between about 150 and 85 Ma there was a mid-ocean ridge in the proto-Caribbean between North and South America (Ross & Scotese 1988) which was connected to the Mid-Atlantic Ridge. However this proto-Caribbean ridge was not connected to the ridge system active at that time in the Pacific.

A second connection may have been via the Chile Rise and south of South America. The

Chile Rise has been subducting beneath South America for the last 10 Ma. Prior to this time, it was connected to the very complex mid-ocean ridge and subduction zone system between South America and Antarctica (Fig. 4) and hence to the S America–Antarctic Ridge and so to the Mid-Atlantic Ridge. The history of the Scotia Sea region is very complex and not fully understood (Barker *et al.* 1991). South America and Antarctica were always in close proximity with no major ocean between them: creation of new crust at mid-ocean ridges was balanced by subduction along the west coast of South America and the Antarctic Peninsula. Nevertheless, there could have been a mid-ocean ridge/subduction zone pathway for vent fauna through this region during the Tertiary.

The Mediterranean region could have provided a third connection from Pacific to Atlantic. Prior to the breakup of Gondwanaland in the early Cretaceous, there was a major ocean, Tethys, between India/Africa in the south and Eurasia to the north. As this oceanic lithosphere has long since been subducted leaving only geological fragments in Eurasia and some magnetic anomalies offshore NW Australia, the locations of the mid-ocean ridges that produced it are somewhat conjectural (e.g. Scotese *et al.* 1988). The ridges must, however, have connected to the mid-ocean ridges in the Pacific to the east (see fourth connection below). However there were a series of major continental transform faults in the western Mediterranean region (Gealey 1988) so that any westward ridge connection through to the Atlantic seems unlikely.

The fourth connection reflects the current one. Today, connection between Pacific and Atlantic is south of Australia and Africa,

Table 1. *Numbers of taxa known from well-collected vents, seeps and whale skeleton (sites are noted on Fig. 1)*

Site	Abbreviation	No. of species	No. of genera	No. of families
Vents				
Galapagos	GAL	74	57	38
11–13°N, EPR	13N	99	78	48
21°N, EPR	21N	88	77	37
Guaymas Basin	GMS	41	36	26
Northeast Pacific	NEP	75	64	45
Mid-Atlantic Ridge	MAR	34	28	25
Mariana, Japan	MBJ	35	33	24
Lau, Fiji, Manus	LFM	65	51	37
Vent total		375	220	107
Seeps				
Sagami Bay, Japan	Jps	20	17	14
Kaiko Trench, Japan	Jpd	9	5	5
Oregon Margin	Ore	8	8	8
Gulf of Mexico	Lsp	30	28	26
Florida Escarpment	Fla	17	15	16
Barbados Prism	Bar	8	8	8
Laurentian Fan	Lfn	9	9	8
Seep total		76	78	53
Whale				
Whale skeleton	WAL	45	37	29

through the Indian Ocean. This pathway has been possible since about 95 Ma when Australia, Africa and India had all separated from Antarctica (Lawver *et al.* 1985; Veevers 1986; Powell *et al.* 1988). An alternate route north of Australia would have been possible prior to about 100 Ma (Audley-Charles *et al.* 1988). The magnetic anomalies west and north of Australia indicate that there was an east–west mid-ocean-ridge there between the late Jurassic and mid-Cretaceous. The ridge itself, and the rest of the oceanic lithosphere formed at it, have been lost down the subduction zones of southeast Asia. However, the ridge was probably connected to the Pacific ridge systems.

Systematic information

As discoveries of vent, seep and whale carcass faunas are relatively recent, systematic descriptions have used current taxonomic hierarchies. For the most part, the same systematists have been consulted thus comparisons and nomenclature tend to be consistent. This feature means that comparison of faunas much more reliable than is usual in biogeography; synonymies and misidentifications are rare. The faunal listing has been assembled from published literature records and consultations with most of the systematists involved. For the analyses here, the

faunal list of Tunnicliffe (1992) is supplemented by newer site lists (Galkin 1992; Desbruyeres *et al.* 1994), taxon publications such as Waren & Bouchet (1993) and numerous individual species descriptions. Only well-described vent sites on spreading ridges (mid-ocean or back-arc) have been included. These sites are indicated on Fig. 1; many other venting areas are known but their biota remains to be described. Information from warm water flows such as volcanic islands is also too limited to assess for inclusion.

Determination of what constitutes a seep is difficult. We chose sites most likely to interact with vent faunas: those with documented sulphide emissions mostly in deeper water. Coastal and shelf seeps have great faunal differences that mask the relations by other seeps to vents. Included are fauna from Sagami Bay, Japan (the one shallow site) (Hashimoto *et al.* 1989), Japan Trench (Sibuet *et al.* 1988), Oregon Margin (Seuss *et al.* 1985), Gulf of Mexico (Carney 1994), Florida Escarpment (Hecker 1985), Bermuda Prism (Jollivet *et al.* 1989) and Laurentian Fan (Mayer 1988) (Fig. 1). The longest list was only 30 species.

Bennett *et al.* (1994) is the primary source of taxa recovered from whale carcasses. These finds are known only from three dredged and one submersible-visited skeleton off the California coast (Fig. 1). Other whale carcass

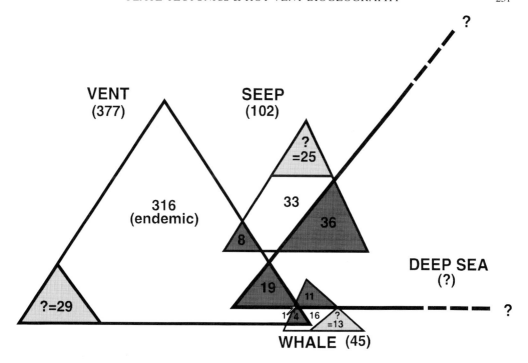

Fig. 5. Overlap of species recorded at vents, seeps and whale carcasses. No species overlaps all three sulphide habitats. A full species listing for the deep-sea is not available thus the triangle is open-sided.

collections are reported from Japan (Fujioka *et al.* 1993), New Zealand (Marshall 1994) and Iceland (Waren 1989) but only the latter two authors give taxon information.

While the fauna of the deep-sea has been studied since the Challenger Expedition of the 1870s, it remains relatively poorly known; many habitats are impossible to sample remotely. Nearly all vent and seep collections have been made by targeted submersible missions. When considering comparisons with the deep-sea, it is important to be aware that collection methods and intensity are not similar. Species described as endemic to sulphide habitats may well emerge in later deep-sea collections.

Table 1 presents numbers of species recovered from the documented sites. The most important interpretation is the need for further information and systematic effort. In particular, the seeps are poorly documented. Effort tends to be concentrated on groups that interest systematists, for instance, the gastropods. Species numbers are, to some degree, a reflection of collection effort: the largest number of biology dives has occurred at the 11–13°N East Pacific Rise site. The number of species known from Galapagos and Northeast Pacific sites (Explorer–Juan de Fuca–Gorda) are comparable

despite many fewer opportunities to collect at the Galapagos than in the Northeast Pacific. The Lau–Fiji–Manus back-arc basins are potentially rich in view of the low sampling effort to date.

Endemism

The vent fauna

The unusual aspects of the vent fauna are discussed elsewhere (Newman 1985; Grassle 1986; Van Dover 1990; Tunnicliffe 1992). For this analysis, the vent animals can be treated as a biogeographic unit. Only 7% of the vent species are known beyond the habitat although a similar fraction is still undescribed (Fig. 5). The degree of endemism is extremely high at the species level and still high at the generic level (50% of described genera). Because there is so little overlap with the surrounding deep-sea fauna, analysis of biogeographic patterns is scarcely compromised by distribution pathways through deep-sea habitats.

Seeps and whales

The hydrocarbon seep fauna probably has a variety of origins. The fauna has the same general composition as the vent fauna (predominantly polychaetes, gastropods, pogonophorans, and crustaceans) but a lower proportion of endemics. The boundaries of the seep habitat are more diffuse than are those of vents. Some deep-sea species may be more tolerant of sulphide and methane fluxes that are low compared to vents. In general, hydrocarbon seeps appear to be more open systems than hydrothermal vents, presumably because physiological barriers for deep-sea species are less severe. Most endemic whale carcass species have deep-sea congenerics; the degree of separation from the deep-sea fauna is less than for vents and seeps.

Shared taxa among vents, seeps and whales

No species is yet known to range from the vents to seeps to whale carcasses. Better documentation of seeps and whales near the whale migration routes is needed: such as between Oregon and Alaska along the Cascadia subduction zone. Nonetheless, there is little overall exchange at the specific level as illustrated in Fig. 5. While acknowledging the proportions in this figure will change with further work, the numbers do not support the notion of dispersal stepping stones (Smith *et al.* 1989) except for a few species (3% of the vent fauna).

The relationship becomes more interesting at higher taxonomic levels. The number of genera and families found only at vents and seeps is eight and five respectively. These numbers do not represent high proportions of the faunas but the fact that whole groups are limited to only these habitats reflects some evolutionary connection through common ancestors who penetrated both vents and seeps. None of these vent/seep taxa is found at whales; whales do share one new family with vents and have one endemic family of their own. There seems less of an evolutionary connection with whales despite the availability of large carcasses since the Mesozoic (Hogler 1994) although further work is needed. Only two genera, the polychaete *Nereis* and the squat lobster *Munidopsis* range among vents, seeps, whales and the deep-sea; both are cosmopolitan marine genera with noted abilities for tolerating low oxygen.

A sulphophilic fauna?

The presence of dissolved sulphide is common to the three habitats. Because of the abundance of sulphide in marine sediments, physiological mechanisms to cope with it are common in marine invertebrates although hydrothermal vents represent the extreme (Vismann 1991). While sulphide is a metabolic poison to metazoans, its potential as an energy source for free-living and symbiotic bacteria that fix carbon is highly attractive to animals that can resist the toxic effects (Southward 1983). The deep-sea sulphide-rich habitats harbour a subset of the major groups present in the deep-sea. They are numerically dominated by polychaete, pogonophoran, gastropod, and crustacean species (Van Dover 1990; Tunnicliffe 1992). Presumably, the physiological constraints induced by the environment are limiting. The whale carcass appears more selective than vent and seep habitats as it is dominated mainly by polychaete species and secondarily by molluscs.

Colonization of vents and seeps is associated with speciation and adaptation; taxonomic differentiation primarily reflects morphological change. Speciation has resulted in the evolution of novel endemic groups of organisms which is especially strong at the vents. Many elements of the seep fauna are endemic to vents and seeps, such as the pogonophoran class Obturata, the polychaete family Nautiliniellidae, the gastropod families Neolepetopsidae and Provannidae, and the crustacean family Dirvultidae, as well as eight shared genera. These commonalities are evidence that the two communities have not evolved independently – novel groups arising in one community could have dispersed to the other. These elements of the vent and seep communities form an evolutionary sulfophilic fauna in the deep-sea. Other elements of this fauna include the animals that have the greatest diversity in these habitats, such as the vesicomyid clams. In this light, another habitat deserves attention: wood and organic falls. Many of the vent/seep non-endemic records, particularly of molluscs and pogonophorans, are otherwise associated with sunken terrestrial remains (Dando *et al.* 1992; Waren & Bouchet 1993) that also generate sulphide.

The bulk of the endemic groups are restricted to vents. The more open seep system may allow more frequent invasion of competitors and predators, thus excluding parts of this sulfophilic fauna. Seep communities may present an opportunity to test the concept of vents as refugia from competition and predation as opposed to refugial survival of mass extinctions (McLean 1981, 1989; Newman 1985; Tunnicliffe 1992). Phylogenetic–geographical analyses could help trace the origination and dispersal of novel sulphophilic groups. Craddock *et al.* (1995) recently illustrated a clear differention of vent and seep mussel lines. Their molecular data supported a seep ancestor for the vent mussels.

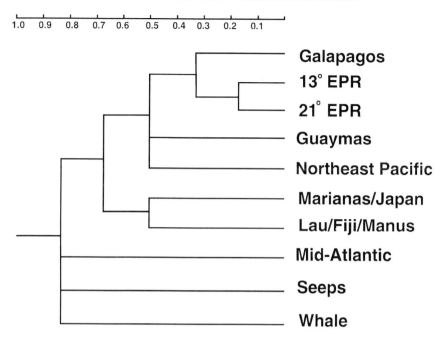

1.0 0.9 0.8 0.7 0.6 0.5 0.4 0.3 0.2 0.1

Galapagos
13° EPR
21° EPR
Guaymas
Northeast Pacific
Marianas/Japan
Lau/Fiji/Manus
Mid-Atlantic
Seeps
Whale

Fig. 6. Cladogram of generic similarities of site fauna from vents, seeps and whale carcasses. Consensus tree from five trees generated by three clustering methods (UPGMA, Complete, Single) used in the Jaccard Similarity Coefficients of the NTSYS© program. Strict consensus option used.

Similarity analyses

There are currently 662 records of species from the vent, seep and whale sites examined here. The similarity of the faunal assemblage at each site was examined using clustering techniques. The NTSYS© package allowed the comparison of several clustering algorithms. The UPGMA, Complete and Single methods of grouping the Jaccard Similarity Coefficient were examined. The strict consensus option was chosen to generate a single tree from the results. The Jaccard was chosen simply because it probably has the greatest popularity in this type of analysis. Trials with other coefficients resulted in no major differences.

Figure 6 presents the cladogram from analysis of generic similarities. While the species trees (not shown) have the same general pattern, genera better reflect biogeographic patterning that results from longer time scales. Problems of differential species splitting among taxa also are reduced. Four primary observations resulted from these analyses. Firstly, the seep and whale groups fell outside the vent groups. The matrix was run separating the seep sites: under one cluster method, the shallow Japan site fell out between the Atlantic and western Pacific back-arc basins reflecting some regional similarities.

Otherwise, the seeps came out as a group. Figure 6 presents them as a single unit because most sites have so few described taxa that it is premature to test their inter-relations.

A second observation is that the pattern appears to reflect modern proximity; this issue is addressed further below. Thirdly, similarity levels are not very high except for 13°N and 21°N. While poor sample size at some sites can be implicated, the vent fauna does display distinct provincialism with faunas specific to regions within the context of the global vent fauna. Lastly, the node connecting Guaymas Basin to the East Pacific Rise sites was not very robust. The consensus index that indicates the agreement among the trees generated by various algorithms is is only 0.62. It is not higher because of differences in placement of Guaymas Basin; it can fall out with Northeast Pacific.

With respect to an indigenous sulphiphilic fauna, a second analysis was run using only the genera that are not found in the deep sea. In this way, complications of interactions with other habitats might be lessened. The same pattern was resolved, except that Guaymas fell out with Northeast Pacific on the consensus tree. This pattern appears to relate to the habitat. Middle Valley and Escanaba Trough in the northeast Pacific are sedimented basins with many taxa

Table 2. *Distance (km) between vent sites*

	GAL	13N	21N	GMS	NEP	MAR	LFM	MBJ
GAL	'	2374	3366	3919	6612	16681	11088	14109
13N	3197	'	1059	1680	4439	18516	9614	11779
21N	4719	1523	'	730	3380	19553	9377	10906
GMS	5665	2467	945	'	3884	20192	10088	11518
NEP	8383	5186	3663	5795	'	22902	8960	8234
MAR	43535	43032	44555	45499	48218	'	20618	25024
LFM	20978	20475	21997	22942	25660	36216	'	5062
MBJ	27023	26520	28042	28987	31705	42261	6045	'

Above diagonal is shortest direct distance; below diagonal is distance via ridges and ridge discontinuities. See text for description. Site abbreviations are given in Table 1. The chosen NEP coordinates are on Axial Seamount – the middle of the Explorer/Juan de Fuca/Gorda system.

distinct from the adjacent bare-rock vents (Petrecca & Grassle 1990; Juniper *et al.* 1992). However, several genera are shared with the sedimented Guaymas Basin for which proximity to the East Pacific Rise sites appears less important.

Biogeographic patterns and spreading history

Proximity and dispersal

The similarity cladogram suggests that proximity is an important control on taxon groupings. Given that most marine invertebrates have a pelagic larva during initial life stages, larval dispersal among sites is an important consideration. Genetic studies have confirmed that gene flow can be high within species on ridge segments up to 100 km apart (France *et al.* 1992; Black *et al.* 1994; Southward *et al.* this volume) so propagules must disperse well, perhaps using hydrothermal plumes (Mullineaux *et al.* 1991). To test whether geographic distance was the most important factor, a Mantel test of matrix comparison was run (NTSYS©). Two matrices of distance were generated (Table 2). Shortest distance between sites by direct oceanic route was calculated by a Global Positioning System using paths that avoided continental shelves. The second distance was measured using way-points along ridges, faults and discontinuities; where ridge disjunctions occurred the shortest connecting distance was used.

The two distance matrices were compared with the species similarity matrix (Table 3) using a conservative method of estimating significance (Legendre & Fortin 1989). Shortest distance gave no significant correlation with either the species or genus matrix. Only ridge pathway distance was significant ($p < 0.01$). While long-

range dispersal can extend the ranges of some species, it is less influential in the overall biogeographic pattern. Ridge pathways appear to be the key factor.

Ridge pathways and spreading history

Eastern Pacific. Table 2 illustrates one example where history is probably more important than proximity: the Guaymas Basin site is actually closer to Juan de Fuca Ridge than to Galapagos by direct distance. The 21°N site is equidistant between Juan de Fuca and Galapogas. However, between the East Pacific Rise and the Northeastern Pacific vents, there is no ridge pathway. The species assemblage, taxonomic diversity and vent faunal character at Northeastern Pacific appear to reflect the isolation of an ancestral group isolated by the ridge separation at about 30 Ma (Tunnicliffe 1988). One may postulate a pan-ridge vent fauna that continues on a divergent course.

The case of Guaymas Basin is interesting. The relatively recent nature of both spreading there (12–14 Ma) and turbidite accumulation at Middle Valley on Juan de Fuca suggest these habitats are more recent than either neighbouring bare-rock ridge. Because some of their taxa are similar, an alternative evolutionary origin for sedimented habitats may exist. The Galapagos site has a higher generic similarity with Northeastern Pacific bare rock sites than it does to the Guaymas Basin. More sample analysis is necessary to examine the relationship further.

The Galapagos site separates well from the East Pacific Rise sites. Separation in time and space are important factors. The relative youth of the ridge may explain the lower diversity at this site compared to the East Pacific Rise.

Cross-Pacific. With the first discoveries of vent fauna in the western Pacific back-arc basins, it

Table 3. *Matrix of similarity coefficients generated from the genus and species presence/absence data for vent sites, seeps and whale carcasses*

	GAL	13N	21N	GMS	NEP	MAR	MBJ	LFM	SEEP	WALE
GAL	1	0.328	0.350	0.127	0.050	0.000	0.028	0.020	0.012	0.000
13N	0.406	1	0.403	0.085	0.018	0.000	0.030	0.011	0.005	0.000
21N	0.340	0.510	1	0.093	0.032	0.000	0.024	0.012	0.011	0.000
GMS	0.194	0.153	0.185	1	0.056	0.000	0.013	0.009	0.000	0.024
NEP	0.210	0.154	0.201	0.192	1	0.000	0.009	0.014	0.012	0.026
MAR	0.050	0.071	0.023	0.033	0.058	1	0.014	0.000	0.008	0.000
MBJ	0.164	0.118	0.156	0.166	0.164	0.127	1	0.096	0.007	0.000
LFM	0.105	0.142	0.133	0.094	0.099	0.075	0.256	1	0.000	0.000
SEEP	0.076	0.064	0.073	0.071	0.089	0.043	0.082	0.084	1	0.000
WALE	0.034	0.027	0.021	0.096	0.077	0.016	0.014	0.022	0.030	1

Species similarity above the diagonal, genus below. Data were analysed in the NTSYS-pc Version 1.60 (©1990, Applied Biostatistics Inc.) using the SIMQUAL program and the Jaccard similarity coefficient.

was a surprise that so many taxa were related to those of East Pacific Rise given the distances involved (Hessler & Lonsdale 1991; Desbruyeres *et al*. 1994). In fact, five endemic vent species are known across the Pacific: two polychaetes, a limpet, a mite and a copepod. For these species, the major pathway must lie through the southern Pacific where ridges were active throughout the Cenozoic. Despite the distance between the Lau/Fiji/Manus site and the Marianas/Japan site, regional similarity is evident: a quarter of western Pacific genera are shared. Persistant tectonic activity in the Philippine Sea area has likely fostered the local vent fauna.

The possibility of an ancient connection across the north Pacific via a Pacific–Kula ridge (Fig. 3) is interesting. It might be most evident in the Northeast Pacific to Japan relationships. The present similarity (Table 3) is not conclusive either way; further sample analysis from the Japan region is necessary to clarify the relationship as there are too few taxa described.

Pacific–Atlantic. Plate reconstructions suggest the Atlantic connection occurred either from the western Pacific and Indian Oceans or from the eastern Pacific either around South America or through an open Isthmus of Panama area. At present, the limited information supports the first route primarily because two endemic vent genera and one species, a copepod, are present at both western Pacific and the Mid-Atlantic vents only. Given how well known the eastern Pacific vents are, a higher affinity to the Atlantic should be detectable if it existed. At the moment, the best test would be discovery of a fauna intermediate between western Pacific and Atlantic in the Indian Ocean. This model also predicts that vent fauna on the Chile Rise,

Atlantic/Antarctic Ridge and in the Scotia Sea area would not indicate a connection to the Atlantic around Cape Horn.

It is interesting to note that, based on limited data, seep similarities suggest contact through a Caribbean seaway. In the site-by-site analysis of generic similarity for seeps, the Oregon Margin clustered with Gulf of Mexico and sometimes Florida Escarpment. An additional observation is that the one seep-endemic vestimentiferan genus (*Escarpia*) is known from Gulf of Mexico, Florida and the California slope.

The Mid-Atlantic Ridge could have been colonized from the Pacific through the Indian Ocean either north or south of Australia. However, North and South Atlantic were unconnected until about 90 Ma in the late Cretaceous (Klitgord & Schouten 1986). Ridge spreading developed in the central Atlantic in the late Jurassic by about 170–175 Ma (Klitgord & Schouten 1986) and in the South Atlantic at about 135 Ma. Thus if the faunal pathway was from the south, a decreasing diversity gradient should exist from the Indian Ocean to South Atlantic to North Atlantic. The North Atlantic opened in stages in the Cretaceous. An active mid-ocean ridge between Canada and Greenland extended northwards into Baffin Bay from 55 Ma until about 35 Ma when the ridge ceased spreading. The Reykjanes Ridge has been active since 55 Ma; Iceland will have presented a major barrier to any migration further to the north.

Extinctions

These analyses are predicated upon the assumption that vent faunas have existed throughout the Cenozoic and into late Mesozoic without major interruption. An alternative scenario is to consider the effects of major extinctions during deep ocean warming and anoxia in the Eocene (Kennett & Stott 1991). Reinvasion of the habitat would occur in the Pliocene either from refugia (?seeps) or by novel derivatives of deep-sea or cosmopolitan species. Rapid rates of differentiation are required to produce the large numbers of new families, subfamilies and genera observed today. Information that does not support this model includes (i) that many vent animals show few derived features that would suggest a recent origin, rather they have an antiquated aspect; and (ii) one might anticipate multiple invasions of such a productive habitat at different sites around the world thus greatly reducing the relatedness from region to region. Recent molecular work suggests that at least one vent group (Neomphalid gastropods) represents long term *in situ* evolution. This concept of frequent extinctions and recolonisation however, requires greater consideration, perhaps with the idea that some vent sites were more heavily affected by extinctions than others thus increasing the exchange of taxa into empty niches.

Conclusions

We conclude with a testable model for biogeographic patterns in vent faunas that is based on plate tectonic history. Because of its greater antiquity, the East Pacific Rise is a source for the ancestral vent fauna. The northeast Pacific fauna was split off 30 Ma ago. Further systematic work is required to examine the colonization of the younger sedimented sites on both ridges perhaps from a different source. An ancient northern Pacific connection may be revealed in the familial and generic levels of the Japan area vents. Southern Pacific vents should show increased similarity to the western Pacific. Regional dynamics around New Guinea/Philippines may be complex but subsequent comparison to the Indian Ocean is important to determine any connection to the Atlantic. A faunal gradient should exist through the Indian to southern and northern Atlantic.

Present evidence does not support extensive interchange with seep faunas and even less with whale carcass faunas. Nonetheless, there is a faunal subset that may have common origins but different speciation paths at vents and seeps partly due to different intensities of interactions with the surrounding deep-sea faunas. Subsequent work should also examine hot spring/sulphide faunas that are not associated with ridge crests as well as sunken wood debris.

The analytical approach adopted here is only one of several ways to examine biogeographic patterns. It will be interesting to examine the phylogenetic relationships of several component groups as the current molecular analyses become available: the vestimentiferans, alvinellid polychaetes, vesicomyids and several gastropod groups among others. Indeed, comparison of taxon-area cladograms for several groups may be very useful in identifying common patterns (Patterson 1981).

Finally, the value of systematics in such an undertaking must be underscored. A wealth of information lies in missed opportunities and unsorted collections. Sites in particular need of further systematic work are Guaymas Basin, southern Pacific, Japan, the Atlantic, the seeps, and the Indian Ocean which begs for exploration! Especially problematic is the lack of coherent taxon lists of animals known from the normal deep-sea. Eventually, with a solid understanding of the phylogenetic relationships of the vent faunas, perhaps biogeographers may be able to suggest models for plate tectonic history where such information is lacking.

We thank the BRIDGE DOVE Workshop for helping to coalesce these ideas. Numerous systematists have contributed information. L. Franklin and B. Koop provided technical assistance; NSERC Canada and the University of Victoria provided research funding to V.T. and scholarship funds to A.McA.

References

ATWATER, T. 1989. Plate tectonic history of the northeast Pacific and western North America. *In*: WINTERER, E. L., HUSSONG, D. M. & DECKER, R. W. (eds) *The eastern Pacific Ocean and Hawaii* The Geology of North America, **N**, Geological Society of America, Boulder, Colorado, 21–71.

AUDLEY-CHARLES, M. G., BALLANTYNE, P. D. & HALL, R. 1988. Mesozoic-Cenozoic rift-drift sequence of Asian fragments from Gondwanaland. *Tectonophysics*, **155**, 317–330.

BARKER, P. F., DALZIEL, I. W. D. & STOREY, B. C. 1991. Tectonic development of the Scotia arc region. *In*: TINGEY, G. (ed.) *Antarctic Geology*. Oxford University Press, Oxford, 215–248.

BENNETT, B. A., SMITH, C. R., GLASER, B. & MAYBAUM, H. L. 1994. Faunal community structure of a chemoautotrophic assemblage on whale bones in the deep northeast Pacific Ocean. *Marine Ecology Progress Series*, **108**, 205–223.

BLACK, M. B., LUTZ, R. A. & VRIJENHOEK, R. C. 1994. Gene flow among vestimentiferan tube worm (*Riftia pachyptila*) populations from hydrothermal vents of the eastern Pacific. *Marine Biology*, **120**, 33–39.

BRIGGS, J. C. 1974. Operation of zoogeographic barriers. *Systematic Zoology*, **22**, 248–256.

CARNEY, R. S. 1994. Consideration of the oasis

analogy for chemosynthetic communities at Gulf of Mexico hydrocarbon vents. *Geo-Marine Research Letters*, **14**, 149–159.

CHILDRESS, J. J. & FISHER, C. R. 1992. The biology of hydrothermal vent animals: physiology, biochemistry and autotrophic symbioses. *Oceanography and Marine Biology: An Annual Review*, **30**, 337–441.

CRADDOCK, C., HOEH, W. R., GUSTAFSON, R. G., LUTZ, R. A., HASHIMOTO, J. & VRIJENHOEK, R. J. 1995. Evolutionary relationships among deep-sea mytilids (Bivalvia: Mytilidae) from hydrothermal vents and cold-water methane/sulfide seeps. *Marine Biology*, **121**, 477–485.

DANDO, P. R., SOUTHWARD, A. J., SOUTHWARD, E. C., DIXON, D. R., CRAWFORD, A. & CRAWFORD, M. 1992. Shipwrecked tube worms. *Nature*, **356**, 667.

DESBRUYERES, D., ALAYSE-DANET, A.-M. & OHTA, S. 1994. Deep-sea hydrothermal communities in Southwestern Pacific back-arc basins (the North-Fiji and Lau Basins): composition, microdistribution and food web. *Marine Geology*, **116**, 227–242.

FISHER, C. R. JR. 1990. Chemoautrophic and methanotrophic symbosis in marine invertebrates, C.R.C. *Reviews in Aquatic Sciences*, **2**, 399–436.

FRANCE, S. C., HESSLER, R. R. & VRIJENHOEK, R. C. 1992. Genetic differentiation between spatially-disjunct populations of the deep-sea, hydrothermal vent-endemic amphipod *Ventiella sulfuris*. *Marine Biology*, **114**, 551–559.

FUJIOKA, K., WADA, H. & OKANO, J. 1993. Torishima whale bone deep-sea animal community assemblage – new finding by Shinkai 6500. *Journal of Geography*, **102**, 507–517.

GALKIN, S. V. 1992. The benthic fauna of hydrothermal vents in the Manus Basin. *Oceanology*, **32**, 768–774.

GEALEY, W. K. 1988. Plate tectonic evolution of the Mediterranean – Middle East region. *Tectonophysics*, **155**, 285–306.

GORDON, G. & JURDY, D. M. 1986. Cenozoic global plate motions. *Journal of Geophysical Research*, **91**, 12389–12406.

GRASSLE, J. F. 1986. The ecology of deep-sea hydrothermal vent communities. *Advances in Marine Biology*, **23**, 301–362.

HASHIMOTO, J., OHTA, S., TANAKA, T., HOTTA, H. *ET AL.* 1989. Deep-sea communities dominated by the giant clam, *Calyptogena soyoae*, along the slope foot of Hatsushima Island, Sagami Bay, Central Japan. *Palaeogeography, Palaeoclimatology and Palaeoecology*, **71**, 179–192.

HECKER, B. 1985. Fauna from a cold sulfur seep in the Gulf of Mexico, comparison with hydrothermal vent communities and evolutionary implications. *Bulletin of the Biological Society of Washington*, **6**, 465–473.

HESSLER, R. R. & LONSDALE, P. F. 1991. Biogeography of Mariana Trough hydrothermal vent communities. *Deep-Sea Research*, **38**, 185–199.

HIBBARD, J. P. & KARIG, D. E. 1990. Alternative plate model for the early Miocene evolution of the southwest Japan margin. *Geology*, **18**, 170–174.

HILDE, T. W. C. & LEE, C. S. 1984. Origin and evolution of the West Philippine Basin: a new interpretation. *Tectonophysics*, **102**, 85–104.

——, UYEDA, S. & KROENEKE, L. 1977. Evolution of the western Pacific and its margin. *Tectonophysics*, **28**, 145–165.

HOGLER, J. A. 1994. Speculations on the role of marine reptile deadfalls in Mesozoic deep-sea paleoecology. *Palaios*, **9**, 42–47.

JOHANNESSON, K. 1988. The paradox of Rockall: Why is a brooding gastropod more widespread than one having a plantonic larval dispersal stage (*L. littorea*)? *Marine Biology*, **99**, 503–519.

JOLIVET, L., HUCHON, P. & RANGIN, C. 1989. Tectonic setting of western Pacific marginal basins. *Tectonophysics*, **160**, 23–47.

JOLLIVET, D., FAUGHRES, J.-C., GRIBOULARD, R., DESBRUYERES, D. & BLANC, G. 1989. Composition and spatial organization of a cold seep community on the South Barbados accretionary prism: tectonic, geochemical, and sedimentary context. *Progress in Oceanography*, **24**, 25–46.

JUNIPER, S. K., TUNNICLIFFE, V. & SOUTHWARD, E. C. 1992. Hydrothermal vents in turbidite sediments on a Northeast Pacific spreading centre: organisms and substratum at an ocean drilling site. *Canadian Journal of Zoology*, **70**, 1792–1809.

KENNETT, J. P. & STOTT, L. D. 1991. Abrupt deep-sea warming, palaeoceanographic changes and bethnic extinctions at the end of the Palaeocene. *Nature*, **353**, 225–229.

KLITGORD, K. D. & SCHOUTEN, H. 1986. Plate kinematics of the central Atlantic. *In*: VOGT, P. R. & TUCHOLKE, B. E. (eds) *The Western North Atlantic region. The Geology of North America*, **M**, Geological Society of America, Boulder, Colorado, 351–378.

LAWVER, L. A., SCLATER, J. G. & MEINKE, L. 1985. Mesozoic and Cenozoic reconstructions of the South Atlantic. *Tectonophysics*, **114**, 233–254.

LEGENDRE, P. & FORTIN, M. 1989. Spatial pattern and ecological analysis. *Vegetatio*, **80**, 107–138.

LE PICHON, X., HUCHON, PH. & BARRIER, E. 1985. Pangea, geoid and the evolution of the western margin of the Pacific ocean. *In*: NASU *ET AL.* (eds) *Formation of Active Ocean Margins*. Terrapub, Tokyo, 3–42.

LONSDALE, P. 1988. Paleogene history of the Kula plate: offshore evidence and onshore implications. *Geological Society of America Bulletin*, **100**, 733–754.

MARSHALL, B. A. 1994. Deep-sea gastropods from the New Zealand region associated with recent whale bones and an Eocene turtle. *Nautilus*, **108**, 1–8.

MARSHALL, L. G., WEBB, S. D., SEPKOSKI, J. J. JR. & RAUP, D. M. 1982. Mammalian evolution and the great American interchange. *Science*, **215**, 1351–1357.

MAYER, L. A., SHOR, A. N., CLARKE, J. H. & PIPER, D. J. W. 1988. Dense biological communities at 3850m on the Laurentian Fan and their relationships to the deposits of the 1929 Grand Banks earthquake. *Deep-Sea Research*, **35**, 1235–1246.

McLEAN, J. H. 1981. The Galapagos rift limpet *Neomphalus*: relevance to understanding the evolution of a major Paleozoic-Mesozoic radiation. *Malacologia,* **21**, 291–336.

—— 1989. New archaeogastropod limpets from hydrothermal vents: new family Peltospiridae, new superfamily Peltospiracea. *Zoologica Scripta,* **18**, 49–66.

MULLINEAUX, L. S., WIEBE, P. H. & BAKER, E. T. 1991. Hydrothermal vent plumes: larval highways in the deep sea? *Oceanus,* **34**, 64–68.

NEWMAN, W. A. 1985. The abyssal hydrothermal vent invertebrate fauna: a glimpse of antiquity? *Bulletin of the Biological Society of Washington,* **6**, 231–242.

NICHOLSON, C., SORLIEN, C. C., ATWATER, T., CROWELL, J. C. & LUYENDYK, B. P. 1994. Microplate capture, rotation of the western Transverse Ranges, and initiation of the San Andreas transform as a low angle fault system. *Geology,* **22**, 491–495.

OSOZAWA, S. 1992. Double ridge subduction recorded in the Shimanto accretionary complex, Japan, and plate reconstruction. *Geology,* **20**, 939–942.

—— 1994. Plate reconstruction based upon age date of Japanese accretionary complexes. *Geology,* **22**, 1135–1138.

PATTERSON, C. 1981. Methods of paleobiogeography, *In:* NELSON, G. & ROSEN, D. E. (eds) *Vicariance Biogeography: a critique.* Columbia University Press, New York, 446–489.

PARKER, T. & TUNNICLIFFE, V. 1994. Dispersal strategies of the biota on an oceanic seamount: implications for ecology and biogeography. *Biological Bulletin,* **187**, 336–345.

PETRECCA, R. & GRASSLE, J. F. 1990. Notes on fauna from several deep-sea hydrothermal vent and cold seep soft-sediment communities. *In:*, McMURRAY, G. R. (ed.) *Gorda Ridge.* Springer-Verlag, New York, 278–284.

POWELL, C. MC. A, ROOTS, S. R. & VEEVERS, J. J. 1988. Pre-breakup continental extension in east Gondwanaland and the early opening of the eastern Indian Ocean. *Tectonophysics,* **155**, 261–283.

ROSS, M. I. & SCOTESE, C. R. 1988. A hierarchial tectonic model of the Gulf of Mexico and Caribbean region. *Tectonophysics,* **155**, 139–168.

SCOTESE, C. R., GAHAGAN, L. M. & LARSON, R. L. 1988. Plate tectonic reconstructions of the Cretaceous and Cenozoic ocean basins. *Tectonophysics,* **155**, 27–48.

SENO, T. & MARUYAMA, S. 1984. Paleogeographic reconstruction and origin of the Philippine Sea. *Tectonophysics,* **102**, 53–84.

SIBUET, M., JUNIPER, S. K. & PAUTOT, G. 1988. Cold-seep benthic communities in the Japan subduction zones: geological control of community development. *Journal of Marine Research,* **46**, 333–348.

SMITH, C. R., KUKERT, H., WHEATCROFT, R. A., JUMARS, P. A. & DEMING, J. W. 1989. Vent fauna on whale remains. *Nature,* **341**, 27–28.

SOUTHWARD, A. G. 1983. Poisons of delight. *Marine Pollution Bulletin,* **14**, 321–322.

SOUTHWARD, E. C., TUNNICLIFFE, V., BLACK, M., DIXON, D. & DIXON, L. 1996. Ocean ridge segmentation and hot vent tubeworms in the northeast Pacific. *This volume.*

STOCK, J. & HODGES, K. V. 1989. Pre-Pliocene extension around the Gulf of California and the tansfer of Baja California to the Pacific plate. *Tectonics,* **8**, 99–115.

—— & MOLNAR, P. 1987. Revised history of early Tertiary plate motion in the south-west Pacific. *Nature,* **325**, 495–499.

SUESS, E., CARSON, B., RITGER, S. D., MOORE, J. C., JONES, M. L., KULM, L. D. & COCHRANE, G. R. 1985. Biological communities at vent sites along the subduction zones off Oregon. *Bulletin of the Biological Society of Washington,* **6**, 475–484.

TUNNICLIFFE, V. 1988. Biogeography and evolution of hydrothermal-vent fauna in the eastern Pacific Ocean. *Proceedings of the Royal Society of London,* **B233**, 347–366.

—— 1991. The biology of hydrothermal vents: ecology and evolution. *Oceanography and Marine Biology: An Annual Review,* **29**, 319–407.

—— 1992. Hydrothermal vent communities of the deep-sea. *American Scientist,* **80**, 336–349.

VAN DOVER, C. L. 1990. Biogeography of hydrothermal vent communities along seafloor spreading centers. *Trends in Ecology and Evolution,* **5**, 242–246.

VEEVERS, J. J. 1986. Breakup of Australia and Antarctica estimated as mid-Cretaceous $(95 \pm 5 \text{ Ma})$ from magnetic and seismic data at the continental margin. *Earth and Planetary Science Letters,* **77**, 91–99.

VISMANN, B. 1991. Sulfide tolerance: physiological mechanisms and ecological implications. *Ophelia,* **34**, 1–27.

WAREN, A. 1989. New and little known mollusca from Iceland. *Sarsia,* **74**, 1–28.

—— & BOUCHET, P. 1993. New records, species, genera, and a new family of gastropods from hydrothermal vents and hydrocarbon seeps. *Zoologica Scripta,* **22**, 1–90.

WEISSEL, J. K., HAYS, D. E. & HERRON, E. M. 1977. Plate tectonic synthesis: the displacements between Australia, New Zealand and Antarctica since the late Cretaceous. *Marine Geology,* **25**, 231–277.

The hydrothermal imprint on life: did heat-shock proteins, metalloproteins and photosynthesis begin around hydrothermal vents?

E. G. NISBET & C. M. R. FOWLER

Department of Geology, Royal Holloway, University of London, Egham, TW20 0EX, UK

Abstract: Molecular evidence implies that the first living community existed around a hydrothermal system. From this setting, life may have colonized a wide variety of hydrothermal environments. These would have included hydrothermal systems on mid-ocean ridges, around komatiite plume volcanoes, and in a variety of shallow-water settings. Hydrothermal systems on ridge segments may have been crucial in providing habitats to sustain the first communities of life.

The early hydrothermal heritage may now be deeply imprinted in all organic life. Biological processes that use essential metals, especially Fe (typically as 4Fe-4S), Mn, Zn, Cu, Mo, Mg, Se and Ni, may all have hydrothermal descent from reactions that first evolved on a natural hydrothermal metal sulphide substrate, and it is possible that the first supply of P, when life began, was also associated with volcanism. The heat-shock proteins and many important 'housekeeping' enzymes of the biosphere, such as urease, hydrogenase, and nitrogenase may all be of hydrothermal origin. Photosynthesis may have originated via infrared phototaxis developed at mid-ocean ridge hydrothermal systems. The modern biosphere may have grown out of bacterial processes originally evolved at mid-ocean ridges.

The evidence is increasingly strong that the first living community existed around a hydrothermal system. Though not proven, this hypothesis is supported by a wide variety of independent clues. All of the deepest branches of life share a common hyperthermophile character (Kandler 1992, 1994a, b), existing at temperatures of 80–110°C. Many of the most ancient proteins appear to have a hydrothermal origin, and key elements that make possible the existence of living communities (S, Fe, Mn, Zn, Mo, and perhaps P, Ni and Mg) may have entered into life processes from an early hydrothermal or volcanic substrate, or from fluids in a volcanic ambiance. In this discussion we explore the basis for the 'hydrothermal-first' hypothesis, and attempt to map out the 'hydrothermal-world' and its imprint and consequences on the modern biosphere.

This discussion is based on the assumption that primitive organisms were *not* highly skilled in extracting chemical species from their environment. Today, the biosphere is immensely sophisticated in its biochemical cycling. In the modern world, after aeons of evolutionary tinkering, rare elements essential to life are extracted by sophisticated biogeochemical processes from the inorganic setting, and then jealously guarded and multiply recycled by the living community. In the earliest Archaean, the first living organisms would have been much less skilled geochemically, and the very first use of any inorganic component (such as Fe-S, and Mn-S, or Ni) is likely to have occurred in a setting where that component was readily and abundantly available as a rock/sediment substrate, or in ambient fluid, almost forcing itself into the primitive organism. If this assumption is correct, and it is the basic axiom of our discussion, then the task of the geologist is to look for those settings where essential chemicals and environmental conditions were available to unskilled primitive organisms.

Although molecular palaeontology, which uses the molecular biology and biochemistry of modern organisms to reconstruct phylogeny, is now well established, the broader use of molecular results to reconstruct palaeoecology is less developed. Here we attempt to reconstruct details of the habitats of the earliest Archaean microbial communities (Kandler 1994a; Knoll & Bauld 1989) by considering deeply conserved aspects of present-day organisms. In this discussion we assume: (1) that deeply conserved properties of life, such as the heat-shock proteins, first evolved to meet specific environmental needs, very early in geological time; and (2) that the earliest organisms were biochemically relatively unskilled.

From: MacLeod, C. J., Tyler, P. A. & Walker, C. L. (eds) 1996, *Tectonic, Magmatic, Hydrothermal and Biological Segmentation of Mid-Ocean Ridges*, Geological Society Special Publication No. 118, pp. 239–251.

The hydrothermal-first hypothesis

There is strong evidence from molecular phylogeny, especially based on the study of ribosomal RNA, that the most deeply rooted modern species of life are the hyperthermophiles, living in temperatures above 75–85°C. A likely, although not proven, inference from this is that the last common ancestral population from which modern life is descended, was hyperthermophile (Kandler 1992, 1994a, b; Achenbach-Richter et al. 1987, 1988).

There are two possible interpretations of the hyperthermophile nature of the most deeply rooted organisms: either (1) that the last common population of prokaryotes existed in a very warm ocean, or (2) that they existed in localized volumes of hot water, specifically either hydrothermal systems that provided local hot sites on the sea floor, or hot evaporative surface salt pools. While the hypothesis that life existed in a very warm ocean is tenable, a hot ocean could have caused massive hydrogen loss from the planet and eventual greenhouse runaway. A rapid transition from an atmosphere with little stratospheric H_2O (as on Earth) to one with abundant stratospheric H_2O (as must have occurred on early Venus, prior to planetary dehydration) takes place when surface temperature exceeds about 340K in a typical model atmosphere (Kasting et al. 1989, 1993). It is possible that a very warm ocean, at 75°C, did exist briefly in the Hadean, but if so the Earth would have been perilously close to the breakdown of the atmospheric 'cold trap' which prevents abundant water vapour from reaching the stratosphere and hence protects the planet's hydrogen from being lost to space. Hot oceans would have meant massive evaporation, immense hurricanes, and mixing of water into the air above the tropopause: in turn, this would allow water into the very high atmosphere and loss of hydrogen to space. Although there is indeed evidence for substantial early H loss from the planet (Yung et al. 1989), by the end of the Hadean global thermal runaway had not occurred on Earth. Thus, although it is possible that the earliest oceans could have been at 75°C for some time, this would not have been a stable state. By the time of life's origin, which would have been after the cessation of heavy meteorite bombardment, several hundred million years after Earth accretion, it is likely that the oceans were somewhat cooler, unless the atmosphere was very different from typical models. Kasting et al. (1989) have suggested a possible early atmosphere with carbon dioxide pressures above 2 bars, and elemental sulphur in the

atmosphere screening off UV radiation, the surface being at about 45°C. If so, the hyperthermophiles could not have lived in the open ocean, which would have been too cool for them.

The implication is that the early hyperthermophile community did *not* live in the open ocean – they must have existed around warmer volcanic vents, or just possibly in hot, salty, evaporative surface pools under sunlight. Given the flux of UV radiation on the surface, and the transience of evaporative pools, the more likely conclusion, then, is that the earliest life lived in the vicinity of hydrothermal systems.

Did hot vents exist under the Archaean ocean? Contrast the Archaean Earth with the presumably cooler (Nisbet et al. 1993) modern planet: in the modern Earth, hydrothermal flow through the oceanic crust is very important. The total heat loss from the Earth today is 4.4×10^{13} W (Pollack et al. 1993). Of the 3.2×10^{13} W lost through the oceanic lithosphere, 1.1×10^{13} W (a quarter of the global total) is lost by hydrothermal circulation in the oceanic crust (Stein & Stein 1994). Some 90% of the heat flux currently discharged along the mid-ocean ridges (ages less than 1 Ma) is due to hydrothermal circulation (Stein & Stein 1994). In the late Hadean and early Archaean (in terminology we adopt the convention that the Hadean aeon ended when life began, and infer that the Archaean began perhaps 4.0–4.3 Ga ago), there are strong arguments that deep oceans existed and that sub-oceanic volcanism occurred. Various authors (e.g. Bickle 1986) have argued strongly not only that mid-ocean ridges did exist, but that they were much more vigorous than today: more heat was lost. In the late Hadean/earliest Archaean it is even possible that the magmas feeding ridges were komatiites (>18% MgO), and much hotter than today. Thus hot vents were probably common. There is also considerable evidence for *deep* oceans (see discussion in Nisbet 1987), possibly 3.5 km or deeper, so vent temperatures probably ranged up to 350°C or more.

From available Archaean igneous rocks, one can imagine a diverse array of Archaean hydrothermal systems. On the continents, plume komatiitic volcanism would have created wide low shields, and smaller alkaline centres would also have existed. At continental margins, and at island arcs, andesitic volcanism occurred. Mid-ocean ridges in the oceans would be connected in places to rifting continental crust, providing potential biological dispersion pathways between subaereal continental volcanic hydrothermal centres and mid-ocean smokers. The ridges would have been segmented, as today, providing a variety of local habitat differences for the microbial ecosystems.

On the modern Earth, black smokers (where water at temperatures up to 350°C is discharged) and other areas of more diffuse cooler flow are globally very important, not only in the Earth's thermal budget but also in its chemical recycling. Presumably, on the Archaean planet, black smokers and other hydrothermal emissions on young oceanic plates were even more vigorous than today. At Archaean ridges, as today, sulphide vent chimneys would have built up, emitting fluids that, compared to seawater, were highly enriched in Fe, Mn and S, and locally in Cu, Zn, Mo, Cd, Co, Ag and Pb (e.g. von Damm et al. 1995; Lowell et al. 1995). Vent gases at modern ridges include CO_2, CH_4, H_2S and H_2, as well as NH_3 that is derived from organic matter. Similar gases occur in varying ratios in shallow-water systems. On the prebiotic Earth, it is difficult to imagine which volatiles were emitted, but it is likely that most of these were present, as well, perhaps, as HCN made by high-temperature reactions in the hydrothermal systems near magma. N would have been present in ammonium minerals. Around land volcanoes and in shallow water, pH is bimodal, typically around 1.8–2.2 in sulphuric acid systems and 7.5–9.0 in carbonate-bicarbonate systems (Brock 1986). In marine systems, pH is also variable, from neutral to very acid. In the late Hadean/early Archaean, in addition to systems on mid-ocean ridges, hydrothermal systems would have also occurred around massive komatiite plume volcanoes (comparable to Hawaii, but lower and much larger), at volcanoes above early subduction zones and around early land volcanoes including carbonatites and kimberlites.

Molecular phylogeny

The evidence for the antiquity of the hyperthermophiles is reviewed by Kandler (1992, 1994a, b). In the discussion we adopt the nomenclature of the three domains proposed by Woese et al. (1990), recognizing as distinct the domains of Archaea, Bacteria (formerly Eubacteria) and Eucarya (formerly Eukaryotes). Both prokaryotic domains of life, the Archaea and the (Eu)Bacteria (Woese et al. 1990) appear ancestrally to have had a hyperthermophilic character, occupying environments where water was hot, in the range 75–110°C (and in many cases acid or salty). Hyperthermophily is distinct from mesothermophily (warm settings, but below 75°C), which may have evolved later (Kandler 1992). Figure 1 shows a phylogenetic tree (after

Kandler 1994a, b) based on 16S rRNA sequence comparisons. It is clear that the hyperthermophile forms occupy the most deeply rooted positions both in the Archaea and in the Bacteria. In the Bacteria, Hydrogenobacter, Aquifex and Thermotogales appear to be the most deeply distinct forms, separately branched most long ago from the phylogenetic tree. These are all hyperthermophiles, implying, but not proving, that the common ancestor of the bacteria was a hyperthermophile (Achenbach-Richter et al. 1987). This was prior to the subsequent divergence that led to the evolution of organisms such as sulphur bacteria, purple bacteria and cyanobacteria. In the Archaea, similarly, the oldest divergence appears to imply hyperthermophilic ancestry. Of the two modern phyla of Archaea, the Crenarchaeota are all hyperthermophiles, dependent on H_2 and sulphur, while most of the members of the other phylum, the Euryarchaeota, also occupy high temperature environments. Only the relatively late-evolved halophiles and some methanogens occupy cooler settings.

The heat-shock proteins

The heat-shock proteins and their analogues, which protect cells by preventing protein denaturation after accidental heat shock (Martin et al. 1992; Welch 1992, 1993; Gething & Sambrook 1992), appear to be of the deepest antiquity. Heat-shock proteins and analogues occur in all three domains of life, and are diversely involved in modern processes as different as photosynthesis, the function of mitochondria, spermatogenesis, and the operation of the nuclear membrane in eukaryotes. They are expressed by organisms in response to heat shock, and appear to help to protect other cellular proteins against denaturation by binding to them during heat shock before they are irreparably damaged by unfolding, and then by refolding and restoring them (e.g. Martin et al. 1992; Parsell et al. 1993). The bacterial heat-shock proteins occur in (eu)Bacteria and as proteins of bacterial descent in eukaryotes. In the Bacteria, for instance, when Escherichia coli is heated to temperatures between 42 and 46°C, the concentration of the heat-shock protein GroEL increases by a factor of 5–10, reaching up to 12% of cellular protein; in yeast the mitochondrial heat-shock protein hsp60 (of bacterial descent, presumably) increases by a factor of 2–3; in mammal mitochondria hsp10 increases tenfold on heat-shock (Martin et al. 1992). In addition, because of their more general role in protein folding (Gething & Sambrook 1992;

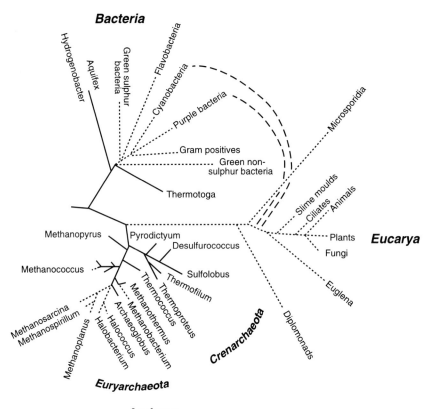

HADEAN   ARCHAEAN PROTEROZOIC PHANEROZOIC

Fig. 1 Phylogenetic tree, based on 16S rRNA evidence (modified from Kandler 1992, 1994, and other sources). Continuous lines represent hyperthermophiles; Dotted lines represent organisms living in cooler conditions; Dashed lines represent symbiotic inclusions of bacterial organisms in Eucarya. The aeon names at the top of the diagram are specifically *not* to chronological scale (especially as the Hadean–Archaean boundary is not yet defined: it may become placed at the origin of life), but instead are broadly illustrative of the development of communities of life. Communities including Bacteria and Archaea existed in the Archaean, but higher Eucarya did not evolve until the Proterozoic. Distances along the lines of the phylogenetic tree from the original ancestor are representative of the evolutionary distance that separates organisms from common ancestry. In the Archaea, the Crenarchaeota are exclusively hyperthermophiles. The Euryarchaeota include some evolved lines that are not hyperthermophiles, including the halobacteria and some methanogens. The Bacteria similarly began with hyperthermophiles and later evolved organisms such as cyanobacteria capable of living in cooler surface conditions. Photosynthesis occurs in the green non-sulphur bacteria, green sulphur bacteria, purple bacteria and cyanobacteria, and appears to have first evolved before the common ancestor of these. Photosynthesis in the plants is derived from symbiotic inclusion of a cyanobacterial-like organism, and likewise the higher Eucarya have derived their mitochondria by the symbiotic inclusion of a bacterial organism. Dashed lines show these symbioses. Halobacteria can also use light, but very differently: this process uses bacteriorhodopsins distantly similar to rhodopsin in the human eye.

Horwich & Willison 1993), they play a wide variety of other roles in all modern life, for instance in folding rubisco, the catalyst of photosynthesis in cyanobacteria and in plants. Rubisco is crucial to the modern biosphere,

providing 'the only quantitatively significant link between the pools of inorganic and organic carbon in the biosphere' (Hartman & Harpel 1994).

Although the heat-shock proteins are best

known in the Bacteria and as proteins of bacterial descent in Eucarya, analogous proteins also occur in the other domains of life, in Archaea and as related proteins in the eukaryotic cytosol (e.g. see Trent *et al.* 1991; Lewis *et al.* 1992; Gething & Sambrook 1992; Horwich & Willison 1993; Yoshida *et al.* 1993). Collectively, the heat-shock proteins and their analogues are likely of the greatest antiquity. Although the heat-shock proteins in Bacteria and their analogues in Archaea and Eucarya do differ, they broadly share common function and structure. This commonality strongly suggests a common inheritance (Trent *et al.* 1991; Horwich & Willison 1993). Parallel evolution is possible, but the conservatism of heat-shock proteins in their later history, and their complexity, argues strongly that they are genuinely primitive, essential 'housekeepers' and thus deeply preserved in nature, and were possessed by the original progenotes prior to the diversification of the three domains from the common ancestor, perhaps around 4 Ga ago. This notion, if correct, implies that protection against heat shock was critically important to the first organisms.

Perhaps the first community was hydrothermal, not dispersed in a warm ocean (Nisbet 1986, 1995). If the early Archaea and Bacteria were living in a warm ocean, with ambient temperatures around 75°C, it is difficult to imagine why protection against heat shock was so important: ocean temperature gradients are small. However, if the early organisms lived in the vicinity of hydrothermal vents, then heat shock and stress would have been frequent and life-threatening, as vents episodically belched out hot fluids. In such a setting with very sharp thermal gradients between hot emissions and ambient water, there would be great advantage to organisms that could survive the dangerously variable thermal and chemical conditions close to a vent. Organisms that had developed heat-shock proteins would have been able to survive closer to the vent, to grow and reproduce rapidly. In contrast, those organisms that were safe from sudden heating, far from the vent, would run the opposite risk of finding themselves in cold, barren conditions where they could not reproduce.

Since the diversification of the domains of life, each prokaryotic domain has exploited heat-shock proteins in a wide variety of roles. The Eucarya have inherited their heat shock proteins both from their ancestral Archaea-like root stock and also from bacteria symbiotically incorporated in the higher Eucarya. Over time, evolutionary tinkering appears to have made use of the antique heat shock proteins in a variety of

different ways, presumably because of their ability to fold other proteins. Among a wide variety of uses, their best known role is as chaperonins (Hemmingsen *et al.* 1988; Ellis 1993) or proteins that assist in the formation of other proteins by folding newly synthesized polypeptides, and by binding to protein surfaces to inhibit incorrect function and allow correct cellular processes. The uses of heat-shock proteins vary widely. Binding to heat shock protein similar to *GroEL* (see above) is an obligatory step in photosynthesis in cyanobacteria and plants (Barraclough & Ellis 1980; Ellis 1993; Hartman & Harpel 1994). In mitochondria, which are bacterially derived organelles in higher Eucarya, the heat shock protein *hsp60* is a critical component of protein maturation (Welch 1993).

In Eucarya, analogues of heat shock proteins that appear to be descended from an ancestor shared with the Archaea also occur, and may be distinct from those of bacterial descent inherited in chloroplasts and mitochondria. The *TF55* family of proteins in Archaea (e.g. *Sulfolobus*) appears closely related to the *TCP1* family in Eucarya (see Horwich & Willison 1993), which occurs in all mammalian cell types, in fruit flies, in yeasts, and in other organisms. *TCP1* was originally discovered in the mouse testis, where it appears to play a major role in spermatogenesis. It has since been found very widely. Just possibly, *TF55* and *TCP1* share a very distant common ancestral protein with the bacterial *hsp60* family of proteins, both protecting a progenote against heat shock. Finally, at the defining core of eucaryote life, *hsp70* is required for protein import into the nucleus of the cell. In its role as gate-keeper of the nucleus, it appears to make the function of the nuclear membrane possible (Dingwall & Laskey 1992).

To summarize, the heat-shock proteins and analogues may be descended from an ancestral protein used to help a very early organism that lived in a setting where heat shock and stress was a major risk. From this organism may be derived all three domains of life. In these domains, heat-shock proteins have been deeply conserved and put to a wide variety of uses, for instance being essential in photosynthesis. Heat-shock proteins and metalloproteins, both part of the postulated early hydrothermal heritage of life, have been extensively exploited by molecular tinkering (Jacob 1983). In the Eucarya, heat-shock proteins have been inherited both from the bacterial line and from an Archaea-like nuclear ancestor. This evidence suggests an early hydrothermal setting for the first living community, and that later organisms tinkered with the hydrothermally evolved equipment to make it perform new functions, such as photosynthesis (for which a mid-ocean ridge origin is argued below).

Inorganic components of life: S, Fe, Mn, Mo, Mg, Ni, etc.

Sulphur, iron, manganese and magnesium, all characteristic of hydrothermal environments, play crucial roles in many functions of life, such as photosynthesis (Barber & Andersson 1994). Sulphur, especially, is a critical ingredient of many polypeptides. Its most likely source would have been around hydrothermal systems, probably around andesitic volcanic centres, though possibly also around komatiitic shield volcanoes.

The roles in life of manganese and iron, especially in association with sulphur, are widespread. Fe, Mn and S are characteristic of hydrothermal vents: in particular, the association of Mn with Fe and S is reminiscent of black smokers on modern mid-ocean ridges, and Mn and Fe of sea floor sediment (e.g. manganese nodules). By inference, perhaps, these elements were first supplied to life from hydrothermal systems, for example around submarine basalt or komatiite eruptions. The importance of these elements to modern life is perhaps best illustrated by their positions at the centre of photosynthesis. In Type II reaction centres, in purple bacteria, water oxidation requires the accumulation of four oxidising equivalents that are necessary to abstract four reducing equivalents ($4e/4H^+$) from two water molecules to produce one molecule of dioxygen (Barber & Andersson 1994). This process involves a cluster of four Mn atoms. A similar mechanism occurs in Type I reaction centres. In the cyanobacterium *Synechococcus*, 4Fe-4S centres occur in the photosystem I reaction centre, which has a two-fold symmetry with the axis passing through the 4Fe-4S centre (Barber & Andersson 1994).

Magnesium is essential in a wide variety of biological processes. MgATP and MgADP are widely significant. For instance, in nitrogenase, electron transfer is co-ordinated by transfer of two MgATP molecules bound by Fe-protein, in a molecular switching mechanism (see Wolle *et al.* 1992). The nitrogenase complex is the enzyme system used in bacteria for the reduction of dinitrogen gas to ammonia. It is crucial to the operation of the nitrogen cycle, and closely linked to the control on nitrogen in the air and in the biota. Mg plays an equally significant role in another important enzyme, hydrogenase in *Desulfovibrio gigas* (Volbeda *et al.* 1995).

Another good example of the centrality of Mg is in light-harvesting in bacterial photosynthesis, where Mg lies at the core of bacteriochlorophyll *a* in purple non-sulphur bacteria (McDermott *et al.* 1995; see also Hartman & Harpel 1994).

A wide variety of trace elements that are most likely of hydrothermal introduction to life is used in the essential 'housekeeping' processes of life. These elements include Zn, Cu, Mo, Se, amongst others – indeed, the list includes a large number of rare elements that would not easily have been accessible to the early, unskilled biosphere, unless life had been sited on an environmental substrate where for some special reason they were more highly concentrated and of higher activity.

Zn, Cu and Mo are widely used elements. Zn proteins are very common, and have wide functions. Zn is the second most abundant metal in higher animals (Coleman 1992), and essential in planktonic communities. In bacteria, Zn is also ubiquitous, for instance, bacterial RNA polymerases contain two Zn ions, as do eukaryotic RNA polymerases, suggesting a feature inherited from the very first life-forms. Alcohol dehydrogenase, alkaline phosphatase and other enzymes also contain zinc (Coleman 1992), sometimes with Mg. Like Zn, Cu proteins are ubiquitous. Mo, rare in most rocks and sediments, but a characteristic hydrothermal element, is also widely used. For example, in nitrogenase (discussed above; Wolle *et al.* 1992) the nitrogenase enzyme is a reversible complex of a Mo-Fe protein and a Fe-protein (Howard & Rees 1994). The Mo-Fe protein contains the site of substrate reduction while the Fe-protein is the exclusive electron donor for the reaction. The electron transfer is co-ordinated by hydrolysis of two Fe-protein-bound MgATP molecules. All manipulation must occur in an atmosphere of less than 1 ppm oxygen, as oxygen is very destructive to the system and presumably was absent at the time, clearly pre-photosynthetic, when it evolved. By analogy with modern hydrothermal systems, those proteins that contain Zn, Cu, Mo etc., and Mg, may have first evolved around hydrothermal systems above subduction zones, possibly in andesitic or andesite-like systems, perhaps in shallow seas on early continental crust of around island arcs.

Hydrogenase in *Desulfurovibrio gigas* has already been mentioned, as it contains vital Mg. Another essential component in this enzyme is Ni. Urease, which is crucial to the biological nitrogen cycle, is also a nickel protein, perhaps the classic metalloprotein. Environmentally, Ni is a relatively rare element, not especially concentrated even in most hydrothermal fluids, and it is not easy to imagine how the early organisms could have obtained it. Possibly Ni was derived from meteorites, but another geologically likely setting is a large komatiitic plume volcano. Such a volcano would form a huge shield structure, comparable to Hawaii but vastly wider and with lower elevation above sea-level. It would produce enormous lava flows, flowing widely and then cooling with vigorous but shallow hydrothermal systems. Komatiite flows often host extensive nickel sulphide deposits, accumulating along the bases of the flows, and potentially exposed on active volcanoes by erosion or earthquakes. Nickel sulphide layers would provide substrates for bacteria. In this setting, the incorporation of Ni and Mg into hydrogenase might be favoured.

To summarize: the hypothesis that the early community of life was hydrothermally based is strongly supported by the components of life. The oldest life-forms appear to have been hyperthermophiles; only on hydrothermal systems would heat shock proteins be so important; only here would S, Mg, Fe and Mn-S all be readily available; and it is here that the incorporation of Cu, Zn, Mo, Se, perhaps Ni, and other elements into basic life processes may have taken place.

The origin of photosynthesis: did it begin on Mid-Ocean Ridges?

The first organisms must have existed with primitive chemotrophic metabolisms. Kandler (1992, 1994) has commented on the metabolisms that sustained the earliest life. Calvin–Benson cycle photosynthesis must have come late; prior to that, organisms were hyperthermophilic chemolithoautotrophs first, then evolved a variety of other pathways. In these, H_2S, H_2, and CH_4 were important chemical species commonly associated with hydrothermal systems. The origin of first anoxygenic and then oxygenic photosynthesis is perhaps the most interesting possible link with the hydrothermal world, especially as it was photosynthesis that may have allowed the living community to escape from hyperthermophile dependence on volcanism, to colonise the planet as a whole.

Recently, it was suggested that photosynthesis as a process may have originated at a hydrothermal system, from infra-red thermotaxis. Nisbet et al. (1995) pointed to the similarity between the wavelengths of light used by bacteriochlorophyll (thought to be the more primitive photosynthetic system) and the wavelengths that would be present in water a short distance from a hydrothermal vent at 350–400°C. They suggested that primitive photosynthesis began with thermotaxis. Bottom-living bacteria that could detect heat stood a better chance of survival and reproduction: by detecting heat they could either move across the mud surface towards it if too cool, or away if too hot, to find optimum conditions. Nisbet et al. noted that bacteriochlorophylls are best able to detect infra-red and far red light, and J. Cann showed that the radiation spectrum absorbed by bacteriochlorophyll matches the wavelengths that are propagated to distances of, say, 50 cm near hot vents at 350–400°C. From this comes the suggestion that bacteriochlorophylls first evolved to play the advantageous but nonessential role of helping thermotaxis; only later were they adapted by evolutionary tinkering to be used for photosynthesis. If correct the hypothesis demands that thermotaxis by infrared detection took place around vents at 350–400°C. Temperatures of this order would imply that the vents were in deep water (say 2.5 km), most likely at a mid-ocean ridge.

Modern ridge bacteria can survive, though not reproduce or grow, in cool water (this may be how they colonise from vent to vent). A bacterium of a species that had evolved near a ridge and was capable of infra-red thermotaxis may later have found itself in shallow water, perhaps in the vicinity of a shallow hydrothermal system. Here, it may have used sunlight as an optional photon source and the species could then have evolved to become an obligate photosynthesiser. Later, bacteriochlorophyll gave rise to chlorophyll, using higher frequency blue-green light to split water and carbon dioxide, and evolving oxygen. One of the attractions of this hypothesis is that it gives a series of small steps, each conferring selective advantage, for the evolution of photosynthesis, rather than demanding an impossibly massive hopeful mutation.

As mentioned above, one important component of photosynthesis is a 4Fe-4S complex; Mg^{2+} also plays a crucial role, as does Mn; and, as its name implies, rubisco is a phosphate compound. Heat shock proteins such as *GroEl hsp60* and *GroES hsp10* are essential to the operation of photosynthesis, in the presence of MgATP. These lines of evidence all support (though certainly do not prove) the hypothesis that photosynthesis evolved out of bacteria initially from a mid-ocean ridge setting, by the accidental association of the need for infra-red detection with the heat shock proteins, an available source of Fe-S and Mn-S in the vent, and Mg perhaps from Mg-clays or fluids around weak or shallow hydrothermal systems (or a komatiite shield flow). This is speculative hypothesis, but perhaps provable by further molecular study of the mechanism of photosynthesis, especially in bacteria.

Curiously, there is a second important class of light-use in the biosphere: that in vertebrate eyes and in *Halobacterium*. There are three possible explanations of this: (1) that a very ancient common ancestor of Archaea and Eucarya possessed this ability, which was later lost by nearly all Archaea except the highly derived *Halobacterium*, and also lost by many Eucarya; or (2) that the two processes are wholly unrelated, and are examples of convergent evolution, perhaps based on two independent cases of the evolution that exploited an antique pre-adaptation common to both the Eucarya and the distantly related Archaea; or, finally, (3) that genetic transfer somehow took place, more likely from *Halobacterium* to an ancestral metazoan or pre-metazoan eukaryote, or possibly reversely to *Halobacterium* from a metazoan.

Archaean ecology: possible locations for early communities of life

In the early Archaean, shortly after the beginning of life, a diverse global set of hydrothermally based biological communities may have developed from an original progenote. In prebiotic synthesis, it is possible that zeolites, with their cage structures and catalytic functions, may have been important, as may have been tubular iron oxides. In the broadest sense, the architecture of heat shock proteins is structurally analogous to the very large cage zeolites, and these proteins may have evolved very early indeed. After the evolution of the Archaea and the Bacteria, a variety of life-processes may rapidly have evolved. On land, complex communities of prokaryotes may have evolved, hosted in hot pools in hydrothermal systems around alkaline and komatitic plume volcanoes, and in pools and fumaroles around andesitic volcanoes above subduction zones. In marine settings, metal- and sulphur-rich muds near hydrothermal vents around subduction zones volcanoes around komatiite plume volcanic islands, and at mid-ocean ridges would have provided sites where metalloproteins could have developed by accidental insertion of hydrothermally derived ions into organic molecules. Eruptions, ocean currents, winds, and rivers would carry dormant Bacteria and Archaea from one hydrothermal system to another, colonizing the available transient volcanic habitats as they were initiated, grew and later became extinct.

Urease may have had a komatiite-based history. It presumably evolved in a system where CO_2 was abundant, as CO_2 is required for the assembly of the nickel metallocentre (Park & Hausinger 1995). One setting in which this could have occurred would have been in a carbonate–bicarbonate hydrothermal system with pH around 8, perhaps, as discussed above, in shallow conditions around an exposed nickel-sulphide substrate in an eroded or fault-broken lava flow on the flanks of a large komatiite plume volcano in a greenstone belt, or in an Iceland-like island on a ridge.

The oxygen content of the early atmosphere and seas remains unknown. It is of great interest that one of the most deeply rooted eubacteria, *Aquifex pyrophilus*, which represents a novel group of marine hyperthermophilic hydrogen oxidizing bacteria (Huber *et al.* 1992), is a microaerophile that today grows chemolithoautotrophically by reducing oxygen, using hydrogen, thiosulphate or sulphur as the electron donor. This lineage of bacteria may be the earliest divergence yet found (Burggraf *et al.* 1992). While much caution should be exercised in drawing inference about conditions from what may be an essentially modern adaptation, *A. pyrophilus*'s need for oxygen suggests the presence of at least small amounts of oxygen (possibly photolytically made, in the upper atmosphere) prior to the evolution of photosynthesis. However, it should be noted that circumstances of nitrogenase evolution perhaps imply a later absence of oxygen (see above).

The simpler of the photosynthetic bacteria may have evolved later than the divergence of *A. pyrophilus* and its relatives, with the evolution of cyanobacteria being still later (see Fig. 1). Cyanobacteria are probably of original hyperthermophile ancestry (Kandler 1994*a*) and remain common in modern hot springs. Perhaps they evolved, from a deep water vent lineage capable of infra-red thermotaxis, via communities in shallow systems around andesitic or komatiitic volcanoes. In early Archaean conditions, ultraviolet radiation may have been a significant problem for surface living organisms; carotenoids may have provided protection from UV-A to early cyanobacteria (Knoll & Bauld 1989). Capable of both photosynthesis and nitrogen fixation, planktonic, shallow-marine and terrestrial cyanobacteria would have immediately reproduced across all accessible habitats, initiating the modern biosphere. Life escaped from its putative hydrothermal birthplace, but carried with it the hydrothermal imprint.

Geological source regions

Hydrothermal systems are ephemeral. Communities of life that were bound to single hydrothermal systems would eventually die. Fortunately, mid-ocean ridge systems are effectively linked (from the bacterial point of view) by ocean floor currents. Even a single prokaryote is enough to start a new colony. Because the Archaean ridges were likely much more active than today, they would have had more closely juxtaposed active segments. Consequently, the Archaean mid-ocean ridge lengths could easily sustain a prokaryotic population, even of individuals less robust than today, and allow a diverse ecology to build up and survive by sporadically colonizing new active segments.

Komatiite plume volcanoes provide a possible setting for the evolution of many early characteristics of life. Around 4 Ga ago, komatitite volcanism would have been extremely extensive, perhaps even with komatitiic mid-ocean ridges, as well as in mantle plumes. Plume volcanoes would have produced huge flat shields, some perhaps sited on the mid-ocean ridges (as Iceland is today). The volcanoes would have produced huge flows covering large areas, sustaining shallow hydrothermal systems producing fluids rich in S, Mg etc. Eroded or fault-broken komatiite flows would locally provide extensive substrates of Ni-sulphide. Although Ni is widely present in basalts, and thus available to life from basalts, it is possible that an exposed nickel sulphide horizon could over a few years or decades support selective evolution of Ni-containing proteins. Around komatiites, deposits of magnesio-pelites, very Mg-rich clays, occur. As mentioned above, perhaps the nitrogen cycle using urease and nitrogenase evolved here, and perhaps the earliest association of Mg with ATP and ADP could also have occurred in such a setting. Fluids around komatiite volcanoes would, however, be poor in phosphorus.

Island arc volcanoes are different, and sustain hydrothermal systems rich in Cu, Zn, Mo, etc. Here hydrogenase and the Zn proteins may have evolved. Greenstone belts offer an interesting association of komatiites and andesites: for example, in Zimbabwe over a few hundred kilometres and within a few tens of Ma widespread komatiite volcanism was followed first by the construction of basalt shields and then andesite volcanism (e.g. see Nisbet 1987).

Connections between ridge, plume island, island arc and terrestrial hydrothermal habitats may have been possible by a variety of processes. Clearly, as today, oceanic currents can move prokaryotes from one active system to another, colonizing new volcanic centres as they arise. Large volcanic eruptions can lift cells into the air, and widely disperse them. Very large eruptions in shallow seas, as may have taken place in the Archaean, could have caused massive hurricanes ('hypercanes'; Emanuel *et al.* 1995) capable of spreading prokaryote cells on a planetary scale.

Conditions on mid-ocean ridge habitats can be extreme, with temperatures up to 400°C (but depending on water depth) and pH as acid as 2.5. Archaea and Bacteria colonizing this habitat could have spread from initial settings around shallow-water centres, such as plume volcanoes on ridge axes, and gradually colonized deeper environments. Possibly it was here, however, that infra-red phototaxis began, as postulated by Nisbet *et al.* (1995), around a hot deep-water vent. Then, later, the thermotactic bacteria could have recolonized shallow water environments and begun sunlight-driven photosynthesis. Ultimately, it was this process that allowed life to escape the hydrothermal systems completely and remake the global atmosphere. Perhaps it was the mid-ocean ridges that gave birth to the modern solar-dependent biosphere.

Origins

The discussion above has pointed to a variety of possible source regions in which important biological processes may have evolved (Figs 2 & 3). The early progenotes may have lived in an environment susceptible to heat shock, with available phosphate and Mg. From this, inferences can be drawn about geological constraints on the site of the birthplace of life. Nisbet (1987) argued that a shallow-water hydrothermal system, with conditions neither especially acid nor alkaline, was the most likely first setting. However, this logic was influenced in part by the optimization of many modern biological processes close to 40°C, and around pH 7; more recently, in contrast, the antiquity of the extreme hyperthermophiles has become apparent (Kandler 1992, 1994*a, b*), implying that the birthplace of life was hotter. As argued above, the temperature of the late Hadean/early Archaean ocean is not known, but is unlikely to have been much above 50°C, or substantial H loss would have occurred to space, either wholly dehydrating the planet or leaving an ocean even richer in D/H ratio than is currently observed. To recapitulate, unless life is very very ancient indeed, and began in the first few hundred million years or so after the accretion, somehow

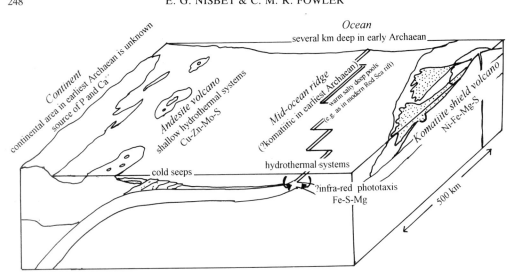

Fig. 2. Possible settings of the evolution of various biological processes. (1) Mid-ocean ridge (possibly komatiitic in the earliest Archean) - ?infra-red thermotaxis leading to photosynthesis. (2) Komatiite shield volcano (possibly where a komatiitic plume intersected a ridge) - ?incorporation of Mg in biochemistry, evolution of urease and hydrogenase in *Desulfovibrio gigas*, with Ni. (3) Andesitic island arcs - ?evolution of nitrogenase, Zn, Cu and Mo proteins. The extent of continental crust in the earliest Archaean remains a matter of vigorous debate: continental crust clearly existed, but it is possible (though not proven) that its area was small.

surviving massive meteorite impacts that could have vapourized the ocean, the hyperthermophile community must have been hydrothermal. By inference, then, the birthplace of life was also hydrothermal, although not necessarily at high temperatures.

The geochemical improbabilities of the origin of life are well known. Here, we confine ourselves to discussion of the phosphorus problem. The backbone of both RNA and DNA is sugar-phosphate: by implication, the first self-replicating nucleic acid (perhaps in an RNA world (Gilbert 1986) must have had ready access to phosphate. Geologically, the incorporation of phosphorus in life is puzzling. Phosphorus is a widespread but relatively inaccessible element. It commonly occurs in continental rocks, or as a minor element in most igneous rocks of the ocean floor, and as apatite. Even in the modern ocean, where phosphorus is widely carried by life, access to P is not easy, and P-limitation restricts the abundance of life. Marine apatite is widely precipitated, and seawater concentrations are low. On a pre-biotic Earth, Moizsis & Arrhenius (1995) have pointed out that instead of apatite, an amorphous protonated magnesium carbonate phosphate would be precipitated. Possibly this was the substrate for early assembly of nucleotides.

In the Archaean ocean, P-access must have

been an immense problem. Sources of P may have included erosion of volcanoes and fluids derived from hydrothermal systems. However, in the earliest Archaean, although the sea water was probably saturated in phosphorus, the concentration of the element was probably low. It is hard to imagine how P could have been gathered by the geochemically unskilled earliest organisms, and even harder to imagine how it was assembled into the first sugar-phosphate chain. In the late Hadean and earliest Archaean, continents did exist, but P-rich intrusions may not have been common on continents and above oceanic crust (in general, alkaline igneous rocks are unusual in the Archaean record, although they occur). Even in a geochemically 'strange' early ocean, 'accessible' phosphate may have been very uncommon. Hydrothermal systems do extract P from the rock, but typically precipitate it rapidly as apatite within the system; thus although it is possible to imagine P-rich fluids, they would not have been common except in very local circumstances.

Unless special chemical events took place, it is arguably unlikely that an ancestral RNA-world could have become established without an abundant local supply of phosphate. This puzzle demands the difficult task of imagining a geological setting in which high concentrations of available phosphate could have occurred to

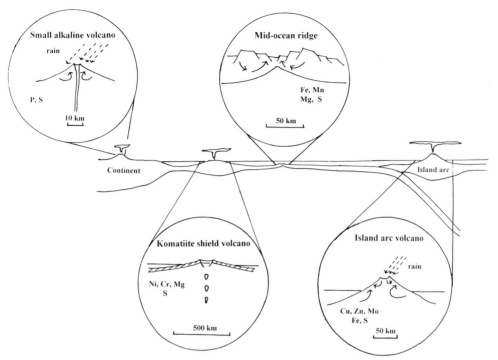

Fig. 3. Settings in which early metalloproteins and heat-shock proteins may have evolved in the early Archaean. Fluid movement indicated by arrows. On land, small alkaline volcanoes may have hosted hydrothermal systems in phosphate rock. At mid-ocean ridges, hydrothermal vents would have been surrounded by highly metalliferous sediments, rich in sulphides of Fe, Mn, Cu, Mo, Cd, etc. Komatiite plume volcanoes would have formed islands as wide low shields, locally hosting hydrothermal systems and exposing substrates rich in Ni sulphides and with Mg-clays produced by alteration. At island-arc and continental-margin volcanoes above subduction zones, Cu, Zn, Mo and Fe sulphides would have been common in sediments around hot pools both in subaerial hydrothermal systems and in shallow-marine settings.

allow the formation of polynucleotides: any nucleotides that did form would have surely been broken up long before they could assemble into nucleic acid. Perhaps the only location on a pre-biological world where phosphorus was abundant would have been in hydrothermal systems that developed around volcanoes erupting extremely alkaline lavas, for instance around a carbonatite (Hogarth 1989). Alkaline volcanics are rare today (Oldoinyo Lengai is the only active modern carbonatite volcano), and were rare in the Archaean but did exist (Nisbet 1987). Diamonds exist in Archaean sediment, implying that there were eruptions of Archaean kimberlites (Nisbet 1987). If so, other alkaline volcanoes may have occurred, some possibly carbonatite complexes in which phosphate rock occured, and where hydrothermal fluids rich in carbon gases and nitrogen circulated through phosphate. Furthermore, very large-cage zeolites, such as faujasite, occur in these settings, possibly providing both catalytic surfaces for the assembly of ancestral molecules and also protective enclosures to allow collections of nucleotides to accumulate and interact, and to save early polynucleotides from being broken up by light or chemical assault. Here, for instance at a fumarole where both liquid and vapour phase reactions could occur on zeolite catalysts, carbon-gas and nitrogen-rich fluids that had reacted with rock at temperatures of, say, 500°C, rising from depth, would interact with phosphorus-rich rock, water and air. Possibly it was above an Archaean phosphatic carbonatite that the warm little pond existed. From this start, colonization of hydrothermal systems around volcanoes in greenstone belt rifts could bring life to the mid-ocean ridges, where a diverse hydrothermal community may have developed the biogeochemical tools that allowed photosynthesis to evolve.

Many people have helped in this widely cross-disciplinary paper. In particular, we would like to thank the anonymous referees for their comments, and P. A. Tyler for helpful advice. J. Cann, V. Tunnicliffe, and C. van Dover, as well as many of the BRIDGE participants, are thanked for their interest and stimulating discussion, and D. Hall, J. Bowyer, J. Ellis and J. Prebble are thanked for comments. None of these is responsible for any error of misunderstanding we may have made, in subjects far from our own.

References

ACHENBACH-RICHTER, L., GUPTA, R., STETTER, K. O. & WOESE, C. R. 1987. Were the original eubacteria thermophiles? *Systematics and Applied Microbiology*, **9**, 34–39.

——, ——, ZILLIG, W. & WOESE, C. R. 1988. Rooting the archaebacterial tree: the pivotal role of *Thermococcus celer* in Archaebacterial evolution. *Systematics and Applied Microbiology*, **10**, 231–240.

BARBER, J. & ANDERSSON, B. 1994. Revealing the blueprint of photosynthesis. *Nature*, **370**, 31–34.

BARRACLOUGH, R. & ELLIS, R. J. 1980. Protein synthesis in chloroplasts. IX Assembly of newly synthesised large sub-units into ribulose bisphosphate carboxylase in isolated intact chloroplasts. *Biochemica Biophysica Acta*, **608**, 19–31.

BICKLE, M. J. 1986. Implications of melting for stabilisation of lithosphere and heat loss in the Archaean. *Earth and Planetary Science Letters*, **80**, 314–24.

BROCK, T. D. 1986. Introduction: an overview of the thermophiles. *In*: BROCK, T. D. (ed.)*Thermophiles: General, Molecular and Applied Microbiology*. Wiley-Interscience, New York. 1–17.

BURGGRAF, S., OLSEN, G. J., STETTER, K. O. & WOESE, C. R. 1992. A phylogenetic analysis of *Aquifex pyrophilus*. *Systematic and Applied Microbiology*, **15**, 351–356.

COLEMAN, J. E. 1992. Zinc proteins: enzymes, storage proteins, transcription factors and replication proteins. *Annual Review of Biochemistry*, **61**, 897–946.

DINGWALL, C. & LASKEY, R. 1992. The nuclear membrane. *Science*, **258**, 942–947.

ELLIS, R. J. 1993. The general concept of molecular chaperones. *Philosophical Transactions of the Royal Society, London*, **B339**, 257–261.

EMANUEL, K. A., SPEER, K., ROTUNNO, R., SRIVASTAVA, R. & MOLINA, M. 1995. Hypercanes: a possible link in global extinction scenarios. *Journal of Geophysical Research*, **100**, 13755–66.

GETHING, M.-J. & SAMBROOK, J. 1992. Protein folding in the cell. *Nature*, **355**, 33–45.

GILBERT, W. 1986. The RNA world. *Nature*, **319**, 618.

HARTMAN, F. C. & HARPEL, M. R. 1994. Structure, function, regulation and assembly of D-ribulose-1,5-bisphosphate carboxylase/oxygenase. *Annual Review of Biochemistry*, **63**, 197–234.

HEMMINGSEN, S. M., WOOLFORD, C., VAN DER VIES, S. M., TILLY, K., DENNIS, D. T., GEORGOPOULOS, C., HENDRIX, R. W. & ELLIS, R. J. 1988. Homologous plant and bacterial proteins chaperone oligomeric protein assembly. *Nature*, **275**, 416–420.

HOGARTH, D. D. 1989. Pyrochlore, apatite and amphibole: distinctive minerals in carbonatite. *In*: BELL, K. (ed.) *Carbonatites: genesis and evolution*. Unwin Hyman, London, 105–148.

HORWICH, A. L. & WILLISON, K. R. 1993. Protein folding in the cell: functions of two families of molecular chaperone, hsp 60 and TF55-TCP1. *Philosophical Transactions of the Royal Society, London*, **B339**, 313–326.

HOWARD, J. B. & REES, D. C. 1994. Nitrogenase: a nucleotide-dependent molecular switch. *Annual Review of Biochemistry*, **63**, 235–264.

HUBER, R., WILHARM, T., HUBER, D., TRINCONE, A., BURGGRAF, S., KONIG, H., RACHEL, R., ROCKINGER, I., FRICKE, H. & STETTER, K. O. 1992. *Aquifex pyrophilus* gen.nov.sp.nov., represents a novel group of marine hyperthermophilic hydrogen oxidising bacteria. *Systematic and Applied Microbiology*, **15**, 340–351.

JACOB, F. 1983. Molecular tinkering in evolution. *In*: BENDALL, D. S. (ed.) *Evolution from Molecules to Men*. Cambridge University Press, 131–144.

KANDLER, O. 1992. Where next with the Archaebacteria? *Biochemical Society Symposia* **58**, 195–207.

—— 1994a. The early diversification of life. *In*: BENGTSON, S. (ed.) *Early life on Earth*. Nobel Symposium s. **84**, Columbia University Press, New York, 152–161.

—— 1994b. Cell wall biochemistry and the three-domain concept of life. *Systematic and Applied Microbiology*, **16**, 501–509.

KASTING, J. F., ZAHNLE, K. J., PINTO, J. P. & YOUNG, A. T. 1989. Sulfur, ultraviolet radiation and the early evolution of life. *Origins of Life and Evolution of the Biosphere*. **19**, 95–108.

——, J. F., WHITMIRE, D. P. & REYNOLDS, R. T. 1993. Habitable zones around main sequence stars. *Icarus*, **101**, 108–128.

KNOLL, A. H & BAULD, J. 1989. The evolution of ecological tolerance in prokaryotes. *Transactions of the Royal Society of Edinburgh: Earth Sciences*, **80**, 209–223.

LEWIS, V. A., HYNES, G. M., ZHENG, D., SAIBIL, H. & WILLISON, K. 1992. T-complex polypeptide-1 is a subunit of a heteromeric particle in the eukaryotic cytosol. *Nature*, **358**, 249–252.

LOWELL, R. P., RONA, P. A. & VON HERZEN, R. P. 1995. Seafloor hydrothermal systems. *Journal of Geophysical Research*, **100**, 327–352.

MARTIN, J., HORWICH, A. L. & HARTL, F. U. 1992. Prevention of protein denaturation under heat stress by the chaperonin Hsp60. *Science*, **258**, 995–998.

McDERMOTT, G., PRINCE, S. M., FREER, A. A., HAWTHORNTHWAITE-LAWLESS, A. M., PAPIZ, M. Z., COGDELL, R. J. & ISAACS, N. W. 1995. Crystal structure of an integral membrane light-harvesting complex from photosynthetic bacteria. *Nature*, **374**, 517–521.

MOIZSIS, S. & ARRHENIUS, G. 1995. Phosphate-graphite association in early Archaean metasedimetary precipitates. (abstract). *European Association of Geochemistry meeting: Earliest history of the Earth. Cambridge, 18–21 Sept. 1995*, 24.

NISBET, E. G. 1986. RNA and hot water springs. *Nature*, **322**, 206.

—— 1987. *The Young Earth*. Allen and Unwin, London.

—— 1995. Archaean ecology: a review of evidence for the early development of bacterial biomes, and speculations on the development of a global-scale biosphere. *In*: COWARD, M. P. & RIES, A. C. (eds) *Early Precambrian Processes*. Geological Society, London, Special Publications, **95**, 27–51.

——, CHEADLE, M. J., ARNDT, N. T. & BICKLE, M. J. 1992. Constraining the potential temperature of the Archaean mantle: a review of the evidence from komatiites. *Lithos*, **30**, 291–307.

——, CANN, J. R. & VAN DOVER, C. L. 1995. Origins of photosynthesis. *Nature*, **373**, 479–480.

PARK, I.-S. & HAUSINGER, R. P. 1995. Requirement of carbon dioxide for *in Vitro* assembly of the Urease Nickel metallocenter. *Science*, **267**, 1156–8.

POLLACK, H. N., HURTER, S. J. & JOHNSTON, J. R. 1993. Heat loss from the Earth's interior. *Reviews in Geophysics*, **31**, 267–280.

PARSELL, D. A., TAULIEN, J. & LINDQUIST, S. 1993. The role of heat-shock proteins in thermotolerance. *Philosophical Transactions of the Royal Society, London*, **B339**, 279–286.

STEIN, C. A. & STEIN, S. 1994. Constraints on hydrothermal heat flux through the oceanic lithosphere for global heat flow. *Journal of Geophysical Research*, **99**, 3081–3095.

TRENT, J. D., NIMMESGERN, E., WALL, J. S., HARTL, F.-U. & HORWICH, A. L. 1991. A molecular chaperone from a thermophilic archaebacterium is related to the eukaryotic protein t-complex polypeptide-1. *Nature*, **354**, 490–493.

VOLBEDA, A., CHARON, M.-H., PIRAS, C., HATCHI-kian, E. C., FREY, M. & FONTCILLA-CAMPS, J. C. 1995. Crystal structure of the nickel-iron hydrogenase from *Desulfovibrio gigas*. *Nature*, **326**, 580–587.

VON DAMM, K. L., OOSTING, S. E., KOZLOWSKI, R., BUTTERMORE, L. G., COLODNER, D. C., EDMONDS, H. N., EDMOND, J. M. & GREBMEIER, J. M. 1995. Evolution of East Pacific Rise hydrothermal vent fluids following a volcanic eruption. *Nature*, **375**, 47–50.

WELCH, W. J. 1992. Mammalian stress response: cell physiology, structure/function of stress proteins, and implications for medicine and disease. *Physiological Review*, **72**, 1063–1081.

—— 1993. Heat shock proteins functioning as molecular chaperones: their roles in normal and stressed cells. *Philosophical Transactions of the Royal Society, London*, **B339**, 327–333.

WOESE, C. R., KANDLER, O. & WHEELIS, M. L. 1990. Towards a natural system of organisms: proposal for the domains Archaea, Bacteria and Eucarya. *Proceedings of the National Academy of Sciences, USA*, **87**, 4576–4579.

WOLLE, D., DEAN, D. R. & HOWARD, J. B. 1992. Nucleotide-Iron-Sulfur cluster signal transduction in the nitrogenase iron-protein: the role of Asp^{125}. *Science*, **258**, 992–5.

YOSHIDA, M., ISHII, N., MUNEYUKI, E. & TAGUCHI, H. 1993. A chaperonin from a thermophilic bacterium, *Thermus thermophilus*. *Philosophical Transactions of the Royal Society, London*, **B339**, 305–312.

YUNG, Y., WEN, J.-S., MOSES, J. I., LANDRY, B. M. & ALLEN, M. 1989. Hydrogen and deuterium loss from the terrestrial atmosphere: a quantitative assessment of non-thermal escape fluxes. *Journal of Geophysical Research*, **94**, 14971–89.

Index